FIFTY KEY THINKERS ON DEVELOPMENT

Fifty Key Thinkers on Development is the essential guide to the world's most influential development thinkers. It presents a unique guide to the lives and ideas of leading contributors to the contested terrain of development studies from both North and South. David Simon has assembled a highly authoritative team of contributors from different backgrounds and disciplines to reflect on the lives and contributions of fifty leading development thinkers from around the world. These include:

- Modernisers like Kindleberger and Rostow
- *Dependencistas* such as Frank, Cardoso and Amin
- Progressives like Hirschman, Prebisch, Helleiner and Streeten
- Political leaders enunciating radical alternative visions of development, such as Mao, Nkrumah and Nyerere
- Progenitors of religiously or spiritually inspired development, such as Gandhi and Ariyaratne
- Development-environment thinkers like Blaikie, Brookfield and Shiva

This invaluable reference is a concise and accessible introduction to the lives and key contributions of development thinkers from across the ideological and disciplinary spectrum.

David Simon is Professor of Development Geography and Director of the Centre for Developing Areas Research at Royal Holloway, University of London. He is the co-editor of *The Peri-Urban Interface: Approaches to Sustainable Natural and Human Resource Use* (Earthscan, 2005).

Also available from Routledge

Economics: The Basics
Tony Cleaver
0-415-31412-7

The Routledge Companion to Global Economics
Edited by Robert Benyon
0-415-24306-8

Fifty Major Economists
Steven Pressman
0-415-13481-1

Fifty Major Political Thinkers
Ian Adams and R.W. Dyson
0-415-22811-5

Fifty Key Thinkers on the Environment
Joy Palmer
0-145-14699-2

FIFTY KEY THINKERS ON DEVELOPMENT

Edited by

David Simon

LONDON AND NEW YORK

First published 2006
by Routledge
2 Park Square, Milton Park, Abingdon, Oxon OX14 4RN

Simultaneously published in the USA and Canada
by Routledge
270 Madison Ave, New York, NY 10016

Routledge is an imprint of the Taylor & Francis Group

Typeset in Bembo by
HWA Text and Data Management, Tunbridge Wells
Printed and bound in Great Britain by
MPG Books Ltd, Bodmin

British Library Cataloguing in Publication Data
A catalogue record for this book is available from the British Library

Library of Congress Cataloging in Publication Data
Fifty key thinkers on development/edited by David Simon
p. cm.
Includes bibliographical references and index.
1. Planners—Biography. 2. Economic development—Planning.
3. Economic development—Cross-cultural studies.
4. Economic policy—Cross-cultural studies.
I. Title: Key thinkers on development.
II. Simon, David, 1957–

HD87.55.F53 2006
338.9—dc22 2005012690

ISBN 0-415-33789-5 (hbk)
ISBN 0-415-33790-9 (pbk)

CONTENTS

INTRODUCTION

The invitation to edit this book proved irresistible, despite several other simultaneous editorial commitments. First, Routledge's confidence in a market for a development studies title in this well-established and successful series that spans many fields of study and research was pleasing. More particularly, though, I perceived the opportunity to address a long-felt gap in the development studies literature, namely a good quality biographical reference work that brings together leading figures from the various constituent disciplines. Precisely because of its inter- (and to some extent but still inadequately multi-) disciplinary nature, as well as the very contested nature of both theory and policy, both of which have been evolving rapidly in this young field, development studies still has a remarkably underdeveloped centre or core. Consequently, there is less sense of shared heritage or of a widely agreed set of leading figures and personalities than in longer-established fields.

Moreover, most academic research and teaching still take place within traditional disciplines, a feature recently reinforced in the UK and several other European countries as a result of the current discipline-based research assessment and audit cultures alongside years of cutbacks and rationalisation in higher education as a whole. Multidisciplinary institutes or centres of development studies were established in various countries amid the optimism of the 1960s and early 1970s but their subsequent fortunes have been mixed. In the USA, the primacy of conventional academic disciplines has never been seriously challenged. At another level, the development field was long beset by a veritable 'town versus gown' divide between academics and practitioners. Michael Edwards' (1989) polemic both reflected and exacerbated the sense of division but did focus attention on the problem. Although he moderated his views within a few years (Edwards 1993, 1994), suggesting that rapid progress towards *rapprochement* had been made, the last decade or so has certainly seen more widespread interaction and collaboration, resulting in a considerable narrowing of the divide.

Attesting to the vitality of the field, numerous recent textbooks and compendia have sought to keep pace with the rapidity of change in the post-Cold War world and the profound debates over the very meaning and future of 'development'. Yet, despite harbingers of doom predicting its demise, and whether we prefer to denote contemporary perspectives by means of prefixes like anti- or post-, or to refashion and reinvigorate the term 'development' itself, there is certainly much ongoing theoretical debate, conceptually rigorous research and dynamism in policy-driven practice.

As such, this volume aims to make a substantial contribution through informed reflection on the life and work of seminal thinkers and actors in the broadly defined field of development studies. The fiftieth anniversary of President Truman's 'Point Four' speech in 1999 was marked by several retrospective and prospective works. This book will continue that trend, given additional impetus by the recent deaths of several prominent, if in some cases highly controversial, figures like Walt Rostow, Charles Kindleberger and James Tobin. As some of the last survivors of the post-World War II impetus to 'develop' the newly independent states, their passing also symbolises the generational change that has been taking place within development studies.

The most difficult challenge in producing this book came at the outset: how to narrow down the long list of nearly 200 names that I jotted down in a very short time to only fifty? It quickly became clear that no universal consensus would be reached, regardless of the final choice. Indeed, this was confirmed through a process of consultation with friends, colleagues, the publisher and the subsequent suggestions of referees of the proposal for this title. In reaching the final list, I have undoubtedly made a few decisions that some people will find surprising. Several contributors have even admitted to not knowing of all the fifty themselves, perhaps illustrating our individual disciplinary and regional biases. A bit of controversy in this diverse and contested field may thus be both good and necessary.

Let me explain the selection process. One helpful factor was that a number of leading lights whose contributions to development have been made as practitioners or activists rather than as 'thinkers' could be excluded. This applies for instance to Chico Mendes, the champion of the indigenous Brazilian Amazonian rubber tappers, who was brutally murdered in 1988 by agents of the powerful ranchers and logging firms whose destruction was being challenged. As it happens, he is included in *Fifty Key Thinkers on the Environment* (Palmer 2001). Interestingly, five thinkers do appear in both the above title and this volume (Thomas Malthus, Karl Marx, Mohandas Gandhi, E.F. Schumacher and Vandana Shiva), while Marx and Gandhi also appear in *Fifty Major Political Thinkers* (Adams and Dyson 2003). In such cases, the respective entries have been written by different authors and bring out different aspects or interpretations of their contributions, so readers could usefully consult more than one.

Overall, the choice of Thinkers is intended to enhance the relevance and contribution of this book by having a group who are broadly representative of the diverse currents and movements in the world of development. This has meant ensuring that the pervasive Anglo-American dominance of development theory and discourse (not least by economists) is challenged through the inclusion of people working in different disciplines, over different periods of time and – crucially – hailing from different walks of life and geographical regions. This volume is by definition therefore a very different undertaking from the self-reflective studies by fifteen eminent Northern development economists (with commentaries by younger economists) commissioned by the World Bank some twenty years ago, when economic development and development were still commonly conflated (Meier and Seers 1987), although several of those economists are also featured here.

Readers will also notice the sharp gender imbalance among the Thinkers featured here; this reflects the strong male dominance in most areas of development thought. Economics is a partial exception but efforts to balance the various disciplines and Northern versus Southern representation made for a few difficult choices. One other possible woman contender, Gro Harlem Brundtland, chair of the UN's World Commission on Environment and Development, is featured in the Environment title in this series (Palmer 2001).

A few of the voices (both 'Southern' and 'Northern') included in this book are far less well-known 'internationally' than they should be, precisely because of the linguistic, cultural, disciplinary and geographical limitations under which many of us labour and which – however unwittingly – sometimes serve to perpetuate the very simplifications and abstractions that we claim to challenge. Their inclusion is therefore deliberate, as a small contribution to overcoming insularity and promoting polyvocality in development, as befits a postcolonial approach.

Similarly, one or two people like Norman Borlaug, progenitor of the Green Revolution, are included because of their vision and the profound impact of their work, even though this might arguably have been somewhat more 'technical' than 'conceptual'. Inevitably, though, some difficult choices have had to be made in view of the artificial limit of fifty thinkers. For instance, despite the undoubted influence of his 'world-systems theory', Immanuel Wallerstein has been omitted, in part because his ideas integrate less directly with development issues than the closely related work of Jim Blaut. Another important criterion for inclusion has been that, to allow 'mature' reflection on the significance of the lives and contributions of the Thinkers, they should be close to the end of their active careers, if not yet retired or dead. Given the relative youthfulness of Development Studies – as evidenced by Malthus and Marx being the only Thinkers included who predate the twentieth century – this still leaves plenty of scope for topicality, while avoiding the temptation for authors to make premature judgements or risk incurring personal embarrassment.

Personally, the process of editing the essays and melding them into a hopefully coherent book has been fascinating beyond my most optimistic expectations. Indeed, it would be no exaggeration to say that it has been the most rewarding editorial challenge I have undertaken to date. First and foremost, it has brought me into direct contact with two diverse sets of people, the authors and the Thinkers, some of whom I had never previously met or had contact with. Some of these new relationships will doubtless endure. Finding contributors willing to write about the identified Thinkers was in itself a challenge, particularly as I was soliciting only one entry per author. In most cases, the task proved remarkably easy, perhaps because the idea of the book caught their imagination too. In a few instances, it was difficult to recruit an author for a particular Thinker or to match a willing contributor to an appropriate Thinker not already 'taken'. Nevertheless, I must acknowledge the efficiency with which commitments were turned into well-crafted essays, and mostly by the agreed deadlines, as a result of which it has proved possible to produce the book on schedule. Many of the authors also found their research and writing most illuminating. Some were actually able to communicate with, and occasionally even interview, the figure on whom they were working. The manuscript has benefited greatly as a result. Inevitably, trying to capture the essence of full and fascinating lives, as well as assessing their lasting contributions, in about 2,000 words has been a considerable challenge.

My second source of inspiration as editor has been how much I have learned about the Thinkers and their lives and times, regardless of how much I knew of them and their work beforehand. This reflects the formula adopted for the book, namely to interweave their biographical 'stories' and an appreciation of their work and legacies.

Third, however, a fascinating set of insights emerged as the essays were edited and integrated into the book manuscript. Two in particular stand out and exemplify the 'added value' of this compendium. The first was the tremendous impact of the Nazi regime on the subsequent evolution of Development Studies through the emigration or escape of many young European Jewish (and some non-Jewish) refugees to the UK and USA, where they later emerged through universities and political office as influential contributors to the emerging ideas and approaches. Holocaust survivors and escapees are surprisingly numerous among the Thinkers represented here. While many other implications of World War II for what was to become Development Studies, such as the stimulus to colonial liberation struggles and formulation of the Marshall Plan, for instance, are quite well known, this one has seemingly not previously been documented.

Second, and somewhat more broadly, the geopolitical shifts and ruptures of the 'end of empire' represented by decolonisation opened up the possibility of interesting, indeed important, new interregional and experiential connections that had important influences on the subsequent thinking and work of the Thinkers concerned. This is exemplified by Arthur Lewis and Walter Rodney, two natives of the Caribbean who studied at the University of London (the former in the 1930s and the latter in the 1960s) and then worked for a time in Ghana and Tanzania respectively. There are better-known examples too. Decades earlier, Gandhi's experience fighting racism in South Africa had proved seminal to his subsequent strategy of non-violent direct action and resistance (*Satyagraha*) in India. Similarly, most of the Northern voices in this collection were profoundly influenced by growing up, travelling and/or working in parts of the South at an early age.

The third important insight was the often close and influential interconnections among some Thinkers whom most people today do not particularly associate with one another (e.g. Boserup, Lipton, Myrdal and Streeten with respect to the *Asian Drama* study in the 1960s) – something that may help explain the processes of interactions that spawned and popularised key ideas and theories at momentous times in the history of development thinking.

Hopefully readers will find this volume both stimulating as a good read and useful for reference purposes. Perhaps some will even seek to explore the lives and contributions of other key figures in the field.

References

Adams, I. and Dyson, R.W. (eds) (2003) *Fifty Major Political Thinkers*, London and New York: Routledge.

Edwards, M. (1989) 'The Irrelevance of Development Theory', *Third World Quarterly* 11(1): 116-36.

—— (1993) 'How Relevant is Development Studies?', in F. Schuurman (ed.), *Beyond the Impasse; New Directions in Development Theory*, London: Zed.

—— (1994) 'Rethinking Social Development: The Search for Relevance', in D. Booth (ed.), *Rethinking Social Development*, Harlow: Longman.

Meier, G.M. and Seers, D. (eds) (1984) *Pioneers in Development*, Oxford: Oxford University Press.

Palmer, J.A. (2001) 'Chico Mendes 1944–88', in Palmer, J.A. (ed.), *Fifty Key Thinkers on the Environment*, London and New York: Routledge, pp. 302–7.

David Simon
Egham, Surrey
March 2005

Note on cross-references

All cross-references between essays are indicated in the text by the relevant Thinker's name cited in bold.

FIFTY KEY THINKERS ON DEVELOPMENT

ADEBAYO ADEDEJI (1930–)

Born in 1930 in Ijebu-Ode in southwestern Nigeria, Adebayo Adedeji received his doctorate in Economics from the University of London in 1967, following initial training in Economics (BSc Hons, London, 1958) and Public Administration (Diploma, University College, Ibadan, 1954 and MPA, Harvard, 1961). After working initially as a civil servant, he joined the University of Ife (now Obafemi Awolowo University, Ile Ife) in 1963, becoming Nigeria's first Professor of Public Administration in 1967 and, concurrently, Director of the Institute of Administration. Between 1971 and 1975 he served as Nigeria's (post-civil war) Minister of Economic Planning and Reconstruction, before joining the United Nations as Executive Secretary of the Economic Commission for Africa (UNECA) with the rank of Assistant Secretary-General in 1975. He was promoted to Under-Secretary-General in 1978. He resigned from the UNECA in 1991 to return to Ijebu-Ode in Nigeria, where he established an independent think-tank, the African Centre for Development and Strategic Studies (ACDESS), which he continues to head as Executive Director.

Adedeji was clearly influenced by his training in economics at a time when development thinking was dominated by notions of teleology and growth. However, this influence appears to have operated in a 'malign' way, in the sense of, first, instilling and, subsequently, reinforcing a seemingly unshakeable belief in both the inability of inherited development policy and the inappropriateness of conventional development practice to respond adequately to the dynamic complexity of Africa's post-independence circumstances and realities. It is therefore probably in his articulation of a vision of a progressively 'nationalist' but increasingly (and variously) integrated continent-of-regions, an Africa secure in itself and valued as an integral member of the global community, that Adedeji's most significant contribution to development thinking lies. There is considerably more to his contribution, however, for in promoting the case for the elaboration of indigenous African development strategies, Adedeji has also illustrated the vital demonstration effect that the skilful deployment of development administration can contribute to policy and planning, to the rethinking of practice and to the elaboration of strategy. Thus in the only comprehensive and coherent analysis of Adedeji's contribution to development thinking and practice to date, Asante (1991) acknowledges the contribution of like-minded colleagues at the UNECA to the formulation of the ideas about African development which came to be closely associated with the person of Adedeji. Asante also notes how Adedeji's tenure as UNECA Executive Secretary provided an indispensable platform for the dissemination of these ideas, pointing out that the start of Adedeji's sustained challenge to orthodox development thinking from an African perspective was concurrent with the emergence of the radical Latin American-based critiques by **Andre Gunder Frank** and Michael Todaro, while Adedeji's role at the UNECA paralleled that played by **Raúl Prebisch** at ECLAC/CEPAL (the Economic Commission for Latin America and the Caribbean).

Adedeji's thinking on African development had its origins in the ferment of decolonisation, the emergence of post-war development as planned socio-economic change and the growing lack of evidence for significant early post-independence trickle-down benefits (Adedeji 1977, 1981). It was also undoubtedly influenced by debates taking place in and around ECLAC, and drew inspiration from the work of both Raúl Prebisch and, to a lesser extent, **Arthur Lewis**, although he differed from both in important respects and has almost certainly never considered himself a *dependentista* (Asante 1991). Not surprisingly, he early on questioned the wisdom of conflating economic growth and material change (readily quantifiable indices – or *means* – of development) with the less-easily measurable (because largely qualitative and continuously evolving) process – or *end* – of development which, for him, was always complex, holistic and people-centred, in addition to encompassing economic *as well as* social, cultural, environmental, political and other forms of non-economic change and transformation. 'Development', he is quoted as saying, 'is a collective responsibility in which all have to share in the labour as well as the fruits', for when 'people become the end and the means of development, their interests, values and aspirations necessarily determine the content, strategies and modalities of [such] development' and, in the process, serve to ensure that development remains anchored to its socio-cultural, political and historical bearings (Asante 1991: 6). For him, too, Africa's *actual* post-war experience of development diverged markedly from this ideal, partly because, their various descriptive labels notwithstanding, the policies and tactics adopted were inherited rather than home-grown, and thus overwhelmingly imitative rather than creative or locally responsive in nature. Second, development efforts failed to alter the dependent status of, or indeed to stimulate wealth or opportunity redistribution within, African economies significantly despite the varied but always heavy reliance of favoured strategies on external sources for inputs of all kinds and as markets for the continent's primary commodities (Adedeji 1981). Development also routinely failed to consider people as subjects rather than objects, he suggested, and, in so doing, neglected to engender a sense of ownership among the vast majority of Africans who were supposed to be the targets and beneficiaries of development intervention (Adedeji 1977).

By the mid-1970s, therefore, and in Adedeji's view, what was sorely needed was a route to 'economic decolonisation', involving, among other things, the 'indigenisation' of national and continental economic development, and the forging of an Africa which,

> [h]aving inherited or borrowed development policy as well as political theory, [would subsequently be able] to revive its own economic assumptions and design its own orientations, just as it ha[d] come to reject much of its neo-colonial legal and organisational legacy.
> (Adedeji and Shaw 1985: 3)

It is this vision, driven in part by a firm belief that economic growth was merely a means to the more desirable end of economic restructuring, societal reform and national/continental transformation for increased self-reliance (Adedeji 1977), which informed his insistence on the need for an African-formulated alternative strategy for African development. Notable among the elements of such a strategy, as detailed in the 1981 Lagos Plan of Action (LPA), were its privileging of increased self-reliance (requiring capacity building and the development of competence) and self-sustainment (involving both rural–urban, sectoral and national market integration); its questioning of the role of foreign trade as an engine of growth (as a prelude to the promotion of internal demand stimuli at the expense of external market demand); its attempt to separate, at least conceptually, internal socio-economic change from export market performance; and its repudiation of the assumption of a positive – rather than negative – correlation between the expansion of advanced and developing country economies (Adedeji 1985; OAU 1980).

Even though it did not advocate autarchy, the LPA was still tantamount to development 'heresy' in its defiant foregrounding of its political-economic purpose. Not only did it represent 'a political declaration, a development strategy, a set of priorities, sectoral programmes of action, and a blueprint for regional and subregional integration', but it also argued for 'a complete departure from the past ... substitut[ing] ... an inward-looking development strategy for [an] inherited externally oriented one [and] put[ting] the development of the domestic market rather than dependence on foreign markets at the heart of ... development ...' (Adedeji 1985: 15). Such defiance mattered little; a key goal of the LPA was, after all, for Africa and its peoples to recover, through the transformative power of a democratised/popularised development, a sense of self-confidence long undermined by slavery, colonialism and neo-colonialism (Adedeji 1977). Furthermore, by striking a dynamic balance between autarchy and vulnerability, the LPA was thought to possess sufficient flexibility and potential to indicate ways out of economic crisis (Adedeji and Shaw 1985). The LPA stimulated widespread debate, generated critiques, and may even have provided a framework for policy formulation and the implementation of strategies; it does need to be understood as part of a wider process of constructing a philosophy of development for Africa, however, one which also highlighted the UNECA's role 'in the international market-place of ideas about Africa in particular and development in general' (Asante 1991: 46).

Nonetheless, despite its capacity for addressing existing constraints and potential for contributing to long-term sustainability (Adedeji and Shaw 1985), implementation of the LPA was slow, hampered in part by the economic crisis of the 1980s but partly also by the gradual nature of the policy change/reform processes envisaged (Adedeji 1985). Confronted by the crises of the 1980s and 1990s, African policy-makers, economies and societies concentrated on immediate survival rather than long-term transformation, with key macro-economic policy being dictated by interpretations and recommendations contained in the Berg Report (World Bank 1981), the very antithesis of the LPA, with which it stood in fundamental contradiction, not least in its opposition to self-reliance and self-sustaining development and in its preference for export-orientation, the market principle and the minimal state (Adedeji 1985). In response, Adedeji and his colleagues at the UNECA produced the African Alternative Framework to Structural Adjustment Programs for Socio-Economic Recovery and Transformation (AAF-SAP), incorporating diagnoses of the root causes of Africa's economic underdevelopment and combined socio-economic and political crises which had initially been set out in the LPA (UNECA 1989). Like the LPA the AAF-SAP was considered crucial to Africa's development future; it was a 'launching pad into the ne[w] decade and beyond' (Adedeji 1990: 112). Adedeji used the AAF-SAP as the basis for challenging the logic, wisdom and ethics of orthodox reform/adjustment, particularly its mantra that 'There Is No Alternative' (TINA), by reiterating that the AAF-SAP *did* represent an alternative which was human-centred, consistent with the development objectives identified in the LPA, and advocated combining 'the short-term objectives of stabilisation and structural adjustment [with] the requirements of long-term restructuring' (Adedeji 1990: 71). He was adamant that it would be inadvisable for the orthodox structural adjustment measures of the 1980s to be continued into the 1990s, and warned that this would plunge the continent into a downward spiral that would be extremely difficult to recover from. Taken together, the LPA and AAF-SAP can, by extension, be considered among Adedeji's most seminal contributions to development thinking; indeed, he would (and probably did) approve their joint message: 'no programme of adjustment or development makes sense if it makes people indefinitely more miserable'. It is not surprising, then, that Adedeji (2002: 4) was less than sanguine that both the LPA and AAF-SAP were 'opposed, undermined and jettisoned by the Bretton Woods institutions' who in this way impeded Africans 'from exercising the basic and fundamental right to make decisions about their future'.

Adedeji's departure from the UNECA in 1991 did not mark an end to his contribution to development thinking and administration in Africa. Certainly, in ACDESS, which, according to the late **Julius Nyerere**, founding president of its Board of Trustees, was established in response to a perceived 'lack of opportunity to express, debate and test ... ideas in an open environment, in an African context, and under African leadership ... dedicated to thinking about and for Africa's future' (http://www.acdess.org/), Adedeji has created a vehicle for the continuing pursuit of his interest in 'prospective and strategic thinking in and about Africa' (Adedeji 1993). Just as significantly, perhaps, it is a vehicle which, like the UNECA, and via its research/training programme, consultancy/advisory operations and services, and its periodic national and international conferences, facilitates continuing interaction with (and between) politicians, policy-makers, bureaucrats, planners, technocrats, NGOs, (sub-)regional organisations and international institutions, universities and academic research institutions (http://www.acdess.org/), thereby allowing Adedeji to continue to 'combine theory with practical experience' (Asante 1991), and research with policy application (Adedeji 1999).

His recent reflections on the road travelled by Africa from the LPA to NEPAD (New Economic Partnership for Africa's Development) are instructive in this regard (Adedeji 2002). Following a favourable representation of NEPAD as a renewed pan-Africanist attempt to reactivate intra- and inter-African integration as well as rejuvenate Africa's partnerships with the international community, he cautions against pursuing NEPAD's goals at the expense of the LPA's principles of structural transformation and socio-economic diversification. NEPAD, he suggests, should be about 'the resources and policies of ... international partners being devoted to achieving Africa-determined development goals' (Adedeji 2002: 17). In this as in so many of his earlier interventions he (re-)focuses attention on (collective) self-preservation and political will. But he also reiterates another of his enduring messages: that Africa's pursuit of sustainable development needs to begin with success in the long-running struggle to 'indigenise' the paradigms, strategies and agendas which must, of necessity, guide the nature, pace, direction and dynamics of any such development. It is a struggle which he will, hopefully, continue to wage for a while yet.

Major works

Adedeji A. (1977) *Africa: The Crisis of Development and the Challenge of a New International Economic Order*, Addis Ababa: Economic Commission for Africa.

—— (1989) *Towards a Dynamic African Economy: Selected Speeches and Lectures 1975–1986*, London: Frank Cass.

—— (1990) *Structural Adjustment for Socio-Economic Recovery and Transformation: The African Alternative; Selected Statements*, Addis Ababa: United Nations Economic Commission for Africa.

Economic Commission for Africa (ECA) (1989) *The African Alternative Framework to Structural Adjustment Programs for Socio-Economic Recovery and Transformation (AAF-SAP)*, Addis Ababa: ECA.

Organisation of African Unity (OAU) (1980) *Lagos Plan of Action for the Economic Development of Africa, 1980–2000*, Addis Ababa: OAU, http://www.uneca.org/itca/ariportal/docs/lagos_plan.pdf.

Further reading

Adedeji A. (1981) 'General background to Indigenisation: The Economic Dependence of Africa', in Adedeji A. (ed.), *Indigenisation of African Economies*, London: Hutchinson.

—— (1985) 'The Monrovia Strategy and the Lagos Plan of Action: Five Years After', in Adedeji A. and Shaw, T.M. (eds), *Economic Crisis in Africa: African Perspectives on Development Problems and Potentials*, Boulder, CO: Lynne Rienner.

—— (1993) 'Marginalisation and Marginality: Context, Issues and Viewpoints', in Adedeji A. (ed.), *Africa within the World: Beyond Dispossession and Dependence*, London: Zed Books.

—— (1999) 'Comprehending African Conflicts', in Adedeji A. (ed.), *Comprehending and Mastering African Conflicts: The Search for Sustainable Peace and Good Governance*, London: Zed Books.

—— (2002) 'From the Lagos Plan of Action to NEPAD and from the Final Act of Lagos to the Constitutive Act: Whither Africa?', keynote address to the African Forum for Envisioning Africa, Nairobi, April 26-29, available at http://64.233.187.104/search?q=cache:AEoDAGQW7l0J:www.worldsummit_Hlt98338960_Hlt98338960 2002.org/texts/AdebayoAdedeji2.pdf+adebayo+ade_Hlt107369836d_Hlt107369836ej_Hlt107369839i_Hlt107369839&hl=en&ie=UTF-8.

Adedeji A. and Shaw, T. (1985) 'Introduction: Africa's Condition and Projections for the Future', in Adedeji A. and Shaw, T.M. (eds), *Economic Crisis in Africa: African Perspectives on Development Problems and Potentials*, Boulder, CO: Lynne Rienner.

Asante, S.K.B. (1991) *African Development: Adebayo Adedeji's Alternative Strategies*, Borough Green, Sevenoaks, Kent: Hans Zell.

Onimode, B. (ed.) (2004) *African Development and Governance Strategies into the 21st Century: Looking Back to Move Forward. Essays in Honour of Adebayo Adedeji at 70*, London: Zed Books.

Onimode, B. and Synge, R. (eds) (1995) *Issues in African Development: Essays in Honour of Adebayo Adedeji at 65*, Ibadan: Heinemann.

Sanmi-Ajiki, T. (2000) *Adebayo Adedeji: A Rainbow in the Sky of his Time. A Biography*, Lagos: Newswatch Books.
World Bank (1981) *Accelerated Development in Sub-Saharan Africa: An Agenda for Action*, Washington, DC: World Bank.

Reginald Cline-Cole

ANIL AGARWAL (1947–2002)

Activist, journalist and scholar, Anil Agarwal was a prominent Indian environmentalist who redefined environmental problems through the eyes of poor people, and who was not afraid to challenge powerful organisations and governments in order to do so. During the rise of global environmentalism in the 1960s and 1970s, it became common to blame poor people for environmental problems through acts such as population growth and deforestation. Agarwal was one of the first critics to challenge these generalisations, and to focus instead on questions of international justice in environmental politics and the choices and risks faced by poor people. Agarwal left various legacies. He founded the Indian think-tank, Centre for Science and Environment, which today remains one of the foremost centres of critical thinking about environment and development. More conceptually, however, Agarwal was a pioneer in debates that are today called political ecology and science and technology studies. Rather than accepting environmental explanations from large organisations as scientifically and politically neutral, Agarwal sought to expose the politics underlying each statement of causality, and to show how such science legitimised or delegitimised different policies. He demonstrated how justice, as a concept, could be integrated into environmental policy between North and South. Agarwal also brought his own style of influencing politics, through a combination of scholarly work, acerbic journalism and careful political campaigning.

Agarwal was born in Kanpur in Uttar Pradesh in 1947, the son of a local landowner. He attended the Indian Institute of Technology in Kanpur, where he studied mechanical engineering, and learnt information about technology that was to characterise his later writings. In a change of career direction, in 1973, Agarwal became a science correspondent at the *Hindustan Times*. In 1974, he wrote about the Chipko movement in the Indian Himalayas, where local villagers opposed logging, and which has more recently become an icon for local environmental struggles in the South. His writing attracted international attention, and in 1979 he won the first A.H. Boerma Award given by the United Nations' Food and Agricultural Organisation in Rome.

In 1980, Agarwal founded the Centre for Science and Environment (CSE) in New Delhi. The CSE was new because it was a non-governmental organisation that focused on environmental matters, and which sought to influence the Indian government and transnational corporations, a role it continues to play today. At the time, mainstream environmental groups in India tended to focus on conservation, and especially conservation of wilderness and wildlife, as their main concern. The CSE, however, highlighted environmental risks faced by poor people in India at a time when livelihoods were being challenged by the decline in traditional biomass-based rural economies and when industrialisation was growing. Agarwal communicated these views widely by editing the CSE journal, *Down to Earth*, which included a supplement for children known as the *Gobar* (or Cowdung) *Times*. Much of the writing was translated into Hindi, Kannada and other Indian languages.

The approach adopted by Agarwal and the CSE began to influence wider debates about the meaning of 'sustainable development'. His reports on *The State of India's Environment*, written with colleagues at the CSE from 1982, challenged the elitist basis of environmentalism, and sought to portray the environment as a political problem partly reflecting international and class-based divisions of power and wealth. Analysts have described this approach as 'red–green environmentalism' – which acknowledges both resources and livelihoods – rather than just the 'green' approach, which highlights conservation alone. Agarwal also believed that orthodox development thinking was wrong to place faith in rapid economic growth as the chief means of achieving social development. He proposed that a new concept of 'gross nature product' should replace 'gross national product' in order to express the impact of growth on environment and livelihoods. Agarwal was also sensitive to the roles of women in protecting resources, and in being vulnerable to environmental hazards. He argued that poverty and environment are interrelated, but that poor people were commonly more protective of resources than commonly thought, and that economic policy should be tailored more closely to address poverty.

Because of such writings, both Agarwal and the CSE quickly developed international reputations. From 1983 to 1987, Agarwal chaired the Environmental Liaison Centre International (ELCI), a Nairobi-based network of environmentalists. His work was reported in the UK-based *New Scientist* and *Economist* magazines, as well as the broadsheets, *Le Monde* (France) and *Asahi Shimbun* (Japan). In 1986, the then Indian Prime Minister, Rajiv Gandhi, invited him to address the Union Council of Ministers, and honoured him with the Padma Shri Award. Agarwal was later asked to address all twenty-seven Parliamentary Consultative Committees in India to educate MPs about his concerns, and to initiate discussions to identify solutions. In 1987, he was elected to the Global 500 Honor Roll of the United Nations Environment Programme.

Much of Agarwal's writing included a critical stance on environmental science, and especially statements that blamed poor people for causing environmental degradation. Instead, he urged a more holistic appreciation of the social and political conditions that make environmental changes problematic, and how proposed solutions may aggravate social injustice. Describing the oft-cited belief that upland deforestation causes lowland flooding in the Himalayas, for example, Agarwal argued that the phenomenon of floods was caused by various factors including lowland water demand, rather than simply deforestation in the uplands. Consequently, policies need to consider how resources (and access to resources) have changed, and for whom, rather than apply simple mechanistic controls on water flow or forest use. He wrote in *Down to Earth* in 1987,

> Floods and shifting of river courses is ... inevitable. Deforestation can aggravate the problem but afforestation cannot get rid of it. Embankments and dams have become an important cause of floods. We need better flood plain management, rather than flood control.

This criticism of popular scientific statements and concern about social justice also affected Agarwal's work in international environmental politics. In one of his most famous works, *Global Warming in an Unequal World* (co-authored with Sunita Narain in 1991), Agarwal criticised the tendency for some analysts to assume that anthropogenic climate change should be addressed by controlling deforestation in developing countries. In particular, Agarwal and Narain condemned a report issued by the Washington DC-based think-tank, World Resources Institute, which allocated national responsibilities for greenhouse gas emissions based on an index largely dependent on current rates of deforestation and methane emissions from wet rice and livestock. The report put the three developing countries of Brazil, India and China among the top six emitting countries.

Agarwal and Narain contested the report on various grounds. First, the report was based on total national emissions, rather than on per capita emissions, which, of course, were smaller in developing countries than in developed countries. Second, the index used highly simplistic estimates for both deforestation and methane emissions. For example, estimates of wet-rice methane emissions were extrapolated globally from Italian figures; deforestation was treated uniformly, with no distinction made between export-led logging and smallholder food production; and no account was taken of the impacts of vegetation that might replace forest. Third, the index focused chiefly on current tropical deforestation, and did not consider historic deforestation in developed countries (which is important as greenhouse gases can exist for many years). Fourth, and perhaps most importantly, the index did not refer to questions of social justice in greenhouse gas emissions, such as acknowledging that much deforestation in developing countries may occur because of poverty and food production, whereas in developed countries burning fossil fuels may be linked to affluence. Agarwal and Narain's criticisms of this index were a watershed in international environmental politics, and demonstrated that scientific reports about environmental problems should not be considered politically neutral, but contain deep political implications about which activities are considered damaging or not, and which countries or people may be considered responsible. Agarwal worked on this theme during the approach to the 1992 Rio Earth Summit (the United Nations Conference on Environment and Development), by advising both the Indian Prime Minister, P.V. Narasimha Rao, and the former Tanzanian President, **Julius Nyerere**, at the South Centre in Geneva, and by joining India's official delegation to the Rio conference. The Rio Summit contained much discussion of sustainable development, and facilitated the signing of the United Nations Framework Convention on Climate Change, and Convention on Biological Diversity.

Agarwal's work after Rio involved a new attention to urban environmental problems, and to the justice of economic globalisation. In particular, he studied how trade and government policy can encourage the provision of clean technology to poor people in cities. In 1996, the CSE published a report on vehicular pollution in Indian cities, which blamed petroleum companies, car manufacturers and regulators and planners. The report was followed by a media campaign, and eventually by government action to phase out polluting cars. In a typically acerbic editorial in *Down to Earth*, Agarwal wrote (1996):

> The western economic dream is a toxic dream. And don't listen to the typical tripe from Indian scientists and officials that India's consumption and production of toxic substances per capita is zilch compared to Western countries. This is utter scientific nonsense trotted out to make you apathetic. It is the exposure levels that matter, which can be very high in India, because of among other causes, high pesticide residue in our food and low quality of drinking water.[1]

Agarwal wrote a series of editorials and writings urging greater global democracy in how environmental problems were solved, and in the processes of globalisation. For Agarwal, it was unacceptable that trade should be used as a means to control environmental misbehaviour by richer countries when poorer countries who suffer pollution or rising sea levels because of these richer countries cannot impose trade sanctions. Yet, globalisation – if conducted with attention to political conflicts and alliances between campaigners in North and South – could also bring opportunities for strengthening the political role of developing countries in international affairs.

Following some of his earlier writings, Agarwal and the CSE also continued to seek ways to demonstrate decentralised rural governance via village communities. Under a campaign entitled 'Making Water Everyone's Business', the CSE supported experiments in water harvesting and land management in Sukhomajri in Haryana, Ralegan Siddhi in Maharashtra and the Tarun Bharat Sangh in Rajasthan. But despite these actions, Agarwal was criticised by some for offering only muted support for the Narmada anti-dam movement in western India, and for allegedly losing some of his initial radical stances by becoming an adviser to the state, thus raising the question as to whether it is possible for a recognised environmentalist to remain radical. Many did not share these criticisms. In 2000, he was given an Environment Leadership Award by the Global Environment Facility – the multilateral funding agency for global environmental problems. In 2001, the Government of India bestowed on him the Padma Bhushan Award, a status reserved for people who have performed distinguished service of a high order to the nation.

Anil Agarwal died in 2002 at just 54. He had experienced a long battle with cancer, and had written about cancer care in India as another example of inadequate attention to social welfare. He left an important legacy through the creation of the CSE, and his personal writings pioneered current thinking about poverty and environment and the hidden politics of environmental scientific assessment. Agarwal made it clear that local questions of environment in developing countries were inherently linked to international political economy, and argued that creating knowledge about environmental problems should not be left to experts in developed countries. He also achieved these aims through establishing a system of campaigning and communication that both harnessed and educated many in poorer countries. Anil Agarwal was one of the most influential thinkers and writers on questions of environment and development because he fought to increase the representation of poor people in both the definition and solution of environmental problems.

Note

1 From online source: http:www.cseindia.org/aboutus/anilji/anilji-book2.htm.

Major works

Agarwal, A. (1982) *The State of India's Environment: A Citizens' Report*, with Sunita Narain, New Delhi: Centre for Science and Environment.
—— (1989) *Towards Green Villages: A Strategy for Environmentally Sound and Participatory Rural Development*, with Sunita Narain, New Delhi: Centre for Science and Environment.
—— (1991) *Floods, Flood Plains and Environmental Myths*, edited with Sunita Narain, New Delhi: Centre for Science and Environment.
—— (1991) *Global Warming in an Unequal World*, with Sunita Narain, New Delhi: Centre for Science and Environment.
—— (1992) *Towards a Greener World: Should Global Environmental Management be Built on Legal Convention or Human Rights?*, New Delhi: Centre for Science and Environment.
—— (1997) *Homicide by Pesticides: What Pollution does to our Bodies*, edited, New Delhi: Centre for Science and Environment, State of the Environment Series 4.
—— *Down to Earth*, journal published by Centre for Science and Environment, http://www.downtoearth.org.in/.

Further reading

There are, to date, no books specifically describing the life of Anil Agarwal, but information about his life can be obtained from the publications and websites of *Down to Earth*, and the Centre for Science and Environment (CSE), and the obituaries below.

CSE: http://www.cseindia.org/.
CSE official biography: http://www.cseindia.org/html/au/_Hlt70072329_Hlt70072330aBM_1_BM_2_nilji/anilji.htm.
Baviskar, A. (2002) 'An activist–environmentalist, Anil Agarwal, 1947–2002', *Frontline* 19: 2, February 1, 2002. http://www.flonnet.com/fl1902/19021210.htm.
Jupiter, T. (2002) 'Anil Agarwal: India's leading environmental campaigner', *Guardian*, 11 January, http://www.guardian.co.uk/obituaries/story/0,3604,630852,00.html.

Tim Forsyth

ELMAR ALTVATER (1938–)

As a representative of the school of Critical Political Economy, Altvater has analysed the limits of the dominant (capitalist) mode of production studiously and creatively. He stresses tirelessly that Fordist production structures and patterns of consumption cannot be translated into a universally applicable avenue for the social development of all. Unusually among social scientists, he has opened himself to basic natural science laws in searching for viable explanatory models for a future-oriented social development. He insists that the entropy principle sets limits.

Born on 24 August 1938, Altvater studied economics and sociology in Munich and wrote a dissertation on 'Environmental Problems in the Soviet Union'. Teaching political economy (or rather the critique of political economy) since 1971 as professor in the Political Science Department of the Freie Universität in (West) Berlin, he has influenced whole generations of students. Altvater's diverse interests have not sustained any damage from the turmoil of the later stages of a 'student rebellion' in decline. As a representative of undogmatic Marxism, he filled his seminars to capacity. Declining profit rates, the internationalisation of capital, and theories of the world market featured as much among his courses as introductions to *Das Kapital* by **Karl Marx** and seminars on state theories. The wide thematic range of his work has always been focused on the creative application of Marxist theory.

Altvater's analyses of international financial capital, monetary and fiscal policy as well as state interventionist questions regarding regulation of markets and societies have gained prominence since the early 1970s. During the 1980s, the intensified debt crisis in the countries of the periphery made him a powerful critic of development on loan (Altvater *et al.* 1991). Inspired by a sabbatical in the eastern Brazilian Amazon region, he devoted a rare case study to Brazilian issues but generalised them in a relevant way by presenting a pioneering study on 'the world market as a force of circumstances' (*Sachzwang Weltmarkt*). In this groundbreaking analysis, he understands space as a real substratum of economic and social processes and their structuring and probes the systemic and systematic connection between regional identity, state development policy and economic crisis tendencies in the world market. Given the latter's dominance over national development strategies and local conditions of ecologically adjusted production, a Fordist catch-up industrialisation strategy must come to nothing. In a chapter (in *Sachzwang Weltmarkt*) on 'The Economy and Ecology', he explores the thermodynamic laws, for him an increasingly central argument for the natural limitation of social development as processes of material conversion, production and consumption are subject to increasing entropy. Hence the economic system and its tendencies cannot be imagined without limitation by natural laws.

Altvater's approach benefited from the fundamental works of Nicholas Georgescu-Roegen (1906–94), a Romanian mathematician, who also studied at the Sorbonne and was apprenticed in economics at Harvard to Joseph Schumpeter, one of the leading names in economic theory and in particular the internationalisation of capitalism. Georgescu-Roegen, who finally settled at Vanderbilt University in the USA, has gained increased influence in recent economic theory because of his ability to combine a new biological or evolutionary approach with economic theory. In his *magnum opus* (Georgescu-Roegen 1971), he emphasised the limits to growth in an economy on the basis of the Second Law of Thermodynamics (which can be translated as 'useful energy gets dissipated') (see also Georgescu-Roegen 1976). Largely ignored by mainstream economics, Georgescu-Roegen's new approach to economic theory had lasting effects on the likes of Altvater, who discovered its relevance from the perspective of an environmentally sensitive school of thought, opening new discourses in the field of evolutionary economics.

Another pioneer inspiring a few social scientists – and even fewer economists – like Altvater in the application of economic theories to development studies was the Russian-born Ilya Prigogine (1917–2003), the Founding Director of the Centre for Statistical Mechanics and Thermodynamics at the University of Austin/Texas since 1967 (now the Ilya Prigogine Centre for Studies in Statistical Mechanics and Complex Systems). He was awarded the Nobel Prize in chemistry in 1977. Like Georgescu-Roegen, he contributed substantially to a new combination of natural and social sciences in theories of relevance for environmental and economic aspects in the reproduction of societies. His major works also highlighted the relevance of dissipative structures and contributed significantly to the understanding of irreversible processes, especially in systems far from equilibrium (see Kondepudi and Prigogine 1998; Prigogine and Stengers 1986, 1997).

Altvater derived environmental problems from the specific dynamic of current capitalism, which in his view is not 'the end of history' in Fukuyama's sense of immanent future development but a demand for a continuing search for sound alternatives for the reproduction of social systems. In the course of this demand, Altvater harks back to the integration of the entropy laws into social science theory as documented in his two works, *Die Zukunft des Marktes* (The Future of the Market) and *Der Preis des Wohlstands* (The Price of Prosperity). 'Thermo-balance' is a key concept in his argument. Critics reproach him for a 'thermo-dynamisation of the social sciences'. Others scold him for utopian sandcastle games in which he now urges a solar instead of a socialist revolution without stressing the immanent contradictions of (post-)Fordist capitalist production or utilising the 'transformist' renewal of the capitalist world order. These objections can be countered by pointing to the merit of his introduction of a new dimension into development discourse that considers the criterion of conscious metabolism in the sense of social conduct and does not downgrade the environment and nature to mere objects of boundless exploitation (Köhn 1995). To some of the objections raised by a critic (Hein 1993), Altvater responded directly by pointing out that ten human generations have appeared since the 'Promethean' industrial revolution of the world market. Nature is finite and has reached the limits of encumbrance and, as Altvater maintains, theoretical approaches that do not strive to integrate the ecological limit in their conceptual world are not up to date (Altvater 1994: 104). He challenges socio-economic theory to take into account the global range of the capitalist social formation and the (similarly global) ecological threat. He rejects the reproach for borrowing from thermodynamic physics the categories of 'entropy' and 'syntropy' that are out of place in socio-economic debate. Even if the cosmic process is understood as open and the earth as an energetic open system, a thermodynamic restriction remains. The existing material energy (fossil or nuclear) fuels industrial production cycles and thereby contributes to a warming up (by emission) of the natural environment (*ibid.*: 108). As he argues further, a consideration of the categories of thermodynamic physics in socio-economic analysis is helpful in revealing the consequences for production. Hence he insists on an expansion of socio-economic concepts since conventional theories today are inadequate for comprehending the ecological problems and he refers to the necessity for compatibility between the production and energy systems in humanity's history. His underlying thesis is that the epoch of production based on fossil sources of energy will come to an end within a few decades, that the transition to a solar society awaits and that the exchange of primary sources of energy cannot occur without radical social changes in the social systems of material and energy transformation (*ibid.*: 110). He describes the process of a necessary social, economic and political restructuring as a 'solar revolution' without any substantive concretion but vital as a future-oriented vision.

Together with his lifetime companion, Birgit Mahnkopf, Altvater illuminates the present *Grenzen der Globalisierung* (Limits to Globalisation) and takes this further into the *Globalisierung der Unsicherheit* (Globalisation of Insecurity). He analyses the worldwide socio-economic processes that maintain competition between states and regions in an increasingly deregulated world market, and at the same time lead to a global 'club society of the owners of financial assets', through the globalisation of financial markets and loss of state autonomy in the sense of its own regulatory possibilities. His reference to the limited character of the Fordist model of development is also central to his argument. Neither the political nor the socio-economic qualities of industrialised countries can be generalised worldwide. He formulated the quintessence of this thematic approach in a concise essay (Altvater 1996) by warning that, in a fatal way, the world is finite and that catch-up development by Third World countries by means of Fordist-type industrialisation of the sort undertaken by Western industrial countries is inconceivable. For Altvater, local social coherence is proven in global economic restrictions through the overlapping of functional spaces (see *Sachzwang Weltmarkt*, esp. chaps 2–5: 56–194). Development is determined twice. The observance of restrictions can succeed only when political institutions set limits. Development proposals must both obey general rules and be very specific. He recognises that the resulting political challenge consists in the lack of material prerequisites for local decisional freedom. Interdependence cannot arise as dependence.

As the editors of the *Festschrift* for his sixtieth birthday conceded in their introduction to the volume (Heinrich and Messner 1998), Altvater's holistic approach might be controversial but accomplishes orientation and integration in a very differentiated scientific world where we increasingly understand details but less and less the comprehensive connections. His ongoing work has so far been most influential in the debates concerning the reproduction of capitalism and the world market, ecological constraints, the theory of the state and more recently global public goods. Consequently, the contributions in honour of his sixty-fifth birthday centre on this topical issue (Brunnengräber 2003). Altvater pleads for the regulation of energy consumption, working conditions and capital movements at the supra-national level. In one of his few texts published in English, he summarised his position thus:

> In an era of globalisation, the conventional paradigm of economic policy is in need of radical rethinking. Such a paradigmatic shift, however, will necessarily have to be accompanied by practical efforts to re-embed the global economic system in qualitatively new social relations and forms of political regulation, on both local and global levels.
>
> (Altvater 2002: 88)

Altvater's contributions help us to keep sight of the essentials, including the political-moral ethics of a theory of development, and provide motivation to remain engaged. 'At some time or other, we must begin,' he said, in justification of his own beliefs:

> This sounds idealistic and the idealism reproach has been very harsh since Marx. Nevertheless, a process is initiated through which the fatal monetarisation and short-term materialism are revoked. Hopefully we will have this time. This is not certain. Unfortunately, catastrophe cannot be excluded. However, no leftist project for catastrophe is developed but rather a project for avoiding catastrophe.
>
> (Altvater 2004: unpaginated)

Major works

Altvater, E. (1987) *Sachzwang Weltmarkt. Verschuldungskrise, blockierte Industrialisierung, ökologische Gefährdung – der Fall Brasilien*, Hamburg: VSA.

—— (1991) *Die Zukunft des Marktes. Ein Essay über die Regulation von Geld und Natur nach dem Scheitern des 'real existierenden Sozialismus'*, Münster: Westfälisches Dampfboot.

—— (1992) *Der Preis des Wohlstands oder Umweltplünderung und neue Welt(un)ordnung*, Münster: Westfälisches Dampfboot.

—— (2002) 'The Growth Obsession', in Panitch, L. and Leys, C. (eds), *Socialist Register 2002: A World of Contradictions*, London: Merlin Press, pp. 73–92.

Altvater, E. and Mahnkopf, B. (1999) *Grenzen der Globalisierung. Ökonomie, Ökologie und Politik in der Weltgesellschaft*, Münster: Westfälisches Dampfboot (4th, revised and enlarged edn; originally 1996).

—— (2002) *Globalisierung der Unsicherheit. Arbeit im Schatten, schmutziges Geld und informelle Politik*, Münster: Westfälisches Dampfboot.

Further reading

Altvater, E. (1994) 'Tchernobyl und Sonnenbrand oder: Vom Sinn physikalischer Kategorien in den Sozialwissenschaften. Replik auf die Kritik von Wolfgang Hein', *Peripherie* 14 (54): 101–12.

—— (1996) 'Von möglichen Wirklichkeiten. Hindernisse auf der Entwicklungsbahn', *Entwicklung und Zusammenarbeit* 37 (2): 44–9.

—— (2004) 'Time Needs Radicalism and Radicalism Needs Time', translated from the German original of 1998, posted to the indymedia website on 10 January 2004 and accessed on 6 February 2004 (http://sf.indymedia.org/news/2004/01/1671226.php).

Altvater, E., Hübner, K., Lorentzen, J. and Rojas, R. (eds) (1991) *The Poverty of Nations: A Guide to the Debt Crisis – From Argentina to Zaire*, London: Zed Books (originally published in 1987 as *Die Armut der Nationen. Handbuch zur Schuldenkrise von Argentinien bis Zaire*, West Berlin: Rotbuch).

Brunnengräber, A. (ed.) (2003) *Globale Öffentliche Güter unter Privatisierungsdruck. Festschrift für Elmar Altvater*, Münster: Westfälisches Dampfboot.

Georgescu-Roegen, N. (1971) *The Entropy Law and the Economic Process*, Cambridge, MA and London: Harvard University Press.

—— (1976) *Energy and Economic Myths: Institutional and Analytical Economic Essays*, New York and Toronto: Pergamon Press.

Hein, W. (1993) 'Elmar Altvater: Entropie, Syntropie und die Grenzen der Metaphorik', *Peripherie* 13(51/52): 155–70.

Heinrich, M. and Messner, D. (eds) (1998) *Globalisierung und Perspektiven linker Politik. Festschrift für Elmar Altvater zum 60. Geburtstag*, Münster: Westfälisches Dampfboot.

Köhn, R. (1995) 'Gesellschaftliche Grenzen der Entropie. Wider die Thermodynamisierung der Sozialwissenschaften', *Peripherie* 15 (59/60): 180–93.

Kondepudi, D. and Prigogine, I. (1998) *Modern Thermodynamics: From Heat Engines to Dissipative Structures*, New York: Wiley.

Prigogine, I. and Stengers, I. (1986) *Order Out of Chaos*, New York: Bantam.

—— (1997) *The End of Certainty: Time, Chaos, and the New Laws of Nature*, New York: Free Press.

Henning Melber

SAMIR AMIN (1931–)

Samir Amin is the best-known Egyptian/Arab thinker in the field of '**Marx**ist' development theory, formidable critic of capitalism, radical political economist, and one of the ferocious champions of anti-globalisation activism. His contribution to political theory is comparable only to those of contemporary Marxist critics of capitalism, including Paul Baran, **Andre Gunder Frank** and Immanuel Wallerstein. Commonly, they applied Marxist development theories in an effort to explain the consequences of capitalist economic, political, cultural and military development and expansion to the developing countries.

Amin's earlier academic writings preceded the emergence of what is now commonly known as development studies. As an economist, Amin began his research (1957–70) with the study of the economies of individual countries: Mali, Congo Brazzaville, Egypt, Senegal, Ghana, Côte d'Ivoire and the Maghreb countries. Since then, Amin's work has been informed by these early theoretical developments, which originated in historical Marxism.

Influenced by the Cold War economic and ideological rivalry and a disappointing first decade of development, Amin's four empirically based works, *The Maghreb in the Modern World* (1970), *Neo-colonialism in West Africa* (1973), *Unequal Development* (1976) and *The Arab Nation: Nationalism and Class Struggle* (1976), echoed the frustrations of a generation of intellectual leftists in developing countries who witnessed the obliteration of the optimism of the decolonisation euphoria. Describing 'independence' as neocolonialism, the relationship between the newly independent Maghreb and West African states and the economically dominant former colonial powers, Amin was able to foresee the overriding patterns of future economic development of these countries and the developing world in general. This realisation influenced his intellectual development for decades, as reflected in his development theorising. His doctoral dissertation (Institute of Economics, University of Paris, 1957) was entitled 'The Structural Effects of the International Integration of Precapitalist Economies: A theoretical study of the mechanism, which has engendered the so-called "under-developed" economies'.

Amin's first major contribution to development theory was so powerful that it continues to surface in most of his current intellectual works. The two-volume work, *Accumulation on a World-Scale* (1974), was instantly recognised as a major contribution, probably not for its originality, but for its Marxist orientation that found resonance in the Cold War ideological schism. In this study, Amin argues that one of the anomalies of capitalist development theory is that it confuses underdevelopment with poverty (*ibid*.: 261–2). Because it describes bare economic characteristics of developing countries, capitalist development theory artificially separates economics from the sphere of social and political organisation, and by doing so it ignores that underdevelopment consists of more than the outward appearances of poverty. Underdevelopment is described as a cumulative sum of the whole history of capitalist expansion structurally constructed as a world system with centre and periphery. Amin calls this his 'theory of capitalist social formation', which exhibits certain characteristics of underdevelopment (*ibid*.: 15–20).

Amin explains underdevelopment as an outcome of three factors. (1) Unevenness of productivity between spheres or 'sectoral unevenness of productivity' between centre and periphery. (2) The disarticulation, astructuration or distortion of the underdeveloped economies of the periphery made up of non-integrated sectors, with less flow of internal exchanges in the bid to satisfy the external demands imposed on them by the economies of the centre. (3) The centre that dominates economically, socially and politically over the periphery.

If dependence and integration into the world capitalist system engenders underdevelopment, Amin's alternative to disarticulation is delinking. In essence, Amin's theorising development is closely associated with dependency theory and its intellectual heritage.

One of Amin's major contributions to development theory is his ability to reinvent the insights of Marxism–Leninism and **Mao**ism (*ibid*.: 63 and 112–16) on the national question in order to explain the origins of underdevelopment. Methodologically, he exhumed Marxism from its Western antecedence and redeployed it to explain underdevelopment as an ultimate result of past colonial experiences or the hegemony of the ex-colonial powers, the centre over their ex-colonies – which constitute the periphery. As a historical process, underdevelopment is the cumulative result of unequal exchange, unequal development and imperialism (*Unequal Exchange*, 1973; *Unequal Development*, 1976; 1977). The central premise is that while economic growth characterises the economic sectors at the centre, it contributes to the development of underdevelopment and the disarticulation of the social formations of the periphery.

For Amin,

> Delinking has nothing to do with exclusion or autarkic withdrawal. It is a matter of subjecting the mutual relations between the various nations and regions of the whole world of the planet to the varying imperatives of their own internal development and not the reverse.

Delinking is therefore a manifesto for change, a rejection of the idea that the expansion of capitalism, and subsequently the currently dominant neo-liberal paradigm, is inescapable and therefore leaves no possibility for national autonomy. He offers polycentricity (i.e. one planet with several competing systems) (Amin 1990: xii) as an alternative to neo-liberal exclusionary and polarising development, also referred to as mal-development (*ibid.*: 80, 94–7, 129–36).

Amin's theoretical strand, ideological bent and activism are best described as a search for praxis or a theory of practice. Much of his writings during the closing decade of the twentieth century and until today have been devoted to the critique of globalisation, which he described as a peculiar new form of managing the international economy. The implications of globalisation for development and development theorising are horrendous. In *Capitalism in the Age of Globalisation* (1997), Amin revisits the idea of polycentric globalisation and argues that there is the need for 'renewing the perspective of global socialism' (*ibid.*: 6) in order to usher in an era of an 'alternative humanist project of globalisation' (*ibid.*: 10). Such an agenda cannot be realised and the crisis of development resolved until popular, democratic forces capable of dominating society get together again (*ibid.*: 135). In this scheme, the intelligentsia is assigned a vanguard role to establish bonds between its own productive thinking and the aspirations and actions of the popular classes, making them social partners for change.

Obsolescent Capitalism: Contemporary Politics and Global Disorder (2003) is Amin's ultimate commentary on the expansion of capitalism from the Thirty Years' War (1914–45) to the current age of neo-liberal globalisation. Here, Amin draws much from his earlier writings, including *Maldevelopment: Anatomy of a Global Failure* (1990); *Transforming the Revolution: Social Movements and the World System* (1990); *Empire of Chaos* (1992) and *Re-reading the Post-war Period* (1994). The future of capitalism is 'prophesied' with the main aim of answering one important question, 'Are the current developments of world capitalist systems permanent or transient' or 'rather, signs of obsolescence of a system that must be overcome if human civilisation is to survive?' (Amin 2003: 1). Theoretically, the volume offers confirmation of Amin's **Marx**ist interpretation of the state of the World, claiming that 'Marx's law of pauperisation resulting from capitalist accumulation has been ever more strikingly confirmed on a world scale over the past two centuries' (*ibid.*: 3).

The realisation that former communist countries such as Russia and China succumbed to the economic power of global capitalism (*Empire of Chaos*, 1992) created the necessity for Amin to rethink development theorising. In his latest writings, he defines development as a concept that does not connote 'catching up', but instead it 'involves the project of a very different (alternative) society', whose twofold aim would be: (a) to free humanity from economistic alienation; and (b) to end the legacy of polarisation on a world-scale. These objectives, according to Amin, cannot be realised without the active participation of the whole world population because the problems facing humanity have an ever-deeper global dimension (*ibid.*: 3–4).

Capitalism is obsolete because it is an ageing system that has entered a state of permanent disorder, leading to a long transition to socialism or catastrophe and the suicide of humanity. Capitalism is obsolete because: (1) the current scientific and technological revolution shows that capitalism has exhausted its reliance on labour, thus hindering the possibility of continual accumulation; (2) the collective Triad[1] of imperialism operating on a world-scale no longer allows the pursuit of dependent capitalist development in the peripheries. In Amin's view, if:

> the collective triad imperialism, especially in its American centre of centres, no longer functions as exporter of capital to the peripheries, but [is] dependent on surplus generated throughout the world, the triad is no longer a significant and dependent on manufacturing media discourse to survive, 'does this not symbolize the obsolescence of a system that has nothing to offer 80 percent of the world population?'
>
> (2003: 93–4)

Because the US hegemony and its new right neo-liberal project thrive on obsolescence capitalism, its alternative is a non-American twenty-first century whose basic requirements include: (1) The dislocation of the current unipolar world system with a multipolar (democratic and regionalised) one, implying that obsolescence capitalism corresponds to 'the dictatorship of transnational capital, attacks any idea of "self-reliance," and treats "delinking" and "national construction" as regressive protectionism' (*ibid.*: 30). (2) The twenty-first century should be more radical than the twentieth and instead of looking back to historical **Marx**ism, historical Keynesianism and national populism, neo-Marxism, neo-Keynesianism and post-capitalism became the counter responses to globalised liberal capitalism (*ibid.*: 136–8). (3) A non-American twenty-first century requires the building of convergences of social and political movements that give expression to the victims of global neo-liberal capitalism, a task which in turn demands respect for diversity. Unlike the twentieth century, the twenty-first century requires building a Left with alternative strategies and tactics for a united front in support of social and international justice (*ibid.*: 140–7).

Amin's contribution to development theory could be summarised in four points: First, critical and innovative application of classical Marxist theory to explain underdevelopment or maldevelopment as phenomena shaped by the capitalist system, implying that the fate of the underdeveloped countries is inseparable from developments in the capitalist world. Second, underdevelopment is a product of neocolonialism as transformed during the late decades of the twentieth century into what he describes as neoliberal global capitalism, which carries the insignia and genesis of obsolescent capitalism. Third, delinking in a polycentric, regionalised world is based on co-operation between 'autocentric' countries capable of directing their economies not for 'catching up' with obsolescent capitalism but for creating alternatives to its hegemony at best and assured demise at worst. Fourth, transforming the current global order is possible only by building convergence within diversity in a global justice society where social movements create spaces for people's participation (Amin 1987). The overall aim of solidarity and engagement with social justice is to engender a 'long transition to world socialism, implying the delinking of a system of criteria of economic rationality from the system of criteria derived from the submission to the globalised world of value' (Amin 2003: 159). Evidently, Amin is still an 'optimist Marxist', hoping that development's ultimate aim is the long transition to world socialism, which would eventually displace obsolescent capitalism.

Note

1 Triad refers to the Marxist dogma that human history is predetermined, evolving in three basic stages (triads), with the two final stages being socialism and communism.

Major works

Amin, S. (1970) *The Maghreb in the Modern World*, Harmondsworth: Penguin.
—— (ed.) (1972) 'Modern Migrations in Western Africa', Studies presented and discussed at the Eleventh International African Seminar, Dakar, April.
—— (1973) *Neo-colonialism in West Africa*, Harmondsworth: Penguin Books.
—— (1973) *Unequal Exchange, Imperialism and Underdevelopment: An Essay on the Political Economy of World Capitalism*, Ranjit Sau: Calcutta and Oxford University Press.

—— (1974) 'Accumulation and Development: A Theoretical Model', *Review of African Political Economy* 1: 19–26.
—— (1974) *Accumulation on a World-Scale: A Critique of the Theory of Underdevelopment*, New York: Monthly Review Press.
—— (1976) *The Arab Nation: Nationalism and Class Struggle*, London: Zed Press.
—— (1976) *Unequal Development: An Essay on the Social Formations of Peripheral Capitalism*, Hassocks: Harvester Press.
—— (1977) *Imperialism and Unequal Development*, Brighton: Harvester Press.
—— (1978) *The Law of Value and Historical Materialism*, New York: Monthly Review Press.
—— (1980) *The Arab Economy Today*, London: Zed Press.
—— (1981) *The Future Maoism*, New York: New Left Review.
—— (1987) 'Democracy and National Strategy in the Periphery', *Third World Quarterly* 9: 1129–56.
—— (1990) *Delinking: Towards a Polycentric World*, London: Zed Books.
—— (1990) *Maldevelopment: Anatomy of a Global Failure*, London: Zed Books.
—— (1990) *Transforming the Revolution: Social Movements and the World System*, New York: Monthly Review Press.
—— (1991) 'The Ancient World System *versus* Modern Capitalist World System', *New Left Review* 14(3): 349–85.
—— (1992) 'Thirty Years of Critique of the Soviet System', *Monthly Review* 44(1): 43–50.
—— (1992) *Empire of Chaos*, New York: Monthly Review Press.
—— (1993) 'Historical and Ethical Materialism', *Monthly Review* 45(1): 44–56.
—— (1994) *Re-reading the Postwar Period: An Intellectual Itinerary*, New York: Monthly Review Press.
—— (1997) *Capitalism in the Age of Globalisation*, London: Zed Books.
—— (1998) *Specters of Capitalism: A Critique of Current Intellectual Fashions*, New York: Monthly Review Press.
—— (2003) *Obsolescent Capitalism: Contemporary Politics and Global Disorder*, London: Zed Books.

M.A. Mohamed Salih

A.T. ARIYARATNE (1931–)

A.T. Ariyaratne's distinctive contribution as a key thinker lies in his life-long efforts to follow a new path to development, independent of both capitalism and socialism. He is the leader of the Sarvodaya Shramadana movement, the largest non-governmental organisation (NGO) engaged in development and poverty alleviation in Sri Lanka today. Sarvodaya's vision for a new society with 'no poverty' and 'no affluence' was based on **Gandhi**'s philosophy of 'truth', 'non-violence', and 'self-sacrifice'. The term 'Sarvodaya' comes from two Sanskrit words 'sarva' (universal) and 'udaya' (awakening). Ariyaratne uses the word 'sarvodaya' in two ways, to mean the awakening of all people and the awakening of individuals in all spheres – psychological, moral and spiritual, as well as social, economic and political. The term 'shramadana' is also derived from two Sanskrit words, 'shrama' (labour) and 'dana' (gift), that is, the gift of labour.

In the Sinhalese language the two words 'sarvodaya shramadana' have come to mean 'the sharing of one's time, thought, and energy for the awakening of all' (*Dana*, Feb. 1987: 15). Sarvodaya believes that development involves more than material growth. It involves psychological, moral and spiritual dimensions as well as social, economic and political ones. Shramadana or gift of labour implies both physical and mental labour. Ariyaratne's shramadana draws on social networks, donated labour, skills and co-operation, showing how social capital can create material wealth.

Dr Ariyaratne was born in Unawatuna village, Galle District, and after graduation worked as a teacher in Galle town before attending a teachers' training college. In 1958, as a newly arrived science teacher at Nalanda College, a Buddhist high school in Colombo (where he served till 1972), along with his students he organised the first of several voluntary workshops (a shramadana) in one of the poorest villages on the island. The workshop consisted of digging wells, building pit latrines, planting gardens, and opening up rural roads using the co-operative labour of students and villagers. Performing manual labour alongside poor villagers was a transformative experience for students who came from upper- and middle-class urban homes. This was an early example of what we now call 'service learning'.

Ariyaratne received a Bachelor of Arts general degree from the Vidyodaya University of Sri Lanka, graduating in economics, Sinhalese language and education. Later he received an honorary doctorate from the same university, and a doctor of humanities from Amelio Aguinaldo University in the Philippines. As the founder and leader of the Sarvodaya Shramadana movement, Ariyaratne has received a number of international awards including the Raman Magsaysay Award for Community Leadership from the Philippines (1969), Feinstein World Hunger Award from Brown University in Rhode Island (1986), Niwano Peace Prize from Japan (1992) and the Mahatma Gandhi Peace Prize from India (1996).

Although there are many publications under his name (Ariyaratne 1988, 1999) and many more about him (Bond 2004; Macy 1985), Ariyaratne is not widely regarded as a development theorist. According to his own homepage (www.sarvodaya.org), 'he was not guided by theory. He wanted to practice first and enunciate theory later. And the practice should be meaningful; the theory should only follow it.' Today at the age of seventy-four, Ariyaratne has remained active in community development, but increasingly has turned his attention to one of the central problems of Sri Lanka, political violence and the long-drawn-out military conflict with the Liberation Tigers of Tamil Eelam (LTTE). The Liberation Tigers have been fighting since 1983 for a separate state for minority Tamils on the grounds that they have suffered discrimination at the hands of the Sinhalese majority.

Since its inception in 1958, Sarvodaya seems to have evolved through four phases (Bond 2004: 7–42). During the period 1958–67, Sarvodaya was primarily a volunteer work camp movement. The work camp begun by Ariyaratne and his students in 1958 was very successful and it launched a larger social movement that quickly spread to other high schools and villages. At this time resources came entirely from local donations and voluntary labour. During the second period (1967–83), Sarvodaya became a formal NGO. In 1972 the movement was recognised by an Act of Parliament and incorporated as a legal body. It began to attract generous foreign funding, became a fully fledged NGO with a large portfolio of projects for village economic development, and adopted methods of cost accounting, monitoring and evaluation. As it grew, Sarvodaya moved away from the ideology of social revolution, co-operated more closely with the government, and acted in the capacity of an extension agency. Apart from local village schemes, Sarvodaya also undertook several well-funded national projects for enterprise development, alternative technology and child care. During the third period (1983–97), the conflict between the government and the Liberation Tigers intensified and spread; civilian life for both Sinhalese and Tamils was severely disrupted. With funding from overseas, Sarvodaya operated a large programme offering rehabilitation and relief work in villages most affected by the Tiger insurgency.

During this period the close co-operation with the government came to an end because Ariyaratne opposed the policy of seeking a military solution to the LTTE problem. Drawing on **Gandhi**an and Buddhist principles of non-violence, Sarvodaya laid out a plan for the peaceful resolution of the conflict through spiritual means but the larger conflict continued. In the final period (1997–), we see clear evidence of Sarvodaya taking an even stronger stand against a military solution. Sarvodaya publicly declared that neither the government nor the LTTE could 'win' the war; all they can do is to draw out the conflict (www.sarvodaya.org). On the other hand, Sarvodaya claimed it knew how to help transcend the war. To that end it organised a large-scale peace movement, announced an alternative framework of power for conflict resolution, and held several well-attended public peace meditations. It also started a programme of 'sister villages' with villagers from the south travelling to the war-ravaged villages of the north to do rehabilitation work repairing houses, wells, tanks, schools, toilets and places of worship. Since Ariyaratne remains central to Sarvodaya, this brief history shows the potential of his philosophy to achieve both personal empowerment and national reconciliation through non-violence, spirituality, compassion and Buddhist principles of 'right livelihood'.

The Sarvodaya philosophy has a strong moral and spiritual foundation. Ariyaratne was well aware that religion, in its institutional form, historically has not played a progressive role in the material transformation of society. And yet he chose this vehicle for his programme of social advancement, a path not unlike that taken by **Gandhi** in India, and by the worker priests of the Catholic based communities in Latin America. Sarvodaya philosophy is founded on Buddhist teaching; Ariyaratne's contribution has been to show that Buddhism can be used to address two principal problems facing contemporary Sri Lanka – poverty and violence. The Sarvodaya philosophy is vast but for the purpose of this exercise I shall present it under three headings.

Personal agency in structural change: Sarvodaya holds that structural change in society must begin with personal change. According to Buddhism, the principles of virtuous living include loving kindness, respect for all living beings, compassion for others, sharing joy in the completion of projects intended for someone else's benefit and even composure in the face of both joy and sadness. All Sarvodaya workers are urged to practise these in their routine everyday engagements. Beginning with personal awakening (puroshodaya), Sarvodaya expands outwards to community or village awakening (gramodaya), and beyond that to the awakening of the country (deshodaya), and finally to the awakening of the world (vishovodaya). All social change must radiate out from the agency of the individual living and practising a righteous life.

Buddhist theory for overcoming suffering: central to Sarvodaya philosophy is the Buddhist notion of four 'noble truths'. The first states that the normal condition of existence is suffering (Dukkha). Second, the root cause of suffering is greed, craving and desire (Thanha). Third is the claim that suffering can be overcome (Niradha). Finally is 'Marga', the Buddhist path to overcome suffering in the world. 'Marga' contains eight elements: right understanding, right thought, right speech, right action, right livelihood, right effort, right mindfulness and right concentration. The 'Eightfold Path' points to the centrality of personal agency in both Buddhist and Sarvodaya philosophy. Ariyaratne uses the first of the 'Four Noble Truths' (there is suffering in the world) by starting from an actual village with great poverty. This becomes the basis for analysis and reflection on the realities of the village. The second principle (craving as a cause of suffering) is presented to villages by describing how distrust, competition, enmity and egocentricity sap the energy of villagers. The third principle (suffering can cease) is translated into a discourse on affection, compassion, kind words, sharing and mutual self-help. Finally, the eightfold path of 'Marga' recommends the avoidance of two extremes – the pleasure of sensual indulgence and the pain of self-mortification. Ariyaratne adapted this Buddhist concept of the 'Middle Way' by recommending a development path of 'no poverty, no affluence'. Sarvodaya does not believe that poverty can be eradicated through creation of wealth and economic development. Sarvodaya begins this task by stating a manifesto of ten basic human needs which includes: a clean and healthy physical environment, water, clothing, food, housing, health care, communication, fuel, education, and spiritual and cultural needs (www.sarvodaya.org). These needs can be satisfied by using local resources, self-reliance, and shramadana. Sarvodaya believes that the dominant models of economic development constitute obstacles to meeting basic needs because such models emphasise increased production, expansion of desire, unlimited growth and open markets.

Non-violence in conflict resolution: a core principle of Buddhism and Hinduism is non-violence which advocates respect for all sentient beings. Following the examples of **Gandhi** and Vinoba Bhave in India, Ariyaratne argued that fundamental social change can be achieved through non-violence. The adoption of the eightfold path of right livelihood automatically excludes the use of physical violence as a means of achieving objectives. Sarvodaya has reached out to the Tamils through membership of the organisation and by doing relief and rehabilitation work in villages of the North. It has organised peace marches, mass meditations, and offered to mediate between the government and the LTTE.

Ariyaratne attributes Sri Lanka's current economic problems to British imperialism which destroyed self-reliant village economies. In that regard his analysis is very similar to that of dependency theorists like **Andre Gunder Frank** (Frank 1966) and **Fernando Henrique Cardoso**. Ariyaratne also believes that unequal global structures and giant transnational organisations are responsible for increasing poverty, armed conflict and ecological destruction in Sri Lanka and throughout the world. Despite that structural analysis, his proposals do not focus on changing global structures but on personal agency and right living. By rejecting the notion that human behaviour is the result of larger forces which actors neither control nor comprehend, Ariyaratne's views are consistent with Western social theories such as structuration (Giddens 1984) and post-structuralism (Foucault 1980).

Sarvodaya's material critique of development arises from its explicit rejection of affluence. The rejection of capital-intensive, high-technology, open economies is clearly at odds with the dominant models of development. Economic growth requires the constant expansion of consumption and desire, which according to the Buddhist theory of craving is what leads to suffering. Economists measure living standards principally by how much people consume. Sarvodaya rejects this path of development.

Even though the concept of no affluence has the potential to yield a serious critique of development, Sarvodaya has weakened the argument by enframing it entirely within the context of Buddhist advocacy of controlling one's craving. Craving is presented in Sarvodaya philosophy as an intrinsic human condition that must be overcome through reflection and moral advancement. But it is not enough to advocate the voluntary limiting of consumption without also addressing the myriad forces of modern capitalism that drive consumption.

Sarvodaya's advocacy of a non-violent solution to the ethnic problems of Sri Lanka is very courageous. A number of Tamil Sarvodaya workers have been killed by the Liberation Tigers, and Ariyaratne himself lives under constant threats to his life. But there are some ironies to Sarvodaya's offering non-violence as a solution to the ethnic problem. Despite its ecumenical outlook, Sarvodaya in Sri Lanka is perceived primarily as a Buddhist organisation. The current conflict in Sri Lanka arises from Tamil reaction to what they perceive as the hegemony of Sinhala Buddhist nationalism. The identity of the Sinhalese and their perception of self, are both intricately tied to Buddhism, its social history and its myths. Conversely, the Sinhalese perception of the Tamil is that of 'the other as non-Buddhist', a dynamic which the Tamils in turn have incorporated into their perception of the Sinhalese and of themselves. Given the signification of Buddhism to the two main parties in the conflict, it is important to reflect on the limits of a social movement that uses Buddhist theology to bring about a non-violent resolution of the protracted ethnic conflict. Nevertheless, Sarvodaya philosophy can contribute to peace in Sri Lanka in two important ways. First, as evidence shows, it provides a means for engaging in peaceful negotiation. Second, the emphasis of the middle path on no affluence and no poverty removes the intensive competition for resources which is one of the driving forces of the ethnic conflict in Sri Lanka.

Major works

Ariyaratne, A. T. (1988) *The Power Pyramid and the Dharmic Cycle*, Ratmalana, Sri Lanka: Sarvodaya Vishva Lekha Press.

—— (1999) *Collected Works*, 7 vols (ed. by N. Ratnapala), Ratmalana, Sri Lanka: Sarvodaya Vishva Lekha Press.

Further reading

Bond, G.D. (2004) *Buddhism at Work: Community Development, Social Empowerment and the Sarvodaya Movement*, West Hartford, CT: Kumarian Press.

Dana: Journal of the International Sarvodaya Shramadana, 1987.

Foucault, M. (1980) *Power/Knowledge*, New York: Pantheon Books.

Frank, A. G. (1966) 'The Development of Underdevelopment', *Monthly Review* 18: 17–31.

Giddens, A. (1984) *The Constitution of Society*, Berkeley, CA: University of California Press.
Macy, J. (1985) *Dharma and Self Development: Religion as Resource in the Sarvodaya Self-Help Movement*, West Hartford, CT: Kumarian Press.

Lakshman Yapa

JAGDISH BHAGWATI (1934–)

Jagdish Bhagwati is currently a University Professor at Columbia University and Senior Fellow at the Council of Foreign Relations. He was born and brought up in Bombay and was educated at Cambridge, Oxford and MIT. Bhagwati has been an advisor to a number of international organisations, including the WTO. His vast published output extends to more than 300 articles and fifty volumes. Jagdish Bhagwati's work has so dominated thinking on a number of economic policy issues in our own time that Paul Samuelson has referred to it as 'The Age of Bhagwati'. This essay provides a brief overview of Bhagwati's work centred on his advocacy of free, non-discriminatory trade, his assault on misguided protectionist policies and his analysis of the so-called phenomenon of the brain drain.

Bhagwati's relentless advocacy of free trade is often referred to as militant and aggressive. This may be so, but his perspective is grounded in economic theory and his analysis and observation of the welfare costs of India's protectionist trade and investment regime.

Students of international economics are familiar with the gains from free trade based on the comparative advantage of trading nations. There is, though, a long list of theoretical exceptions to the case for free trade, including the infant industry argument, arguments based on distortions in the labour and product markets in the domestic economy, and externalities. These and other arguments have so often been invoked for the institution of protection in both developed and developing countries that protection appears to be the rule and free trade the exception. In a seminal article (1963), co-authored with the late V.K. Ramaswami, using the conventional tools of the trade theorist, Bhagwati demonstrated that the best way to address distortions or domestic market failures was through domestic policies while maintaining free trade. The best policy in the presence of various distortions is an appropriate subsidy which rectifies the distortion at its source but leaves the international price unchanged and avoids the loss of consumer surplus associated with a tariff. If the distortion is in the trade sphere, such as monopoly/monopsony power in international trade, it should be rectified with appropriate trade policy instruments such as an export tax or an import tariff.

This message relating to trade carries over to policies regarding foreign capital inflows. It is that capital inflows in the presence of a tariff can be immiserising. Bhagwati enunciated the concept of immiserisation in an article (1958) written whilst he was an undergraduate student reading for the Economics Tripos at Cambridge. The message of the article is that economic growth may so worsen a country's terms of trade that it may actually reduce economic welfare. Extensions of the idea by Bhagwati himself (1968, 1973) and others demonstrate that capital inflows in the presence of a tariff on an importable good in a two-good two-country mode can be immiserising. Whilst the capital inflow induces growth at a constant tariff inclusive of domestic prices, the tariff imposes the familiar production and consumption costs, and the rewards accruing to foreign capital must also be reckoned as a cost.

Yet another of Bhagwati's policy prescriptions relating to trade and capital flows is that countries pursuing an export promotion (EP) policy are likely to attract both a higher volume of foreign direct investment (FDI) and experience substantially higher rewards from it than countries pursuing import substitution (IS) policies (1978). An EP policy here is defined as one which is neutral in terms of the protection it provides for importables and the incentives it affords exportables. In such an environment investment decisions of foreign firms would be based on comparative advantage dictated by market forces and not artificial and uncertain policy inducements. Econometric tests suggest that Bhagwati's twin propositions are robust (Balasubramanyam, Salisu and Sapsford 1996). It is worth noting that Bhagwati's interpretation of the EP policy does not rule out tariffs, it merely requires that investment decisions should be based on comparative advantage and should not be influenced by artificial policy incentives which distort labour and product markets.

Bhagwati is known for identifying a variety of policies which reduce welfare and labelling them in memorable phrases. Immiserising growth, in relation to FDI and Directly Unproductive Profit Seeking (DUP, pronounced DUPE) are all now a part of the lexicon of economics. DUP, as defined by Bhagwati (1982), refers to ways of making profits by undertaking activities which do not produce goods and services. A whole array of activities including tariff-seeking, lobbying, tariff evasion, procuring import licences and seeking premium prices for such licences and smuggling constitute DUP. All such activities, with rare exceptions, decrease welfare by drawing resources away from production-oriented, utility-enhancing activities into non-production-oriented, profit-seeking enterprise. Most DUP activities arise because of various sorts of distortions induced by policy interventions in trade.

An arresting feature of Bhagwati's work is its enduring quality; his insights into trade policy are applicable to the ever-changing scenarios in the world economy. One such scenario, which Bhagwati refers to as 'ironic role reversal', is the growing fear of globalisation in the developed countries, which mirrors the fears expressed by developing countries. The fears of developing countries concerning trade and foreign investment are well known; integration with the global economy would exacerbate income inequalities, perpetuate backwardness and result in dependency. The opposition of developed countries to free trade with the developing countries is based on the relatively low wages and poor standards relating to protection of the environment in the developing countries. Bhagwati has relentlessly questioned the wisdom of these arguments for protection and restated the proposition, which he enunciated in the 1960s, that market failures should be fixed with appropriate domestic policies and free trade should be maintained externally. His provocative latest book on globalisation (2004) denounces many popular fallacies associated with globalisation current in both the developed and developing countries.

During the late 1970s and the early 1980s, Bhagwati produced a series of articles and books analysing the welfare implications of emigration of talented people from developing countries to the developed world (1974, 1977). Here again his analysis questions received wisdom encapsulated in the so-called 'cosmopolitan model' of brain drain, which argues that the emigration of talented people improves welfare all round – of the emigrants, the countries from which they depart and those which receive them. Bhagwati detects a variety of situations in which the emigration of talented people could impose welfare costs on the countries from which they depart. These include situations in which the wage paid to the prospective emigrants is less than their social marginal product and in some cases less than their private marginal product also, and in cases where those who leave have to be replaced by others in the labour market with state-funded subsidies for training and education.

These and other welfare-reducing effects of the brain drain underlie Bhagwati's advocacy of a brains tax or what is now known as the 'Bhagwati tax', which requires skilled migrants to pay a special tax in addition to the taxes they pay in the countries to which they emigrate. The proceeds of the tax are to be transferred to their home countries. The economic rationale for a brains tax is that immigration of skilled people benefits developed countries, there are capital flows imbedded in the flow of skilled people from developing to developed countries and the former deserve to be compensated for this transfer of capital received by the latter. It is worth noting that Bhagwati's advocacy of a tax on skilled emigrants is based on considerations of equity and efficiency and not as a device for restricting emigration. The advocate of free trade is also an advocate of free flows of labour, be they of the skilled or the unskilled variety. He has frequently championed a liberal humanitarian approach to the regulation of illegal immigrants into the USA.

This brief overview of Bhagwati's contribution to the literature on international economics hardly does justice to the breadth of his scholarship, his many insights and his economic philosophy developed over three decades of research and teaching. His crusade for free trade, grounded in the belief that free trade is an important moral force for good, has evolved over long years of theoretical work and analysis of trade policy of both the developed and developing countries. These ideas have influenced economic policy both in India and abroad.

Major works

Bhagwati, J.N. (1958) 'Immiserising Growth: A Geometrical Note', *Review of Economic Studies* 25 (June): 201–5.

—— (1968) 'Distortions and Immiserising Growth: A Generalisation', *Review of Economic Studies* vol. 35, no. 104: 481–5.

—— (1971) 'The Generalised Theory of Distortions and Welfare', in J.N. Bhagwati *et al.* (eds), *Trade, Payments and Growth*, Amsterdam: North Holland, pp. 69–90.

—— (1973) 'The Theory of Immiserising Growth: Further Applications', in Swoboda, A. and Connolly, M. (eds), *International Trade and Money*, London: George Allen and Unwin, pp. 45–54.

—— (1977) 'The Brain Drain: International Resource Flow Accounting, Compensation, Taxation and Related Policy Proposals', Prepared for the Division of Transfer of Technology, UNCTAD.

—— (1978) 'Anatomy and Consequences of Exchange Control Regimes', *Studies in International Economic Relations*, 1(10): 232, New York: National Bureau of Economic Research.

—— (1982) 'Directly Unproductive Profit Seeking (DUP) Activities', *Journal of Political Economy* 90(5): 988–1022.

Bhagwati, J.N. and Hamada, K. (1974) 'The Brain Drain, International Integration of Markets for Professionals and Unemployment: A Theoretical Analysis', *Journal of Development Economics* 1(1): 19–24.

Bhagwati, J.N. and Ramaswami, V.K. (1963) 'Domestic Distortions, Tariffs and the Theory of Optimum Subsidy', *Journal of Political Economy* 71(1): 44–50.

Further reading

Balasubramanyam, V.N. (ed.) (1997) *Jagdish Bhagwati: Writings on International Economics*, Oxford: Oxford University Press.

Balasubramanyam, V.N., Salisu, M.A. and Sapsford, D. (1996) 'Foreign Direct Investment and Growth in EP and IS Countries', *The Economic Journal* 106(434): 92–105.

Bhagwati, J.N. (1998) *A Stream of Windows: Unsettling Reflections on Trade, Immigration and Democracy*, Cambridge, MA: MIT Press.

—— (2004) *In Defense of Globalisation*, Oxford: Oxford University Press.

V.N. Balasubramanyam

PIERS BLAIKIE (1942–)

Piers Blaikie has the reputation of an iconoclast. Frank Ellis, in a tribute when Blaikie retired from the School of Development Studies at the University of East Anglia (UEA) in 2003, wrote of a 'totally over-the-top character, given to extravagant gestures and even more extravagant juxtapositions of ideas'.[1] A second characteristic of Blaikie's academic career is his willingness, even delight, in up-ending and challenging his own assumptions and writings in order to push the research frontier on to new ground.

Blaikie took a first degree in Geography at the University of Cambridge in 1963. In 1964, funded by a Hayter Studentship, he began a PhD at Cambridge under the supervision of Professor Benny Farmer, spending a year in India and undertaking fieldwork in the Punjab and Rajasthan. He successfully defended his thesis on the 'Spatial organisation of some villages in Northern India' in 1967 (see Blaikie 1971). He then went on to join the Department of Geography at the University of Reading as a Lecturer, where he stayed until 1972. From 1972 through to 2003 Blaikie was successively Lecturer, Senior Lecturer, Reader and, from 1990, Professor in the School of Development Studies at UEA. It was here that he made his mark. He has also held visiting positions at the Norwegian University of Science and Technology in Trondheim; the Department of Geography at the University of California, Los Angeles; the Department of Geography at Berkeley; the East–West Center in Hawaii; and the Department of Geography at the University of Hawaii at Manoa. Since he published his first book in 1975, Blaikie has authored or co-authored nine major volumes. His journal output has been, by contrast, relatively modest: just thirty mainstream academic papers since 1970. It would seem, therefore, that Blaikie successfully resisted the publication demands of the UK's Research Assessment Exercise.

Blaikie's contributions to development thinking have been, in order of importance, in three key areas: environment, environmental processes and environmental change; agrarian change, particularly in Nepal; and AIDS and family planning.

Of all Blaikie's books, the one that has had the greatest impact on the most people is his *Political Economy of Soil Erosion in Developing Countries* (1985). Like many influential books, the core idea on which the wider argument rests seems, with hindsight, to be so obvious and unremarkable that it is difficult to believe that nobody had thought of it earlier. It is that soil erosion is not due to mismanagement, overpopulation or environmental context, but due to the tendency for surplus to be extracted from farming households by the nature of the political and economic system, requiring that farmers, in their turn, extract surplus from the land, leading to erosion and degradation. In this way, class relations become centrally implicated in the process. 'A principal conclusion of this book', Blaikie writes, 'is that soil erosion in lesser developed countries will not be substantially reduced unless it seriously threatens the accumulation possibilities of the dominant classes' (*ibid.*: 147).

Another surprising aspect of the book is that it was marketed as an introductory text for 'colleges and universities', is written in clear and accessible prose, and is just 188 pages in length. Despite its lack of pretentiousness, I can remember my own excitement when I read the book having just completed my PhD. Here was a social scientist invading the traditional turf of natural science to argue that a physical process – soil erosion – could only be understood in terms of political economy. Even now students in my lecture classes of 200 first-year undergraduates sit up and think afresh when I suggest to them that physical processes like soil erosion and land degradation require more than technical solutions and are products and reflections of the political, social and economic domains. This, of course, is common currency now but the book was seminal in bringing such views into the mainstream and helping to spawn a whole new subset of intellectual endeavour under the mantle of political ecology. Scholars like Nancy Peluso, Raymond Bryant, James Fairhead, Jules Pretty and Melissa Leach all owe an intellectual debt to this low-key and simply expressed undergraduate textbook. The book has been criticised for its rather essentialist and non-dynamic approach to politics (see Watts in *PIHG* 1997) but this seems to me to be expecting too much of a volume that was marketed as an introductory text for undergraduates and even harboured the hope that sixth-form school pupils would use it! There can be few seminal works (in 1997 it was labelled a 'classic' in human geography, see *PIHG* 1997) that have been so modestly clothed.

Blaikie further developed these ideas, with **Harold Brookfield**, in their part-jointly authored, part-edited *Land Degradation and Society* (1987). In this book the argument that land degradation is an interdisciplinary issue *par excellence* is promoted with vigour and the authors also propose that the task of explanation lies, in major part, with social scientists for the reason that natural scientists have extracted the process of land degradation from its political, social and economic context. The first sentence of the first chapter brazenly lays the authors' cards on the table: 'Land degradation should by definition be a social problem' (1987: 1).

More recently still, Blaikie co-authored with Terry Cannon, Ian Davis and Ben Wisner *At Risk: Natural Hazards, People's Vulnerability and Disasters* (1994). Like the earlier two volumes, *At Risk* arose out of shared dissatisfaction with prevailing views that disasters 'were "natural" in a straightforward way' (*ibid.*: xiii). In the book the authors connect risk to vulnerability and vulnerability to livelihoods and normal living: 'The crucial point about understanding why disasters occur is that it is not only natural events that cause them. They are also products of the social, political, and economic environment because of the way it structures the lives of different groups of people' (*ibid.*: 3). This book was updated and revised in 2003 (Wisner *et al.* 2003).

Less well known, but just as influential among a smaller group of scholars and development practitioners, is Blaikie's work (with collaborators) on Nepal. The main value, as I see it, in this series of books and papers is the longitudinal dimension. By longitudinal I don't just mean in the usual sense that Blaikie and his collaborators have returned to the field to update their work, but longitudinal also in terms of the explanatory framework employed to account for the patterns and processes identified.

In 1973 the Overseas Development Group at UEA were awarded funding from the Economic and Social Committee for Overseas Research (ESCOR) of the Ministry of Overseas Development (MOD) to evaluate the impact of roads in West-Central Nepal. The funders got more than they bargained for. The three project directors wrote a long, three-volume report which, as well as detailing the impacts of roads, also placed this squarely within a broader politico-economic context. The reports were later condensed into a single, more digestible summary, *The Effects of Roads in West-Central Nepal* (1977). Three years on from this, *Nepal in Crisis: Growth and Stagnation at the Periphery* (1980) was published. Neo-Marxist in vision and using the language of dependency, this book was a profoundly pessimistic view of Nepal's present and likely future. Blaikie and his co-authors believed that road-induced and market-led integration would 'not deliver the benefits of increased agricultural production, increased commercialisation, and trade as forecast in the economic appraisal documents' (Blaikie *et al.* 2002: 1256). Rather, the outcome would be a deepening dependency and growing underdevelopment. (One reason why the authorities banned it for a time was that Nepal's leftists scrambled to get hold of a copy and the UK's ambassador in Kathmandu wished it had never been written. It became, as the authors put it with understatement in a 2001 reprint, 'somewhat of a political "hot potato" ' (Blaikie *et al.* 2001: 276).)

Blaikie and his collaborators returned to the hills of Nepal in 1998 to review and update their conclusions (Blaikie *et al.* 2002). While not saying that they were wrong in their 1980 diagnosis (they were correct in predicting that agriculture would not be invigorated and that dependency would deepen), they do admit that events did not evolve in quite the way they anticipated. In particular, they failed to appreciate how the development of non-farm opportunities would make stagnation in agriculture less critical in livelihood terms. They conclude: 'The original model underestimated the capacity of the global labour market to provide work and remittances to sustain rural life and to stave off a more generalised crisis ...' (2002: 1268–9). This admission also, and interestingly, provides a valuable lens for looking again at *The Political Ecology of Soil Erosion*. Just as *Nepal in Crisis* failed to gauge the trajectory of global and national change and the degree to which individuals would be able to benefit rather than classes lose out (but who did?), so too with the soil erosion volume. The dependency perspective, the general pessimism, and the view that individuals as members of classes would be squeezed and controlled looks rather dated in the light of what has actually happened over the intervening years.

The final area where Blaikie has made a contribution to development thinking – although not, in my view, to the same level – is in family planning and AIDS research. Blaikie's interest in family planning can be traced back to 1970 when he took leave of absence to work on the diffusion of family planning (and agricultural) technologies in Bihar (India). Following a paper in *Population Studies* in 1970 (Blaikie 1970), Blaikie published a research monograph on *Family Planning in India: Diffusion and Policy* (1975). With its strong locational, model-building approach it was very much a product of its time. We see, once again, resonances with Blaikie's other work in his pessimistic take on the future and his view that people do not plan for families in the poor, rural Indian context largely for structural reasons. This research also provided the inspiration for a paper in *Progress in Human Geography* on innovation diffusion (Blaikie 1978). Blaikie's main contribution to research on AIDS was a book, co-authored with Tony Barnett, *AIDS in Africa: Its Present and Future Impact* (1992). This book was one of the first to draw on thorough fieldwork (in this instance in Uganda) to map out the likely dynamic of the disease.

Stepping back from the minutiae, it is possible to make three general statements about Blaikie's work and his contribution to scholarship. First, his publications show a clarity of argument and expression which is becoming increasingly rare. It would seem that his confidence that he has something important and valuable to say means he is not scared to say it simply. Second, Blaikie's work is invariably a careful blend of the empirical and the theoretical. He gets into the field, collects data and is involved at the sharp end of development thinking. But, as an academic and not a consultant, he also made it his mission to theorise up (as well as down) and to embed his results in wider considerations and perspectives. The third feature of his work is a common thread that runs through and links his best publications, whether on land degradation or agrarian change: a desire to challenge those who might apply simple solutions of a technical or managerial nature to 'problems' that have complex politico-economic roots. Time and again his work emphasises that 'space is what the political makes it' (1978: 289).

Note

1 Frank Ellis, 'Farewell to Piers Blaikie and Stephen Biggs', www.uea.ac.uk/der/newsletter/ellis.htm.

Major works

Blaikie, P. (1975) *Family Planning in India: Diffusion and Policy*, London: Edward Arnold.

—— (1985) *The Political Economy of Soil Erosion in Developing Countries*, Harlow, Essex: Longman.

Blaikie, P. and Barnett, T. (1992) *AIDS in Africa: Its Present and Future Impact*, London: Belhaven Press.

Blaikie, P. and Brookfield, H. (1987) *Land Degradation and Society*, London: Methuen.

Blaikie, P., Cameron, J. and Seddon, D. (1980) *Nepal in Crisis: Growth and Stagnation at the Periphery*, Oxford: Clarendon Press.

—— (2002) 'Understanding 20 Years of Change in West-Central Nepal: Continuity and Change in Lives and Ideas', *World Development* 30(7): 1255–70.

Blaikie, P., Cannon, T., Davis, I. and Wisner, B. (1994) *At Risk: Natural Hazards, People's Vulnerability and Disasters*, London: Routledge.

Wisner, B., Blaikie, P., Cannon, T. and Davis, I. (2003) *At Risk: Natural Hazards, People's Vulnerability and Disasters*, London: Routledge, 2nd edn.

Further reading

Blaikie, P. (1970) 'Implications of Selective Feedback in Aspects of Family Planning Research for Policy-makers in India', *Population Studies* 26(3): 437–44.

—— (1971) 'Spatial Organisation of Agriculture in Some North Indian Villages', *Transactions of the Institute of British Geographers* 53: 15–30.

—— (1978) 'The Theory of the Spatial Diffusion of Innovations: A Spacious Cul-de-sac', *Progress in Human Geography* 2: 268–95.

—— (2000) 'Development, Post-, Anti-, and Populist: A Critical Review', *Environment and Planning* A 32: 1033–50.

Blaikie, P., Cameron, J. and Seddon, D. (1977) *The Effects of Roads in West-Central Nepal*, Norwich: University of East Anglia.

—— (2001) *Nepal in Crisis: Growth and Stagnation at the Periphery*, New Delhi: Adroit Publishers, 2nd edn.

PIHG (1997) 'Classics in Human Geography Revisited', *Progress in Human Geography* 21(1): 75–80.

Jonathan Rigg

JAMES M. 'JIM' BLAUT (1927–2000)

Jim Blaut was an early and rigorous critic of modernisation theory and development policy. In common with another geographer, David Harvey, Blaut's work has influence far beyond that discipline. However, Blaut's critique of Eurocentrism and diffusionism as essential pillars of Western imperialist ideology and world view is quintessentially geographic. The works he produced during the last ten years of his life (Blaut *et al.* 1992; Blaut 1993, 2000; but see also his earlier embryonic works, 1969, 1970a, 1976, 1987a) were detailed, scholarly, yet polemical treatments of the origins and trajectory of a world system we know today as neo-liberal globalisation. In this he was a pioneer, with such authors as Immanuel Wallerstein, Edward Said, Franz Fanon, **Andre Gunder Frank** and **Samir Amin,** in conceptualising the world from the point of view of the subaltern, the colonised, and the oppressed.

Blaut argues that diffusionists believe that 'independent invention is rather uncommon, and therefore not very important in culture change in the short run and cultural evolution in the long run' (1993: 11). In fact, he finds that many historians and proponents of modernisation-as-development believe that 'only certain select communities are inventive' (*ibid.*: 12). Diffusionism's basic model assumes that Greater Europe has been, and is, the Inside (source of modernity and innovation) and that non-Europe is the Outside. This division and the supposedly innate superiority of Europe are based on the assumption that Europe enjoys a better mode of thinking (rationality) as well as better climate and better soils (Blaut 1994). Diffusionism then makes seven claims (*ibid.*: 14–17; see also Blaut 2000: 3–12):

- Europe naturally progresses and modernises.
- Non-Europe naturally remains stagnant, unchanging, traditional and backward.
- The basic cause of European progress is an intellectual or spiritual factor, a system of values.
- Likewise, the backwardness of non-Europe is attributable to the absence of these values, in particular the lack of rationality.
- Non-Europe progresses by the diffusion of innovation from Europe.
- The historical reverse flow of material wealth from non-Europe to Europe is justifiable as a partial repayment for the priceless gift of rationality, innovation and modernity.
- Non-material reverse flow from non-Europe to Europe is entirely made up of ideas that are ancient, savage, atavistic, uncivilised, or even evil.

This model of the world began to take shape in the fifteenth and sixteenth centuries, flowering fully in the eighteenth and nineteenth centuries. Although these propositions may seem extreme, a look at recent bestsellers such as Ben Barber's (1996) *Jihad vs. McWorld* or Thomas Barnett's (2004) *The Pentagon's New Map* reveals what Blaut exposed as 'inside' and 'outside' alive and well. Barnett calls these the 'core' and the 'gap'.

Blaut's treatment is characterised by an exceptional level of detail and meticulous historical research. Although only two of the planned three volumes in his project to demolish Eurocentrism were finished by the time of his death (Blaut 1993, 2000), he went a very long way toward that goal. For students of development, these two volumes provide essential vaccine against a new, virulent Crusader Virus that began to sweep the world from the US White House and Pentagon shortly after 11 September 2001. Although the notion of 'spatial modernisation' was unique to geography, and therefore a specialist and obscure concept, assumptions about diffusion and innovation run like fat throughout the marbled meat of development theory and practice. Probably the best-known canon of that school is **Walt Rostow**'s (1960) *The Stages of Economic Growth: A Non-Communist Manifesto*. Blaut's critique, therefore, provides as important a corrective lens as, for example, Wolfgang Sachs's archaeology of the concept of 'development' (Sachs 1999) or of Said's deconstruction of the idea of 'orientalism' (Said 1988).

What also sets Blaut's contribution apart is its range and methodological complexity, as well as its unity over nearly fifty years. Beginning his career in the 1950s and 1960s as a cultural geographer concerned with the practicalities of tropical agricultural production, he quickly began to criticise top-down attempts to 'modernise' peasant agriculture by 'tropicalising' the techniques of European and North American farming (Blaut 1967, 1970b). He also appreciated before many others the importance of what has become known as indigenous technical knowledge – and he employed, as did a few other geographers and anthropologists of his generation, techniques that would later become standard for researchers and NGOs alike under its contemporary name, participatory research. A good example is Blaut's study of farmer perceptions of the causes of soil erosion in Jamaica's Blue Mountains (Blaut 1959). Blaut had a life-long preoccupation with methodological and theoretical questions that balanced and informed his 'mud on the boots' fieldwork and approach to the practice of development. Therefore he also theorised his early work with peasant farmers (Blaut 1979).

His experiences with the rich knowledge possessed by farmers in Singapore, Jamaica, St. Croix, Costa Rica and Venezuela led Blaut to more general questions. How do children learn, informally, to become farmers?

How do children learn to map and to navigate about on the surface of the planet? These spawned a series of empirical studies and theoretical debates that began in the 1970s and were still ongoing at his death.

The significance for Blaut's contribution to development studies of this work on place learning is twofold. First, he and his close colleague, David Stea, discovered that mapping skills learned spontaneously very early in childhood are actually 'unlearned' during formal schooling (Blaut and Stea 1974; Blaut 1997; Sowden *et al.* 1997). Second, Blaut built on a series of cross-cultural studies of children's place learning (Blades *et al.* 1998) to develop a theory of mapping as a human universal (Blaut 1991; Stea *et al.* 1996). Eurocentrism and diffusionism – bedrocks of Western imperialism to the present day – necessarily privilege one group's knowledge and skill over another's. Just as those assumptions denigrated and ignored peasant farmers' local knowledge, so they also tend to ignore and, indeed, to destroy early childhood mapping skill. Some of the fiercest academic polemics to involve Jim Blaut centred not on his overt political ideas or anti-imperialism, but over whether (following the Swiss psychologist Piaget) children have built-in stages and limitations in their ability to understand spatial relations. Seeing this as yet another Western, rationalist idea, the roots of which go back to the eighteenth-century philosopher Kant, Blaut coined the phrase 'Can'tianism' – employing a rather esoteric pun (thus, the belief that 'children can't ...') and attempted to refute it, asserting, 'Children Can!' (Sowden *et al.* 1997).

Ultimately, Eurocentrism must deny human universalities such as mapping and spontaneous, local innovation by peasants. Contemporary imperialism may give lip service to the universality of human rights, but its Eurocentric assumptions demand that the West name, enumerate, and protect them. African philosophers – whose work Blaut knew – have also countered the claim by Western academic philosophy that African cultures have produced no 'philosophy' in the strict sense, only 'folk wisdom'. Thus one sees a unity in Blaut's contribution that embraces his studies of tropical agriculture, of place learning, and his critique of Eurocentrism.

Blaut integrated these themes nicely in a conference paper (Blaut 1994), noting that Eurocentric diffusionism leads to certain typical claims about tropical soils and family farmers in the tropics, namely:

- Soils are difficult to manage, and small-scale farmers destroy them when they try.
- European guidance is necessary to bring these soils into sustainable production.

From the 1930s onwards, tropical soil science was producing data that contradicted these claims, yet the dominant assumptions remain and still influence policy. In one striking example, he shows how 'irrational' refusal to drain hillside farms on the contour is based on local knowledge that such drainage is likely to lead to landslides.

Blaut's contribution to development studies is also noteworthy for the way in which he put his ideas into practice. He was a strong advocate of Puerto Rican independence from the United States and wrote extensively on the national question, publishing in both Spanish and English, in aid of that struggle (Blaut 1987b; Blaut and Figueroa 1988). He also wrote a good deal about ghettos as internal colonies, his intellectual efforts again guided by his solidarity with the cause of Puerto Ricans (and other minority, oppressed groups) (e.g. Blaut 1974). Blaut was also active promoting the rights of the Palestinian people and in support of the anti-apartheid struggle in South Africa.

Jim Blaut was born in New York City in 1927 and attended the University of Chicago at the young age of 16 (Matthewson and Stea 2003). He had the unique experience of studying at the Imperial College of Tropical Agriculture in Trinidad during the twilight of the British Empire. He then completed his PhD at Louisiana State University – writing at great length on a one-acre family farm in Singapore (Blaut 1953). While undertaking this fieldwork, he was an instructor at the University of Malaysia (1951–3).

Blaut later held academic positions at Yale (1956–61), Cornell (1960), Clark (1967–71) and the University of Illinois – Chicago Circle Campus (1972–2000). He maintained a strong interest in the Caribbean, often teaching and consulting there. Blaut served as agricultural consultant to the Venezuelan Government (1963–4) and as UNESCO advisor to the Dominican Republic's Planning Board in 1964. He taught at the University of Puerto Rico from 1961 to 1963 and again in 1971–4. He directed the Caribbean Research Institute based in the Virgin Islands from 1964 to 1966 and acted as a consultant to the institute from 1966 to 1967. As late as 1982, Blaut was back in the field, climbing on foot into the mountains of Grenada to interview young *rastamen* who produced charcoal for sale in Georgetown, once again using the gifts he had been developing for three decades for conversational-style interviews.

Blaut says of himself that he was an 'activist as a teenager in the Old Progressive Party (of Henry Wallace), working in Georgia' (Blaut 2005). He was later an activist in the Puerto Rican Socialist Party and was arrested (with a number of his graduate students) for blocking a draft board entrance in Worcester, Massachusetts as a Vietnam War protest (*ibid.*). Blaut was an enormously gifted and effective teacher who has left behind several generations of left-wing intellectual workers, spread throughout the world (Matthewson and Stea 2003; Wisner and Matthewson 2005).

The rewards for scholar-activism in bourgeois society include academic censure. Blaut was denied promotion and tenure at Clark University despite arriving at the decisive faculty meeting literally with a wheelbarrow filled with his publications. The Graduate Student Association at Clark University responded by voting no confidence in the then director of the Graduate School of Geography, Saul Cohen. Later in life Blaut did receive recognition, receiving the Association of American Geography's 'Distinguished Scholar of the Year' award in 1997.

Major works

Blaut, J.M. (1953) 'The Economic Geography of a One-Acre Farm in Singapore: A Study in Applied Microgeography', *Journal of Tropical Geography* 1(1): 37–48.

—— (1983) 'Nationalism as an Autonomous Force', *Science and Society* 46: 1–23.

—— (1987a) 'Diffusionism: A Uniformitarian Critique', *Annals of the Association of American Geographers* 77: 30–47.

—— (1987b) *The National Question: Decolonising the Theory of Nationalism*, Atlantic Highlands, NJ: Zed Books.

—— (1993) *The Colonizer's Model of the World: Geographical Diffusionism and Eurocentric History*, New York: Guilford Press.

—— (1994) 'Eurocentrism, Diffusionism, and the Assessment of Tropical Soil Productivity', paper read at the International Conference on Advances in Tropical Agriculture in the Twentieth Century and Prospects for the Twenty-First Century, University of the West Indies, Trinidad, 4–9 September.

—— (1997) 'Piagetian Pessimism and the Mapping Abilities of Young Children', *Annals of the Association of American Geographers* 87, 1: 168–77.

—— (2000) *Eight Eurocentric Historians*, New York: Guilford.

—— (2005) 'My Legacy (Assessed by Me, Jim Blaut, on 16 September, 2000. Incomplete and Hasty with Some Redundancy)', in Wisner, B. and Matthewson, K. (eds), *The Work and Legacy of J.M. Blaut*, special issue of *Antipode*, 37(5), November.

Blaut, J.M., with contributions by S. Amin, R. Dodgshon, A.G. Frank, and R. Palan, and with an introduction by P.J. Taylor (1992) *Fourteen Ninety-Two: The Debate on Colonialism, Eurocentrism, and History*, Trenton, NJ: Africa World Press.

Further reading

Barber, B. (1996) *Jihad vs. McWorld: How Globalism and Tribalism Are Reshaping the World*, New York: Ballantine.

Barnett, T. (2004) *The Pentagon's New Map: War and Peace in the 21st Century*, New York: Putnam.

Blades, M., Blaut, J.M., Darvizeh, Z., Elguea, S., Sowden, S., Soni, D., Spencer, C., Stea, D., Surajpaul, R. and Uttal, D. (1998) 'A Cross-Cultural Study of Young Children's Mapping Abilities', *Transactions, Institute of British Geographers* NS 23: 269–77.

Blaut, J.M. (1959) 'A Study of Cultural Determinants of Soil Erosion and Conservation in the Blue Mountains of Jamaica', *Social and Economic Studies* 8: 403–20.
—— (1967) 'Geography and the Development of Peasant Agriculture', in Cohen, S.B. (ed.), *Problems and Trends in American Geography*, New York: Basic Books, pp. 200–20.
—— (1969) 'Jingo Geography: Part I', *Antipode* 1(1): 10–13.
—— (1970a) 'Geographic Models of Imperialism', *Antipode* 2(1): 65–85.
—— (1970b) 'Realistic Models of Peasant Agriculture', in Field, A.J. (ed.), *Town and Country in the Third World*, Boston: Schenkman, pp. 213–24.
—— (1974) 'The Ghetto as an Internal Neocolony', in Morrill, R. and Eichenbaum, J. (eds), *New Perspectives in Urban Location Theory*, special issue of *Antipode* 6(1): 37–42.
—— (1976) 'Where Was Capitalism Born?', *Antipode* 8(2): 1–11.
—— (1979) 'Some Principles of Ethnogeography', in Gale, S. and Olsson, G. (eds), *Philosophy in Geography*, Dordrecht: Reidel, pp. 1–7.
—— (1991) 'Natural Mapping', *Transactions, Institute of British Geographers*, NS 16(1): 55–74.
Blaut, J. and Figueroa, L. (1988) *Aspectos de la cuestión nacional en Puerto Rico*, San Juan: Editorial Claridad.
Blaut, J.M. and Stea, D. (1974) 'Mapping at the Age of Three', *Journal of Geography* 73(7): 5–9.
Matthewson, K. and Stea, D. (2003) 'James M. Blaut (1927–2000)', *Annals of the Association of American Geographers* 93(1) (March): 214–22.
Rostow, W.W. (1960) *The Stages of Economic Growth: A Non-Communist Manifesto*, Cambridge: Cambridge University Press.
Sachs, W. (1999) 'Archeology of the Development Idea', in Sachs, W. (ed.), *Planet Dialectics: Explorations in Environment and Development*, London: Zed Books, chap. 1, pp. 3–22.
Said, E. (1988) *Orientalism*, New York: Vintage.
Sowden, S., Blaut, J., Stea, D., Blades, M. and Spencer, C. (1997) 'Children Can', *Annals of the Association of American Geographers* 87(1): 152–8.
Stea, D., Blaut, J.M. and Stephens, J. (1996) 'Mapping as a Cultural Universal', in Portugali, J. (ed.), *The Construction of Cognitive Maps*, Boston: Kluwer.
Wisner, B. and Matthewson, K. (eds) (2005) *The Work and Legacy of J. M. Blaut*, special issue of *Antipode* 37(5), November.

Ben Wisner

NORMAN BORLAUG (1914–)

During the 1960s, rapid increases in agricultural yields, particularly in wheat and rice, in parts of Asia and Latin America were heralded as a solution to the problems of hunger facing millions of the world's poorest people. The use of high-yielding varieties (HYVs), chemical fertiliser, irrigation and machinery was termed 'the Green Revolution'. Debates about the relative successes and failures of these changes in agricultural production continue, but Norman Borlaug, sometimes termed 'the Father of the Green Revolution', remains almost unknown.

Borlaug was born in Cresco, Iowa, USA on 25 March 1914 and grew up on a farm. His future interests were strongly influenced by this environment. During the 1930s he studied at the University of Minnesota, where he gained a BSc in Forest Management in 1937, followed by a Masters in Plant Pathology in 1939 and a doctorate in 1942. His doctoral thesis was on a common fungus, rust, which attacks a wide variety of crops. His work focused on the movement of rust spores and found that they could travel vast distances. He then worked as a microbiologist for the du Pont de Nemours Foundation in Delaware, where his research concentrated on agricultural chemical products such as fungicides and bactericides.

Borlaug's scientific research focused on how technology could improve agricultural yields. His convictions regarding the role that science could play in agriculture were based not only on this work, but also on his observations growing up in the US Midwest. The 'Dust Bowl' of the 1930s is often used as an example of how farming methods inappropriate for a particular physical environment can cause long-term environmental, as well as social, damage. As the economic depression of the 1930s worsened, farmers sought to increase yields by intensifying production. This left large areas of land without appropriate vegetation cover, leading to high levels of soil erosion. Borlaug observed that it was not the use of scientific agricultural techniques which caused these problems, but their misuse or lack of use. He argued that those farmers who adopted appropriate scientifically informed practices did not suffer the same losses.

The 1940s saw the beginning of large-scale 'development assistance' from the global North to global South. In 1943, the Rockefeller Foundation, in conjunction with the Mexican government, set up the Comparative Wheat Research and Production Program in Mexico. Borlaug went to Mexico in 1944 to become head of this Program. This provided him with the opportunity to put his ideas into practice.

The Program was meant to concentrate on teaching Mexican farmers how to improve their agricultural techniques, but under Borlaug's leadership it also developed a very strong focus on innovation. Borlaug was determined to breed new forms of wheat that would help increase yields and reduce risk for poor farmers. These innovations included developing a strain of wheat (*ceredo*) which was insensitive to the number of hours of sunlight in a day and, most famously, varieties of dwarf wheat. Borlaug and his fellow researchers argued that traditional wheat's long stalks limited yields, partly because of the energy expended on growing the inedible long stalks rather than the ears of wheat, and also because tall stalks often got damaged in wind or rain, so making harvesting more difficult. Dwarf strains had much higher yields if grown with appropriate fertiliser and irrigation. Experiments on dwarf rice strains were being conducted at the same time at the International Rice Research Institute in the Philippines and at China's Human Rice Research Institute.

In 1963 the Mexican Program was transformed into a new institution known as the International Centre for the Improvement of Maize and Wheat, often referred to by its Spanish acronym CIMMYT (Centro Internacional de Mejoramiento de Maíz y Trigo). Given the success of the dwarf wheat programme in Mexico, Borlaug was impatient to transfer the technology and practices to other parts of the world where starvation and hunger were much more widespread. This led him to focus on India and Pakistan. At that time, seed distribution was controlled by state companies, so he focused his attention on these organisations. Borlaug's focus on wheat, rather than indigenous crops of the subcontinent, was driven not because he felt that wheat was intrinsically better than lentils or other local staples, but rather because high-yielding varieties of indigenous crops had not been developed, wheat can grow in a wide variety of physical environments, and it provides significant calories. This latter point was key as Borlaug's mission was to address the perceived mismatch between population size and food supply.

Unsurprisingly, there were significant obstacles to his attempts to introduce HYVs to India and Pakistan as it represented a mammoth shift in cultural acceptance and understandings of the role of particular foodstuffs, as well as farming methods. Borlaug would not give up and eventually the governments agreed to limited adoption of HYV wheat because of the widespread famine in parts of their countries. Borlaug often argued that the military hostilities between India and Pakistan in 1965 meant that he was able to introduce his ideas with little government interference once approval had been given. As well as using HYVs, irrigation was important, as was the use of inorganic fertiliser.

Yields increased very rapidly and by 1968 Pakistan was self-sufficient in wheat and by 1974 India was self-sufficient in all cereals. These figures were particularly timely as neo-Malthusian ideas about the 'population timebomb' were becoming increasingly widespread in the global North. India was often used as an example of how rapid population growth was outstripping food supply and would lead to widespread famine, disease and war. Borlaug argued that **Malthus**'s predictions did not take into account scientific advances in food production such as those he promoted. However, Borlaug was, and continues to be, concerned about population growth and its implications for food security.

In 1970, Borlaug was awarded the Nobel Peace Prize. This reflected the awareness that precarious food supply can lead to major tensions and violence between individuals, communities and countries. In his acceptance speech, Borlaug stated that 'food is the moral right of all who are born into this world'. He acknowledged that people had other rights as well, but that 'without it [food] all other components of social justice are meaningless'. The acceptance speech demonstrates his passion for practical and effective research, and his drive to address problems of food supply throughout the poorest regions of the world. His commitment to developing research capacity in the countries of the South is also apparent. For example, in his overview of the early work of the Mexico Program, he stressed the provision of training and fellowship programmes, emphasising that 'researchers in pursuit of irrelevant academic butterflies were discouraged' (Borlaug 1970). Borlaug's own career focus on the practical implementation of scientific developments, rather than the writing of academic papers, perhaps partially explains his low profile among the academic community working on 'development', particularly in the global North.

Borlaug retired in 1979, but this did not represent an end to his work. Since then he has been particularly involved with research and projects to promote improved agricultural practices in sub-Saharan Africa. Despite earlier financial support from charitable organisations such as the Rockefeller and Ford Foundations and multilateral agencies including the World Bank, a significant amount of Borlaug's more recent work has been funded by a Japanese foundation. In 1986 he helped launch the Sasakawa-Global 2000 Program. The work of this Association involves Green Revolution-style projects in numerous countries in sub-Saharan Africa. Borlaug is also Distinguished Professor in the Soil and Crop Sciences Department at Texas A & M University.

Borlaug and supporters of his methods claim that the increasing reluctance of US-based organisations to fund his work reflects the growing power of environmental movements. The use of non-indigenous crops and HYVs, as well as the increased use of irrigation, pesticides and inorganic fertiliser has been blamed for decreasing soil fertility, water pollution, soil erosion and other environmental problems, particularly in marginal environments. In addition, while the Green Revolution may have made dramatic increases in yields at the start, these increases are impossible to sustain. Finally, the social impacts of the Green Revolution have been highlighted. In the vast majority of cases Borlaug is not criticised directly, but rather the changes in agricultural practices which he promoted. For example, given the costs of HYVs, pesticides and inorganic fertilisers, the Green Revolution, it is argued, has often exacerbated existing class and caste divisions. While richer farmers can afford these new inputs and benefit from increased yields, poorer farmers are left behind and may have to sell their land and become landless labourers or urban migrants.

Hunger and starvation remain daily realities for millions of people in the global South today. New agricultural technologies and the possibilities opened up by genetic modification are, some argue, the answers to these problems. Borlaug supports the use of genetic modification, arguing that opponents who claim such processes are 'unnatural' fail to understand the genetic mixing that happens in nature without human interference. He is also passionate in his criticisms of many groups and individuals who, he claims, lobby against the use of pesticides, fertilisers and GM crops from the 'comfort' of Europe or the USA.

These views, however, do not necessarily reflect the criticisms that certain forms of agricultural practice and the use of GM seeds, in particular, have received from people living in the global South. In India and Mexico, for example, despite past crop yield increases, there have been widespread protests against GM crops. Borlaug admits that at times scientists have not been successful in presenting their work effectively, implying that if people were presented with the 'facts' then there would be no protests.

When discussing his current work in sub-Saharan Africa, as in an interview in 2000 (Bailey 2000), Borlaug mentions continued obstacles to meeting individuals' food needs. While agricultural technology may improve yields, there are issues of distribution to consider. This involves not only physical distribution in terms of infrastructure, but also social distribution, encompassed in **Amartya Sen**'s ideas of entitlement. Sufficient food may be produced, but if you do not have enough money to buy it then you will still starve. For some development theorists and practitioners, it is these issues of distribution which should now get far greater attention than the focus on increasing global food supply.

Norman Borlaug's commitment and passion in addressing a key development debate of how to feed an increasing population with a finite amount of land is admirable. His focus on implementing projects, influencing governments and working with local people demonstrates not only his scientific credentials, but also his ability to adapt and work within specific social and cultural contexts. While the enthusiasm that greeted the 'Green Revolution' in the 1960s and 1970s may have been overly optimistic, the Green Revolution's contribution to saving millions of people from starvation cannot be underestimated. Borlaug's Nobel Prize acceptance speech stressed that he was just one part of a large team which deserved recognition. This is certainly true, but without Borlaug's vision and determination the team's results may have been less successful and much less widely implemented.

Major works

Borlaug, N. (1970) 'The Green Revolution: Peace and Humanity', speech on the occasion of the awarding of the 1970 Nobel Peace Prize. Available from www.theatlantic.com/issues/97jan/borlaug/speech.htm.
—— (2002) 'Agriculture and Peace: The Role of Science and Technology in the 21st Century', U Thant Distinguished Lecture Series, United Nations University Tokyo, October. Available from www.unu.edu/uthant_lectures/.

Further reading

Bailey, P. (2000) 'Billions Served', *Reason*, April. Available from reason.com/0004/fe.rb.billions.shtml.
CIMMYT www.cimmyt.org/.
Easterbrook, G. (1997) 'Forgotten Benefactor of Humanity', *The Atlantic*, January 1997. Available from www.theatlantic.com/issues/97jan/borlaug/borlaug.htm.
The Norman Borlaug Heritage Foundation: www.normanborlaug.org.
The Sasakawa Africa Association: www.saa-tokyo.org.

Katie Willis

ESTER BOSERUP (1910–99)

Ester Boserup was a development economist, who worked as a civil servant for two decades and later as an independent researcher and consultant for the United Nations and its agencies concerning issues and problems in developing countries.

Boserup was born in Copenhagen on 18 May 1910, joined the University of Copenhagen in 1929 and graduated in 1935 as 'cand. polit.'. The main emphasis of her studies during the Great Depression was theoretical economics but she also attended lectures in sociology and agricultural policy. Part of her degree work led to a paper comparing **Marx** and Keynes, a shorter version of which was published in the Danish economic journal, *Nationaløkonomisk tidsskrift* (Boserup 1936). She involved herself for a time with a small group of independent socialist intellectuals and later continued to participate in the socialist Danish *Clarté* movement.

After graduation she worked for twenty years as a civil servant, first a decade in the Danish economic administration[1] and then, from 1947, in Geneva with the Research and Planning Division of the Economic Commission for Europe, where her work contributed greatly to the early success of the annual Economic Surveys.

In 1957 she and her husband, Mogens, moved to New Delhi, accepting a proposal from **Gunnar Myrdal** to engage in a joint study of South and Southeast Asian agriculture. She travelled extensively within India, discussing with Indian agricultural experts and advisers and experiencing unfamiliar agricultural systems. She soon came to the opinion that 'the generally accepted theory of zero marginal productivity and agrarian surplus population in densely populated developing countries was an unrealistic theoretical construction' (Boserup 1999). Her and her husband's developing scepticism towards prevailing development theory put them at odds with Myrdal's evaluation of the Asian situation. Unable to continue in the joint study, they returned to Denmark when their contract expired in May 1960, after delivering the agreed chapter to *Asian Drama* (Myrdal 1969).

Thereafter, she accepted short-term consultancies but continued to reflect on her experiences from Asia and consequently wrote the controversial book *The Conditions of Agricultural Growth* (1965). In this book she argued that population growth is the major cause of agricultural change and that the principal mechanism of change is the intensification of land use through increases in the frequency of cropping. While such intensification would not occur without population growth, which reduces labour productivity – a contentious claim that remains unproven (Grigg 1979) – it is, she argues, a necessary collective response to population pressure. Furthermore, such intensification leads to technological advances, such as the adoption of new fallowing systems, which require new tools and techniques and which shape institutions, land tenure systems and settlement forms (Grigg 1979). This theory overturns the direction of causation implicit in **Malthus**ian and neo-Malthusian approaches, which see technological change as an autonomous process inducing rather than proceeding from population growth. Economists were generally unenthusiastic about Boserup's model but it refocused many modern debates and stimulated further thought on the issues (Giovanni 2001). Other responses, such as expansion at the extensive margin, adoption of new crops or out-migration, can also relieve population pressure, and other influences such as urban growth or development of trade can stimulate agrarian change, yet the thesis remains 'a fruitful interpretation of agrarian change' (Grigg 1979). In spite of all its shortcomings, *The Conditions of Agricultural Growth*, which Boserup was invited to discuss at the International Geographical Union in 1967, remains a small masterpiece of enduring importance.

In 1964–65 she spent a year in Dakar, where her husband was director of the United Nations Institut de Développement Economique et de Planification (IDEP). During this period she also became a consultant to the Center for Industrial Development (later UNIDO) and started focusing on the linkages between industry and agriculture and both rural–urban and rural–rural migration. Her earlier work had denied the Malthusian claim that marginal productivity would be reduced to zero on agricultural land and she now emphasised the role of desire for money incomes rather than land shortage in generating the large labour migration from poor subsistence villages to urban, mining and export-producing centres. She was intrigued by the importance of the gender division of labour to the pattern of African migration. Predominantly male migrants could leave wives and children to take care of themselves for long periods, since in traditional villages it was the women and children who supplied nearly all the subsistence food and fetched water and fuel for cooking. Her interest led her to gain funds to gather data on the hours worked by men and women in both paid and unpaid work, including domestic services, the gathering of fuel and water and so on, in addition to agricultural work and non-agricultural production and services.

Many students of gender studies will have first come across Ester Boserup's work through her well-known book, *Woman's Role in Economic Development* (1970), which formed the output of this project and provided intellectual inspiration for awareness of this important topic. Her work sought to make the 'invisible woman' visible by drawing attention to the economic contribution made, for example, by women in agricultural families, virtually all of whom, despite their economic productiveness, were formally classified as outside the labour force, as housewives or 'non-active' persons. She highlighted the role of what she called the 'bazaar and service sector', stressing that what many considered as largely concealed unemployment was in fact production of necessary goods and services with preindustrial technology catering largely for the market of low income groups (1970: 175–86). Such activities were later coined as the 'informal sector' by the United Nations' International Labour Organisation (ILO). This highlights how effortlessly she integrated into theory, analysis and policy prescription the roles of women as both producers and entrepreneurs and as consumers and reproducers, thereby demonstrating her ability to bring the theoretical rigour of a critical economist to bear on case study material.

In 1970 she received the Rosenkjaer Pris of the Danish Broadcasting Company and took the opportunity to popularise her ideas on economic development and the sexual division of labour in six radio broadcasts. She also suggested that improvement of women's education and status might lead to a reduction of family size (Boserup 1970: 224–5). Such ideas have been gradually moving into the mainstream among agencies concerned with development and the empowerment of the poor. This had influence on her being appointed as a vice chairman in 1972 of the UN Symposium on Population and Development in Cairo, in preparation for the UN World Population Conference in Bucharest in 1974.

In 1981 she published the book *Population and Technological Change: A Study of Long-Term Trends*, in which she extended the scope of her earlier theories on population growth and adoption of new tools and techniques in agriculture to include non-agricultural activity. Throughout history, both the invention of new technologies and their spread have been demand-induced. Even the most 'primitive' people speculate and experiment with new methods, materials and weapons, when the motivation for change becomes strong. Here population growth plays a significant role. She suggested that many technologies can be properly exploited only if the population is dense enough. Population growth makes urban civilisation possible. Population size enables labour specialisation and thus allows economies of scale to be realised in industrial production.

Through the following years she continued to research the interrelationships between population, family structure, status of women, national culture and economic development, both in rural and in urban areas, always arguing for recognition of the complexity of the issues taking *all* the relevant determinants into consideration, instead of using simplified demographic models. The role of fertility remained a central concern and she argued that high fertility may result in both higher agricultural production and higher investments and savings in kind.

Her contribution was recognised through the award of three honorary doctorates: in Agricultural Sciences by the Agricultural University of Wageningen, The Netherlands, in 1978; in economics by the University of Copenhagen in 1979; and in Human Sciences by Brown University, Providence, USA in 1983. In 1989 she was elected Foreign Associate of the National Academy of Sciences in Washington on an interdisciplinary vote. She spent the rest of her career as a consultant and independent researcher.

It is important to put her work into the context of the period when she worked. She was in a small minority both as a female civil servant and development economist.[2] Much of her work appeared as discussion papers which informed UN publications and lectures at symposia and invited conferences. She challenged conventional wisdom first with regard to the relations between agrarian systems and population, and later with regard to women in development and technological change. Her models have stood the test of time. Her ideas and arguments were mainly exposed to the academic community through her much later work in the 1970s and 1980s (aged in her sixties and seventies) mainly through her three books (1965, 1970, 1981) and some journal articles. All this was managed without the benefit of a university post. Shortly before her death on 24 September 1999, she published an unpretentious and objective historical account of all her published work (Boserup 1999). Ester Boserup's work provoked multidisciplinary debates and has influenced various disciplines, particularly anthropology, demography, agricultural economics, geography and sociology.

Though she was an economics graduate, she saw the limitations of narrow economic analysis. She emphasised this especially for least-developed countries for which the use of monetised transactions for production, consumption, investment and income (both personal and national income) as proxies for totals including non-monetised transactions led to false conclusions (Boserup 1996). She was at times very critical of classical economics and its assumption of fixed capacities of land, labour and capital, all subject to variable capacity utilisation. She argued for long-term analysis focusing on changes in the capacities themselves and emphasised the need for this to address changes in those structures which economists usually leave to be studied by other scientific disciplines, for instance national cultures. Such models therefore need – like her own – to be based on interdisciplinary understanding and in later papers she attempted to lay out an overarching framework for long-term interdisciplinary modelling (Boserup 1996). Such a framework involves delineation of the flows of pressure between six structures: environment; population; technology level; occupational structure; family structure; and culture.

Her work concentrated on examining human society and its processes of social change, especially in the context of dynamic relationships between natural, economic, cultural and political structures, while evaluating both economic and non-economic factors. Her analyses of rapid technological change are even relevant today in the context of the processes of globalisation and social change in developing countries. Reflecting on technological changes in the 1970s and 1980s, she concluded that rapid technological change has created conflicts with national culture through its radical influence on the way of life: cultural attitudes and behaviour, which may have been rational before, are no longer so. But many groups and governments attempt to bring about economic development and technological change without cultural change, and this leads to serious conflicts within and between countries. The importance of these problems for economic development is overlooked by economists, when they make the assumption that rational behaviour is the rule whatever the circumstances (Boserup 1995). She elaborated on the positive and negative effects of technological change both in the short and long run, and on the need for government control of the private sector, both nationally and internationally, to ensure that it uses sustainable technology.

Notes

1 As a chief of the planning office she was responsible for estimating future export earnings and suggesting the distribution of import licences by commodity and country, discriminating against luxuries and items that could be obtained by

Danish production and against countries which did not buy much from Denmark.

2 Women have probably made greater inroads into rhetoric, but there are still concerns regarding the inclusion of women in the material as well as the personnel of economics. See website of the Royal Economic Society, Women in Economics www.res.org.uk. At the 2002 World Bank ABCDE seminar, Ravi Kanbur pointed out that most of the economics profession still adheres to and teaches the unitary 'black box' model of the household, which effectively makes gender allocations a private matter.

Major works

Boserup, E. (1936) 'Nogle centrale økonomiske spørgsmål I lys af den marxistiske teori' (Some Central Economic Issues in Light of Marxian Theory), *Nationaløkonomisk Tidsskrift* 75: 421–35.

—— (1965) *The Conditions of Agricultural Growth: The Economics of Agriculture under Population Pressure*, London and Chicago: George Allen and Unwin. (Reprinted by Earthscan, London, 1993.)

—— (1970) *Woman's Role in Economic Development*, Earthscan, London.

—— (1981) *Population and Technological Change: A Study of Long-Term Trends*, Chicago: University of Chicago Press.

—— (1995) 'Obstacles to Advancement of Women during Development', in T. Paul Schultz (ed.), *Investment in Women's Human Capital*, Chicago, IL: University of Chicago Press, pp. 51–60.

—— (1996) 'Development Theory: An Analytical Framework and Selected Applications', *Population and Development Review* 22(3): 505–15.

—— (1999) *My Professional Life and Publications 1929–1998*, Copenhagen: Museum Tusculanum Press, University of Copenhagen.

Further reading

Giovanni, F. (2001) 'Review of Ester Boserup, "The Conditions of Agricultural Growth: The Economics of Agrarian Change under Population Pressure" ', Economic History Services 16 April 2001 (URL: http://www.eh.net/bookreviews/library/federico.shtml).

Grigg, D. (1979) 'Ester Boserup's Theory of Agrarian Change: A Critical Review', *Progress in Human Geography* 3(1): 64–83.

Myrdal, G. (1969) *Asian Drama: An Enquiry into the Poverty of Nations*, 3 vols, New York: Twentieth Century Fund.

Schultz, T.P. (1990) *Economic and Demographic Relationships in Development*, Baltimore, MD: Johns Hopkins University Press. (Articles selected, edited and introduced by T. Paul Schultz.)

Vandana Desai

HAROLD BROOKFIELD (1926–)

It is rare for geographers to be seen as critical thinkers on development issues. Indeed, even this collection, edited by a geographer, includes just three. Of these Harold Brookfield is assuredly the doyen. Indeed, one of the others, **Terry McGee**, described Brookfield as 'the acknowledged "guru" of development geography' (1978: 71). His contributions to development studies range from a much lauded book on interdependent development, which emphasised the necessity for development to be seen in a diversity of interlocking ways, a longstanding focus on local studies (tied to the belief that indigenous knowledge is a key to development) and the environment and sustainability. Such diverse interests were shaped around a search for pragmatic solutions and the eschewing of arid theorising, dogma, ever-changing semantics and political correctness. While some contemplate the lives and livelihoods of those they refer to as 'others', Brookfield preferred the language of everyday life, of ordinary people and their problems, and was unafraid to use words such as poverty.

Harold Brookfield grew up in southern England, completing his BA and PhD degrees at the London School of Economics. The topic of his PhD thesis, post-eighteenth-century urban development in coastal Sussex, completed in 1950, may partly have accounted for the peripatetic life that ensued. After briefly being Lecturer in Geography at Birkbeck College, London, he became Lecturer in charge of the infant Department of Geography at the University of Natal. There he first began to engage with development and social justice issues concerning South Africa and Mauritius. After nearly three years he left for the University of New England (UNE) in Australia, the first of many moves to and within Australia that eventually led to him spending over half his life in Australia, and has made him the most distinguished of contemporary Australian geographers. At UNE, in the small town of Armidale, he found attitudes to the indigenous Australian population very similar to those in South Africa (Rugendyke 2005). From there he went on to the Australian National University (ANU), eventually his longstanding academic home, though Chairs soon followed at Pennsylvania State University, McGill University and Melbourne University, with a two-year period as a Fellow at the Institute of Development Studies, University of Sussex. By 1982 he had returned to ANU where he remained, publishing substantial works well into official retirement.

Early work on the historical geography of southern England and Ireland was soon followed by detailed work in South Africa, where he wrote an early critique of apartheid (Brookfield 1957), and in Mauritius where he continued his interest in the evolution of plural societies. The shift to ANU in 1957 brought the start of a lifelong fascination with Papua New Guinea and detailed fieldwork in the highlands of PNG where contact had been relatively recent. Collaboration with the anthropologist Paula Brown brought his first major book on the region (Brookfield and Brown 1963). Other books on Pacific market places, and the extraordinarily innovative and successful text *Melanesia* (Brookfield with Hart 1971), marked his association with the region and emphasised his own commitment to both small-scale local studies and the comparative method, 'characterised by the careful collection of data and intimate knowledge of a region built up through many years' (McGee 1978: 70). In some respects that commitment became the hallmark of work on the Pacific region, celebrated in his edited collection *The Pacific in Transition* (1973) – mainly the work of his own students and colleagues. The focus on PNG was so strong that the prominent American geographer, Marvin Mikesell, identified a 'New Guinea syndrome' where 'this once remote and mysterious island has played a role in the recent history of cultural geography comparable to the influence of Sauer and his students from Mexico' (1978: 8).

As his colleague Oskar Spate observed, this work was valuable because of 'the realisation that other people also had ways of life which were not based on classical economics, but were not irrational and were tied up with their culture and environment and hence the Brookfield style with its close links to anthropology' (Rugendyke 2005). A focus on problems, the comparative method and indigenous authority was innovative. Brookfield himself believed that a key contribution was the 'bringing of geographers and geographical work into contact with that of other disciplines which were concerned with people and land, anthropology and pre-history, in particular' (*ibid.*). Despite a focus on comparison, Brookfield largely avoided quantitative research methodologies. He was critical of the study of patterns and processes of development derived from statistical methods, arguing for 'a process-oriented discipline in which the question "How?" should be at least as important as the question "Why?" '. In addressing the former question, he 'found that it was far easier to seek cause and effect at microscale, where individual decision makers could be seen in action. I therefore advocated this scale of work to my colleagues' (Brookfield 1984: 35). Remarkably, in this sense, his empirical focus at the local scale, on the interactions between people, their cultures and their environments, rather than on quantitative methodologies and spatial models, pre-empted post-modern reflections that 'there is no universal or unique understanding of development or the environment' (Simon 2003: 35).

Brookfield's work in Papua New Guinea especially was with a thoroughly multidisciplinary group of prehistorians, economists and anthropologists and, while at the research school in Canberra, 'transdisciplinary exchange was as important as interdisciplinary discussion' (Brookfield 1984: 27). He has said, 'I became very much associated with anthropologists, so much that at one stage Anthony Forge said to me that "you are an anthropologist who earns his pay masquerading as a geographer" ' (Rugendyke 2005). Indeed, it is not surprising that Brookfield chose to title his recollections of that period of geography the 'Experiences of an Outside Man' – an iconoclast tangling with anthropology and other disciplines and aghast at the irrelevance of the quantitative revolution. Culture is written on the land and Brookfield has said he would most like to be remembered for his work in cultural ecology.

Soon afterwards his work extended to the smaller islands of the Caribbean (though he wrote little directly on this) and later to the smaller eastern islands of Fiji. Perhaps surprisingly he never directly addressed issues related to size other than in one book chapter (Brookfield 1975b). Simultaneously the wide-ranging reflections on development that were evident in *Melanesia* were developed further in the companion volume *Colonialism, Development and Independence: The Case of the Melanesian Islands in the South Pacific* (1972). Together they drew together strands of culture, history and geography, in a broad political, social and economic context, at a time when the Melanesian nations were taking the first faltering steps towards independence. These diverse issues came together at a world scale in his magisterial *Interdependent Development* (1975a) a *tour de force* later to become a 'classic in human geography revisited' (Corbridge 1996; O'Connor 1996). That book began a theoretical backlash against the irrelevance of several geographical traditions, and argued the case for a more historical and environmental perspective, allied, however, to an empirical position that was to become stronger in the future.

Brookfield always emphasised that his own principal long-term interest was the 'adaptation of the use and management of land to variability and change in society, economy and natural environment': the 'soul' of geography (Brookfield 2004: 40). That was evident and two further books established his reputation in this area. The first, largely neglected because of its seemingly peripheral focus, was an intricate account of the lives of villagers in the eastern islands of Fiji which brought together a number of human and physical geographers, while drawing on the preliminary work of economists and others (Bayliss-Smith *et al.* 1988). In some respects that was a local version of the jointly written *Land Degradation and Society* (Blaikie and Brookfield 1987) that had appeared in the previous year. There and elsewhere Brookfield stressed the interdisciplinary analysis of environmental issues, arguing that processes of land degradation cannot be understood if divorced from their political, social or economic contexts. In its turn it too became a classic to be reprinted three times (see **Blaikie** in this volume). By then the focus on the environment, the *leitmotiv* of Brookfield's work, had been drawn into wider debates over development. *Interdependent Development* was described as 'a striking anticipation of green development thinking' (Corbridge 1996: 86) and what sounded 'remarkably like a call for what would now be called "sustainable development" ' (O'Connor 1996: 88).

While Brookfield never lost interest in the Pacific, and especially PNG, his interests began to shift towards Asia, and a long period of work in Malaysia brought a renewed focus on environmental issues. It also heralded work on both urbanisation and industrialisation (Brookfield *et al.* 1991; Brookfield 1994) in a nation that was changing much faster than the island states of the Pacific and elsewhere. Renewed focus on the environment resulted in a book on environmental change in Borneo and Malaysia (Brookfield *et al.* 1995) and a growing emphasis on agrodiversity. That led to Brookfield's commitment to the more than decade-long United Nations University's project on People, Land Management and Environmental Change (PLEC) that has preoccupied him ever since.

The PLEC project, an acronym very close to the Tok Pisin (PNG) word for village, *ples*, began in 1992 – the year of the Earth Summit in Rio de Janeiro – as an initiative to examine relationships between population and environmental change in rural areas. The original focus was on research, centred around five 'clusters' – PNG, Amazonia, West and East Africa and montane south-east Asia – but it evolved into a practical concern for agrodiversity, defined in 1994 as 'the many ways in which farmers use the natural diversity of the environment for production, including not only their choice of crops but also their management of land, water and biota as a whole' (Brookfield and Padoch 1994: 9). Four years later, PLEC joined the Global Environment Facility (GEF) and became a large demonstration and capacity-building project, involving the participation of small farmers as experts in conservation and sustainable development, alongside indigenous researchers and institutions. Again this project invoked the comparative method, to 'bring together "bottom-up" research in different regions' (*ibid.*), and enable hypothesis generation, with an emphasis on indigenous management practices. At a time when others might have retired, or merely gazed from distant offices, Brookfield continued to emphasise and demonstrate the value of fieldwork and commitment. Amidst many papers and reports, three more books emerged from the PLEC project (Brookfield 2001; Brookfield *et al.* 2002, 2003); these are most unlikely to be the last words.

For half a century Brookfield has worked in the people-environment tradition, with one reviewer noting his 'many original and insightful contributions to issues of tropical human ecology and development' which 'furnishes significant conceptual gunpowder for the long-awaited resurrection of the human-environment tradition in geography' (Airriess 1996: 3). As many geographers sometimes uneasily shifted their focus from troubling development issues, Brookfield remained true to the needs of the discipline, and allied disciplines, ever seeking to raise their focus beyond their navels. He emphasised the necessity for theory to be closely linked to practice, triumphed the ethnographic method that gave voice to the participants in development, cherished interdisciplinary approaches, led the way in recognising the relationship between cultural ecology, the environment and development and raised awareness of humanistic values.

Major works

Bayliss-Smith, T.P., Bedford, R.D., Brookfield, H.C. and Latham, M. (1988) *Islands, Islanders and the World: The Colonial and Post-colonial Experience of Eastern Fiji*, Cambridge: Cambridge University Press.

Blaikie, P. and Brookfield, H. (eds) (1987) *Land Degradation and Society*, London and New York: Methuen.

Brookfield, H.C. (1957) 'Some Geographical Applications of the Apartheid and Partnership Policies in Southern Africa', *Transactions of the Institute of British Geographers* 23: 225–47.

—— (1972) *Colonialism, Development and Independence: The Case of the Melanesian Islands in the South Pacific*, Cambridge: Cambridge University Press.

—— (1973) *The Pacific in Transition*, London: Arnold.

—— (1975a) *Interdependent Development*, London: Methuen.

—— (1975b) 'Multum in parvo: Some Questions about Diversification in Small Countries', in P. Selwyn (ed.), *Development Policy in Small Countries*, London: Croom Helm, pp. 54–76.

—— (1984) 'Experiences of an Outside Man', in Billinge, M., Gregory, D. and Martin, R., *Recollections of a Revolution: Geography as Spatial Science,* London: Macmillan Press, pp. 27–38.

—— (1994) *Transformation with Industrialisation in Peninsular Malaysia*, Kuala Lumpur: Oxford University Press.

—— (2001) *Exploring Agrodiversity*, New York: Columbia University Press.

—— (2004) 'American Geography and One Non-American Geographer', *GeoJournal*, 59: 39–41.

Brookfield, H.C. and Brown, P. (1963) *Struggle for Land: Agriculture and Group Territories among the Chimbu of the New Guinea Highlands*, Melbourne: Oxford University Press.

Brookfield, H.C., Hadi, A.S. and Mahmud, Z. (1991) *The City in the Village*, Kuala Lumpur: Oxford University Press.

Brookfield, H.C. with Hart, D. (1971) *Melanesia: A Geographical Interpretation of an Island World*, London: Methuen.

Brookfield, H.C. and Padoch, C. (1994) 'Appreciating Agrodiversity: A Look at the Dynamism and Diversity of Indigenous Farming Practices', *Environment* 36(5): 6–45.

Brookfield, H.C., Padoch, C., Parsons, H. and Stocking, M. (2002) *Cultivating Biodiversity*, London: Intermediate Technology Publishing.

Brookfield, H.C., Parsons, H. and Brookfield, M. (2003) *Agrodiversity: Learning from Farmers Across the World*, Tokyo: United Nations University Press.
Brookfield, H.C., Potter, L. and Byron, Y. (1995) *In Place of the Forest: Environmental and Social Transformation in Borneo and the Eastern Malay Peninsula*, Tokyo: United Nations University Press.

Further reading

Airriess, C. (1996) Review of Brookfield, H.C., Potter, L. and Byron, Y., 1995: *In Place of the Forest: Environmental and Socio-Economic Transformation in Borneo and the Eastern Malay Peninsula*, in *The Geographical Review* 86: 611–13.
Connell, J. (1998) 'Development Studies in Australian Geography', *Australian Geographical Studies* 26: 157–71.
Connell, J. and Waddell, E. (2005) 'Introduction: That Most Remarkable of Outside Men – Harold Brookfield's Intellectual Legacy', *Asia Pacific Viewpoint* 46(2): 192–33.
Corbridge, S. (1996) Classics in Human Geography Revisited, on Brookfield, H., 1975, *Interdependent Development*, Commentary 1, *Progress in Human Geography* 20: 85–6.
McGee, T. (1978) 'The Geography of Development: Some Thoughts for the Future', *Australian Geographer* 14: 69–71.
McTaggart, D. (1974) 'Structuralism and Universalism in Geography: Reflections on Contributions by H.C. Brookfield', *The Australian Geographer* 12: 510–16.
Mikesell, M. (1978) 'Tradition and Innovation in Cultural Geography', *Annals of the Association of American Geographers* 68: 1–16.
O'Connor, A. (1996) Classics in Human Geography Revisited, on Brookfield, H., 1975, *Interdependent Development*, Commentary 2, *Progress in Human Geography* 20: 87–8.
Rugendyke, B. (2005) 'W(h)ither Development Geography in Australia?', *Geographical Research* 43(3): 306–18.
Simon, D. (2003) 'Dilemmas of Development and the Environment in a Globalising World: Theory, Policy and Praxis', *Progress in Development Studies* 3(1): 5–41.

John Connell and Barbara Rugendyke

FERNANDO HENRIQUE CARDOSO (1931–)

Fernando Henrique Cardoso was born in Rio de Janeiro, Brazil in 1931 and trained as a sociologist at the University of São Paulo. He became a prolific author and one of the most distinguished social scientists in the world, but also a successful political leader who served as President of Brazil. His contributions to development thinking are often referred to his study of dependency and development in Latin America during the late 1960s and early 1970s. Less attention has been devoted to the evolution of his ideas around development during his transition from scholar to political leader since the 1980s. Many of those ideas appeared in speeches, interviews, and in the column he wrote regularly in newspaper *Fohla de São Paulo* for many years. They can also be found in the policies he implemented during his administration as president of Brazil. His academic and political careers have generated considerable scholarly commentary. Cardoso's work includes more than 200 articles, books and book reviews, as well as hundreds of speeches, interviews and journalistic notes. His most important speeches are available through the internet and many publications focusing on his work are available in English. A large database of his life and work is available at the Instituto Fernando Henrique Cardoso (http://www.ifhc.org.br/acervo_i.htm).

His best-known book, *Dependency and Development in Latin America*, written with Enzo Faletto in 1969, addresses the problems of economic development through an interpretation emphasising the political character of the process of economic transformation. Cardoso and Faletto (1969) incorporated in their analysis the historical situation in which the economic transformations occurred in order to better understand these changes and their structural limitations. Cardoso argued that the situation of dependency was neither stable nor permanent. Although he recognised that the dynamic elements of international capitalism lie outside the periphery and its control, he argued that dependent development should not be interpreted as a lack of dynamism in parts of the periphery (Resende-Santos 1997). Cardoso recognised the changing nature of international capitalism and that the reconfiguration of the international distribution of labour opened opportunities for the periphery (Cardoso 1973; Goertzel 1999). For Cardoso, dependency was about position and function within the international economy, about how Latin American countries participated in the global economy (Cardoso 1982). He rejected the assertion that dependency produced underdevelopment. For him, dependency was fundamentally a question of power and domination, internationally and locally (Cardoso and Faletto 1969; Cardoso 1973).

The dependency and development thesis has had a significant impact on generations of scholars concerned with development, but as mentioned above, it would be a mistake to base Cardoso's contribution to the study of development only on that thesis. Cardoso developed further elements of his thesis as central elements in his understanding and practice on development. His attention to social as well as institutional analysis allowed him to identify that politically, dependent development in Brazil produced institutional deformities that constituted important obstacles to democracy and development (Font 2001). Brazil's political system was characterised by its clientelism, corporatism and corruption and Cardoso was an outspoken advocate of fundamental institutional restructuring in Brazil. He was convinced of the importance of strengthening democratic practices in order to assure a stable transition to the next stage in Brazil's development. For him, many of the solutions to the defects in Brazil's political and economic system were bound up in his notion of social democracy (Resende-Santos 1997).

Cardoso became progressively drawn to an active role in the pro-democracy movement in Brazil during the 1970s. His commitment to open opportunities for development in Brazil led him to take an active role in the restructuring of political institutions. This approach was conducive to the political career he initiated in 1978 when he was elected an alternative senator, and later a senator in 1983. Cardoso's involvement in understanding and shaping the process of political and structural reform in Brazil led him to become a founder member of the Brazilian Social Democratic Party (PSDB) in 1982 and to a successful career as a senator. Following a short spell as Foreign Minister in 1992/3, his successful performance as Brazil's finance minister in controlling hyperinflation in the early 1990s positioned Cardoso as a popular presidential candidate. Cardoso was elected and served two terms as President of Brazil (1995–2003). His political philosophy stressed change and the need continuously to redefine ideas and political agendas as circumstances change. That philosophy allowed him to maintain an efficient relationship between ideas and policy choices.

It is precisely the bridge between ideas and the practice of development that targeted criticism at Cardoso. For many of his critics, Cardoso abandoned his **Marx**ist roots to become a defender of liberal democracy and capitalist development. One set of criticisms accused him of embracing political institutions he once criticised. This is direct reference to his early work in the 1960s and 1970s when Cardoso allegedly regarded political parties and other political institutions as formal expressions of class domination; he is criticised for becoming involved in those same political institutions since the late 1970s (Packenham 1992).

A second set of criticisms focuses on Cardoso's term as president of Brazil, labelling him as a neoliberal serving the interests of the multinational business elite. These criticisms focused on Cardoso's macro-economic policies, including the privatisation of state-owned enterprises (Lesbaupin 1999). Scholars studying Cardoso's work and life have suggested these criticisms are based on misconceptions of Cardoso's ideas. For Goertzel, Cardoso was interested in Marxism as social theory, not as a political dogma (1999). The analysis of Cardoso's presidential administrations in Brazil documents his achievements in critical areas for national, regional and local development (Amann and Baer 2000; Dantas 2003; Goertzel 2003).

Resende-Santos (1997) suggests that Cardoso's critics had a difficult time reconciling the progressive social philosophy and reformist idealism that in their view defined Cardoso's early years as a scholar with the neoliberal political leader. For Resende-Santos, this is due to a misunderstanding of Cardoso's political philosophy that stresses change and the need for constant redefinition of ideas and political agendas as circumstances change:

> Cardoso has always rejected dogmatic thinking, especially as displayed in the political projects of both the left and the right in Latin America. He has always maintained that ideas must change as circumstances change ... Cardoso is, and has always been, much more a pragmatist than a revolutionary. His is a moderate, gradualist agenda that stresses what is viable, rather than one committed to absolutes.
>
> (Resende-Santos 1997: 146)

Cardoso's scholarly work and political career illustrate his contributions to the study of development and invite reflection on our own understanding of this concept. Cardoso conceptualises development as a dynamic multidimensional process that does not propose a finished model but a road towards the transformation of society. He emphasises three central dimensions of that process: social equity, institutional rebuilding leading to democracy, and alternatives leading to sustainable economic growth. Cardoso's attention to the dynamic interactions among these dimensions illustrates his understanding of the complex reality that development seeks to address.

For Cardoso, the deficiencies in Brazil's political and economic system are central to understanding the problem of social equity. Social democracy makes it possible to correct those deficiencies and improve social equity. But social democracy has only an instrumental value in the sense that it creates a path towards the transformation of society (Resende-Santos 1997). A central element in addressing Brazil's economic and political deficiencies is the role of the state. For Cardoso, the main reason for state reform is not neoliberal ideology, as many critics have suggested, but rather a profound general crisis of the state. The state remains an important actor in development, but it needs to redefine its mission, refocus, and increase its effectiveness according to the changing conditions in the world economy and within Brazilian society (Font 2001). Cardoso realised that in order to compete successfully in the global economy, Brazil required internal reforms to enhance the process of international insertion.

He considers that the periphery is more in danger of being left behind and excluded from participating in the global economy than of remaining dependent as a supplier of raw materials. For him, Brazil's best chances of a successful future lay in a strong market economy, an efficient regulatory state and effective social programmes. The development policies that he implemented during his presidency of Brazil reflect those ideas. Cardoso accepted the leading economic role of the market while maintaining that it does not address all the needs of society, creates problems of its own, and tends to dissolve human solidarity. For him, the state plays a fundamental role in reducing inequalities, poverty and other social problems. But Cardoso understood that development also needed a strong civil society:

> One has to overcome the simplistic notion that what is in the best interest of the citizenry has to originate necessarily in the State. The mechanism for regulating 'privatised public' activities must be guided by the needs of the people, and to that end, the direct participation by representatives of civil society in the bodies concerned is fundamental. The State must be porous and permeable to the needs of the citizens. The identification of the State with national interest is a political construction work that requires major efforts to reach consensus ... In sum, politics is less 'the art of the possible' and more the 'art of making possible that which is necessary'.
>
> (Cardoso 2001: 286)

In summary, Cardoso has made two significant contributions to our understanding and practice of development that have not yet been adequately assessed. In terms of ideas, Cardoso has had remarkable clarity in conceptualising development as a dynamic multidimensional process, clearly identifying its critical dimensions and the interactions among them. There are two principal reasons why scholars have not devoted enough attention to Cardoso's perspective on development. First, he has not taken the time to present his conceptual framework explicitly in an integrated form. Second, the evolution of his critical thinking around development has not appeared in traditional academic publications. The most interesting part of that evolution is when he became a practitioner of development and had the opportunity to confront, first-hand, his ideas with reality. Of course, the best way he had to express his thoughts on development as president of Brazil were his speeches in domestic and international events.

The second contribution is his experience and performance as a development practitioner. The practice of development in the context of crises deserves better attention to enhance our understanding of opportunities and obstacles for social change. Poverty, inequality, deficiencies in the political system, financial crises and the pressure from a challenging range of domestic and international interests created obstacles to putting his ideas on development into practice. Clearly his administration was not perfect, but it is encouraging to see positive results conducive to social change in Brazil.

The lack of attention to Cardoso's legacy in the practice of development is in part due to the focus on the short-term performance of his administration's policies rather than to their contribution to a comprehensive process of development. Time is proving Cardoso right. Despite criticisms of his social and economic policies, many of these have been maintained by Luiz Inacio Lula da Silva, Cardoso's successor as president and leader from the political left.

A final issue worth noting about Cardoso's legacy is his remarkable balance in resisting the corrosive effect of power and pursuing good governance. He regarded the presidency as a political institution with the obligation to account for its acts. This contrasts with many other Latin American presidents. In times when development is contested as a concept guiding transformations of society, Cardoso's legacy should be considered an inspiration to strengthen the ideas and practice of development as a tool for social change. Cardoso himself insists on having his performance judged on its ability to set Brazil on a new course.

Major works

Cardoso, F.H. (1973) *Estado y Sociedad en America Latina*, Buenos Aires: Ediciones Nueva Vision.

—— (1982) *Politicas Sociales para la Decada en America Latina,* Montevideo: Centro Latinoamericano de Economia Humana.

—— (2001) 'Radicalizing Democracy', in Font, M. (ed.), *Charting a New Course; The Politics of Globalization and Social Transformation: Fernando Henrique Cardoso*, Lanham, MD: Rowman & Littlefield, pp. 284–8.

Cardoso, F.H. and Faletto, E. (1969) *Dependencia y Democracia en America Latina*, Mexico: Siglo XXI Editores. A Portuguese version (*Dependencia e Desenvolvimiento na America Latina*) appeared in 1970, while the University of California Press published an updated English version (*Dependency and Development in Latin America*) in 1979. The book has also been translated into French, German and Italian.

Further reading

Amann, E. and Baer, W. (2000) 'The Illusion of Stability: The Brazilian Economy under Cardoso', *World Development* 28(10): 1805–19.

Dantas, F. (2003) 'After Eight Years, Lula Inherits a Better Country. Brazil Advanced During Fernando Cardoso's two terms as President but it is still fragile and unequal', http://www.estadoa.com/eleicoes/governolula/noticias/2003/jan/01/24.htm. Last accessed 20 September 2004.

Font, M. (2001) 'Introduction. The Craft of a new era: the intellectual trajectory of Fernando Henrique Cardoso', in Font, M. (ed.), *Charting a New Course; The Politics of Globalisation and Social Transformation: Fernando Henrique Cardoso*, Lanham, MD: Rowman & Littlefield, pp. 1–34.

Goertzel, T. (1999) *Fernando Henrique Cardoso: Reinventing Democracy in Brazil*, Boulder, CO: Lynne Rienner.

—— (2003) 'Eight Years of Pragmatic Leadership in Brazil', a supplement to Fernando Henrique Cardoso: Reinventing Democracy in Brazil, http://crab.rutgers.edu/~goertzel/fhc.htm. Last accessed 10 September 2004.
Lesbaupin, I. (organizador) (1999) *O desmonte da nacao. Balanco do Governo FHC*, Petropolis: Editora Vozes.
Packenham, R. (1992) *The Dependancy Movement: Scholarship and Politics in Development Studies*, Cambridge, MA: Harvard University Press.
Resende-Santos, J. (1997) 'Fernando Henrique Cardoso. Social and Institutional Rebuilding in Brazil', in Domínguez, J. (ed.), *Technopols: Freeing Politics and Markets in Latin America in the 1990s*, University Park, PA: Pennsylvania State University Press, pp. 145–94.

Roberto Sánchez-Rodríguez

MICHAEL CERNEA (1934–)[1]

'Development anthropology is a contact support,' Michael Cernea tells his students. His career, from Researcher in the Romanian Academy of Sciences in the late 1950s and early 1960s to Senior Social Advisor for Sociology and Social Policy in the World Bank in the 1990s, is testament to this observation. Cernea's has been a professional life characterised by constant high-stakes struggles over social development ideas within different bureaucratic and political settings. His has been a career where the thinking through of ideas, and the acting upon them, have been one and the same process – as a thinker in development, Cernea has to be understood as much in terms of his relationships to particular institutions as to development anthropology and sociology. Indeed, many might argue that Michael Cernea's most critical contribution has not been just the body of intellectual work that has elaborated on the place of culture, social relationships and institutions in development, on participation and local knowledge, or on the inherence of risk in the development process, but rather the embedding of those ideas in that behemoth of development, the World Bank. From the 1970s to the mid-1990s, Cernea contributed more than anybody in pushing the 'social envelope' at the Bank, his mantle now arguably assumed by one-time co-author, co-conspirator and good friend, Scott Guggenheim. He had the ear of Presidents and Vice-Presidents in the process, and from time to time was unrepentant in giving those ears a good chewing. He has been one of those quintessential reformists (and at times small 'r' revolutionaries) inside the Bank recognised by outside commentators and analysts as critical to any form of pro-poor, pro-environment and pro-civil society change in the institution (Fox and Brown 1998).

How far personal history determines subsequent careers is always a matter of interpretation but in Michael Cernea's case there are at least apparent continuities. He grew up as part of the Jewish community in Jassy (Iasi), a Romanian town close to the Russian border. Persecution was early, real and immediate: he was kicked out of primary school for being a Jew, and during the infamous pogrom of Jassy witnessed – albeit as a child – mass killings and his father badly beaten. The experience of persecution and resettlement was immediate and palpable – he was resettled twice himself, first by the pogroms and then in the face of the advancing German army. The experiences left him politically active and sensitised and after the war he became a member of the Socialist Youth Movement and a young journalist. Though initially a supporter of the Romanian Left's rise to government power in the post-war years, he grew more concerned with time at the directions taken by the regime and the society it governed. His PhD thesis, 'About Dialectics in the Socialist Society' (University of Bucharest, 1962), in which he tried to legitimise theoretically the persistence of contradictions, conflicts and tensions within Romania's allegedly homogeneous society and structures, reflected these growing concerns. That he chose such a topic also presaged strategies of enquiry later in life at the World Bank – strategies in which he aimed to change an institution by challenging its foundational ideas from within that same institution.

As would later be the case at the Bank, this strategy generated resistance and so – coupled with the effects of anti-Semitism in the Romania of the late 1950s – he had to wait four years simply to schedule the defence of his PhD. Still, by the time he made his defence, the effects of the Khruschev thaw had changed the political context in Romania again, and Cernea was finally able to address the topic that really moved him, peasant economic rationality, although not without political fallout. This early work (Cernea 1970, 1971) operated at the boundaries of anthropology, sociology and rural economics, exploring cultural dimensions of productive strategies, and the relationships between peasant rationalities and co-operative agriculture. This concern for the socio-cultural foundations of the economy would reappear later in a different guise in his sustained questioning of what he referred to as the 'economic reductionism' of the World Bank (see Cernea 1994 and below).

As the thaw continued, foreign scholars visited Romania in increasing numbers, a process that ultimately came to define a turning point both in Cernea's career (see his Malinowski Award lecture, Cernea 1996) and in the future that the World Bank was yet to have. Cernea was nominated by one of these visitors for a fellowship at the Center for Advanced Studies in the Behavioral Sciences (CASBS) at Stanford, one of the most prestigious academic fellowships in the USA. The politics behind Cernea being able to assume the fellowship were equally convoluted, but shortly following the meeting of Ceauşescu and Nixon (a meeting itself triggered by Ceauşescu's resistance to the Russian invasion of Czechoslovakia), Cernea finally received the visa allowing him to travel to California, albeit only in time for the latter part of the fellowship. His time there (1970–1) led to the distillation in English of his ideas on peasant society, but more importantly, to a friendship with US sociologist, Robert Merton, and a growing reputation in the USA.

The links between the year at Stanford, his initial contacts with the World Bank, and his being interviewed for a position in Washington are neither direct nor without intrigue, but were very real. Whatever the case, in 1974 – following a talk given at Bank headquarters on the role of the family plot in collective farms – he was offered a job. His position was to be the first sociologist to work in McNamara's newly created central Rural Development Division, a division that in many respects occupied the pivotal piloting role in McNamara's rural development-led approach both to poverty reduction (and, if more implicitly, to the taming of rural radicalisation) and to changing the Bank. Cernea took the Bank – much as he did Romania – as simultaneously an institution he worked (and largely lived) in, as an object of analysis and as something he wanted to change. He believed – then and still now – that such change could (only?) come from the injection of sociological knowledge into the very foundations of the institution's way of interpreting and acting on the world.

In time these convictions led him to study the Bank's project cycle in depth and explore the entry points for sociological knowledge in the processes through which the Bank designed and implemented its operations. As part of this project, he seized on the seminar as an instrument of institutional change and began inviting outside social scientists to speak on the different ways in which sociological knowledge and an awareness of the social dimensions of development could be brought into the project cycle. This cycle of seminars culminated in *Putting People First* (Cernea 1985), which – notwithstanding its focus on the World Bank – became one of the early foundational texts on participatory rural development. Cernea's introduction to the collection argued that the World Bank was in the business of financially induced development, but that for any such induced change to be successful it was imperative that the Bank understand the social structures that existed in the area of intervention and that could play the roles required of them in this process of induced development: echoes of Merton's influence (Cernea 1985). Social science knowledge – and social scientists – were thus essential for the Bank (Cernea 1994).

This was a recurrent theme in his work: social science was to be conducted not for its own sake but for development's sake, and one of the key purposes of such social science was to challenge, constantly, the ideas underlying economists' and others' models of development. While the argument would be made intellectually – and Cernea has been a prolific writer – most strategically for him, it had to be made bureaucratically. His analysis was that if arguments about social science (or anything else) were to have any teeth in the Bank, they had to be turned into bureaucratic instruments that made such knowledge a requirement of normal practice in the institution. Thus, another area in which Cernea contributed greatly was in developing and implementing operational directives and operational policies that required projects to have social appraisals. This contribution, if less visible to outside readers, was critical if any institutional change was to derive from his ideas. The easier part of this process was to generate the ideas and write them; the harder part was to get them through the Bank's approval process and then, once approved, keep them alive in the face of the constant pressure from other parts of the Bank to get rid of them.

The culmination of these arguments was a series of events beginning in late 1995 on the occasion of the Bank's formal recognition of Cernea's twenty years working for the institution. At the event, he caught the new Bank President, James Wolfensohn, and managed to persuade him of why it was that social development had to be central to the poverty orientation Wolfensohn was promoting. This led to a series of exchanges in which Cernea convinced Wolfensohn to create a social development task force that would report on the state of social development work in the institution. Again negotiated and contested – 'development anthropology is a contact support' – this was a process that the Development Economics Vice-Presidency and two Managing Directors were able to contain and to some degree capture. But not entirely, for in 1997, and against the advice of the Bank's Chief Economist, Wolfensohn approved the creation of a Social Development Department in the Bank.

Cernea's work has also pushed thinking on the links between culture and development, on cultural patrimony and on the environment, on the concept of social policy and on articulating various social policies. However, of most significance perhaps – both intellectually and also for human welfare – has been his work on involuntary resettlement, risk and vulnerability. Cernea joined the Bank during the period in which the large-scale rural development and infrastructure programmes that would later bring the institution into such criticism, were being hatched. As early as 1978 he began work on guidelines on resettlement, subsequently issued as Bank policy (1980). Not long after came a sharp fight within the Bank over the provisions for resettlement in India's infamous Narmada Dam project. This became the seed for a review of the Bank's overall performance on resettlement (Cernea 1986) and two subsequent documents giving policy guidelines for involuntary resettlement (Cernea 1986, 1988). This latter piece has been translated into Bahasa, Turkish, Chinese, Italian, Spanish and

French and has had over a dozen print runs. If the 1986 review was critical, a later 1993–94 review (Cernea, Guggenheim and Associates 1994) pulled even fewer punches. It systematically drew attention to the failures of the Bank to follow its own policy. It also constituted one of Cernea's key – empirically sustained – statements on the way in which development inherently brings risk to people at the same time as bringing possibilities of opportunity.

The resettlement reviews were not anti-development statements in *any* sense at all, but they *were* statements that made clear – empirically, intellectually and with forceful bureaucratic power – that development processes handled irresponsibly, without adequate social science insight, without adequate consultation and involvement of poor people, are likely to increase vulnerabilities, and very often for the poorest. The intellectual core of this argument was later captured in his 'Impoverishment Risks and Reconstruction Model' (Cernea 1997, 2001, 2002, 2004; Cernea and McDowell 2000) of the eight risks that he came to identify as inherent in the process of displacement: landlessness, joblessness, homelessness, marginalisation, food insecurity, loss of access to common property resources, increased morbidity and community disarticulation (Cernea 1997; also Mahapatra 1999). The model typifies Cernea's intellectual and professional project in that it offers a framework intended not only to help understand risk, but also – through identifying and making explicit such risks – to trigger a response to such risk such that it might be mitigated even before its effects fully manifest themselves. Its purpose is to be at once analytical, predictive and methodologically useful. It has become a leading model used internationally in research and policies about development-triggered involuntary resettlement (Koenig 2002).

Cernea's lasting contributions to development are many, reflected in a long list of publications, advisory roles and academic honours. Perhaps the most important among them, however, will be to have changed an institution through the sustained and forceful insistence on a few ideas: that social science knowledge is critical to development, that induced development will fail in any meaningful sense if ordinary people are not involved in shaping the forms it takes, and that the cherished disembodied concepts of so much development theory cannot be considered separately from the social structures in which they are embedded. Whether and for how long the Bank will continue to reflect these victories on the battlefields of knowledge depends on geopolitics as much as intellectual debate, but *even if* there is a reversal in the shifts that Cernea helped win, to have changed the institution's practices during the three decades he has spent there means that his ideas, and his work, will have affected literally millions of lives forever.

Note

1 I am very grateful to Michael Cernea for the interviews, materials and time he gave me during the preparation of this entry.

Major works

Cernea, M. (1970) *Two Villages, Social Structures and Technical Progress* (in Romanian, senior author, research co-ordinator), Bucharest: Edit. Politica.

—— (1971) *Changing Society and Family Change: The Impact of the Cooperative Farm on the Traditional Peasant Family*, Stanford, CA: Center for Advanced Studies in Behavioral Sciences.

—— (ed.) (1985) *Putting People First: Sociological Variables in Rural Development Projects*, New York: Oxford University Press.

—— (1986) *Involuntary Resettlement in Bank-Assisted Projects: A Review of the Application of Bank Policies and Procedures in FY79-95 Projects*, AGR, Operations Policy Staff, World Bank, February.

—— (1988) *Involuntary Resettlement in Development Projects: Policy Guidelines for Bank-Financed Projects*, Washington, DC: World Bank.

—— (1994) 'A Sociologist's View on Sustainable Development', in Serageldin, I. and Steer, A. (eds), *Making Development Sustainable: From Concepts to Action*, Environmentally Sustainable Development Occasional Paper Series No. 2. Washington, DC: World Bank.

—— (1996) *Social Organization and Development Anthropology. The 1995 Malinowski Award Lecture*, ESD Studies and Monographs Series, No. 6. Washington, DC: World Bank.

—— (1997) 'The Risks and Reconstruction Model for Resettling Displaced Populations', *World Development* 25(10): 1569–88.

—— (2001) 'Eight Main Risks: Preventing Impoverishment During Population Resettlement', in de Wet, C. and Fox, R. (eds), *Transforming Settlement in Southern Africa*, Edinburgh: Edinburgh University Press for the International African Institute, pp. 237–52.

—— (2002) *Cultural Heritage and Development: A Framework for Action in the Middle East and North Africa*, Washington, DC: World Bank.

—— (2004) 'The Typology of Development-Induced Displacements: Field of Research, Concepts, Gaps and Bridges', paper to the US National Academy of Sciences Conference on the Study of Forced Migration, 22–23 September, Washington, DC.

Cernea, M. and Guggenheim, S. (eds) (1993) *Anthropological Approaches to Resettlement: Policy, Practice, and Theory*, Boulder, CO: Westview Press.

Cernea, M., Guggenheim, S. and Associates (1994) *Resettlement and Development. The Bankwide Review of Projects Involving Involuntary Resettlement 1986–1993*, Washington, DC: World Bank.

Cernea, M. and McDowell, C. (eds) (2000) *Risks and Reconstruction: Experiences of Resettlers and Refugees*, Washington, DC: World Bank.

Further reading

Fox, J. and Brown, D. (eds) (1998) *The Struggle for Accountability: The World Bank*, Cambridge, MA: MIT Press.

Koenig, D. (2002) *Toward Mitigating Impoverishment in Development-Induced Displacement and Resettlement*, Oxford: Refugees Studies Centre, University of Oxford.

Mahapatra, L.K. (1999) *Resettlement, Impoverishment and Reconstruction in India*, New Delhi: Vikas.

Anthony Bebbington

ROBERT CHAMBERS (1932–)

Robert Chambers has been at the forefront of attempts since the early 1980s to place the poor, the destitute, the marginalised, the excluded and the powerless at the heart of the processes of development problem identification, decision making, policy formulation and project implementation. He popularised the notions of 'farmer first' and 'putting the last first', and has been highly influential in challenging professionals in all echelons of the development sector to foster critical self-awareness in their approach to their role and activities. In so doing, he has played a catalytic role in the introduction, refinement and dissemination of the participatory development ideology and associated methodologies. His 'new paradigm' has rapidly become the new orthodoxy.

Robert John Haylock Chambers was born on 1 May 1932. He received an Open Scholarship to Cambridge in the natural sciences in 1949, but this was interrupted by National Service from 1950 to 1952, when he was commissioned in the Somerset Light Infantry. Studying at St John's College, he graduated with a Double First in the BA History Tripos in 1955, and was awarded an MA in 1959. His life-long appetite for travel and discovery was whetted in 1955 as a member, and short-term leader, of the first scientific expedition to Gough Island in the South Atlantic – then one of the last places in the world about which little was known. He still lists running, mountaineering and rock-climbing as his principal recreational pursuits.

Chambers was appointed a District Officer in Kenya between 1958 and 1962, in the process gaining valuable insight into both colonial administration and local rural development. He then shifted his emphasis towards training and research, becoming a Lecturer in Public Administration at the Kenya Institute of Administration, and then a Research Officer at the East African Staff College. After receiving his PhD from the University of Manchester in 1967 for a thesis on settlement schemes in tropical Africa, and after short teaching appointments at the Universities of Manchester and Glasgow, he returned to Kenya to become a Senior Research Fellow in the Institute for Development Studies at the University of Nairobi from 1969 to 1971. From 1972 until 1997 he was a Fellow at the Institute of Development Studies (IDS), University of Sussex, where he established an international reputation as a leading advocate for and promoter of appropriate and just forms of development. During the 1970s, Robert Chambers continued to work on the management of rural development in eastern Africa, later extending his areal expertise to include India and Sri Lanka. Also, while at the IDS, he spent a period in the mid-1970s with the UNHCR working with rural refugees in eastern Africa and southeast Asia.

He continues to work at the IDS in the role of Research Associate. He was awarded the OBE in 1995, and has received Honorary Doctorates from the University of East Anglia (1995) and the University of Edinburgh (1998). The two principal influences on his way of looking at the development world were **E.F. Schumacher** and Ivan Illich, although he professes not to have been in total agreement with their ideas.

Although Chambers' journey on the path towards appropriate development probably started in the late 1950s, it was a paper published in the journal *World Development,* in 1981, entitled 'Rural Poverty Unperceived: Problems and Remedies', that, in his words, 'started all this off'. A number of themes and issues that recur in his later work were first introduced in that article. Its central argument, and that of much of Chambers' work, is that history has presented us with particular ways of looking at and approaching the poor and the disadvantaged, and 'development' more generally, which are often far removed from frequently unseen and unperceived realities. 'Doing development' typically falls to outsiders – people who, however well-intentioned, are neither from the target areas nor have directly experienced the problems of poverty, deprivation, insecurity, vulnerability, exploitation and so on. Their visions, interpretations and actions are thus built upon preconceptions and misperceptions about the rural poor, and are often influenced by stereotypical views of their circumstances and capabilities. Their interventions are also frequently constrained by a number of institutionalised biases that are influenced by practical realities (time, access, resources), attitudes (planning arrogance, political correctness, timidity), inequalities (hierarchies, differentials of power), segmentation (sectoralisation, project focus) and so on which determine that those who are in greatest need, or who are less visible and vociferous, remain unseen, unheard and unhelped. The result is that mountains of cash, eons of effort and one of the world's largest 'industries' have collectively failed to make much more than a dent in deprivation and destitution.

The paper sets the challenge to the principal actors in development to retrain and reprogramme themselves. Chambers emphasises the need for a series of 'reversals' where practitioners, academics, administrators, scientists and the like step away from what he pointedly calls 'rural development tourism' and start to bring about a revolutionary change in forms of cognition and behaviour. One of the principal reversals – now a mantra for the burgeoning non-governmental sector – is in the learning process, with the outside expert becoming the unimportant pupil who learns from and with those in need of development assistance, as opposed to the hitherto pervasive 'I talk, you listen' approach and attitude. The challenge is to see rural poverty as it really is, rather than as it is perceived to be, and to tailor actions accordingly in order truly to make a difference. The need is to confront global cores and peripheries of knowledge by looking for, being receptive to and finding means of incorporating local forms of knowledge that, time and again, have been shown to be tried and tested, appropriate but woefully overlooked by outside 'experts' and seriously compromised by the predilection for modern science and technology. Imagination and inventiveness are needed to find ways of reaching the poorest of the poor, the socially excluded, the politically voiceless, the geographically peripheral and the environmentally marginal.

The ideas in the *World Development* article were developed, honed and expanded in a series of accessible books produced during the 1980s and early 1990s, most of which were published via the Intermediate Technology Group founded by **E.F. Schumacher**. *Rural Development: Putting the Last First* (1983) emphasised the need for new professionals and a new professionalism with which to rise to the challenge of an inclusive and emancipatory development. The emphasis was placed squarely on 'the last': the hundreds of millions of largely unseen people in rural areas who are poor, weak, isolated, vulnerable and powerless. The book also exposed the gulf between the two principal cultures involved in the development project - the academic and the practical – and the tension between scientific/technical and local/indigenous knowledge. It looked at tactical means of confronting the 'superiority complex' of the development profession, of bringing about reversals in learning and of underpinning effective practical action.

Farmer First (1989) fleshed out a 'new paradigm' centred on localised, particularistic, contextualised, sympathetic, democratic and participatory means of supporting people in marginal contexts. The idea was not to come up with 'another development' of the kind advocated by postdevelopment writers such as **Manfred Max-Neef**, Arturo Escobar and Wolfgang Sachs, but forms of intervention involving a 'new professionalism' that complemented and enhanced existing science and technology-based approaches. The solutions to farmers' problems lay, in part, within their own capacities and capabilities. The challenge was to find ways of unlocking their development potential by creating a suitable enabling environment. The targets of development were to assume increasing control of the development process. Outsiders were to become the facilitators, not the paternalistic controllers, of local development. New institutions and institutional practices were to become established in order to allow the new paradigm to flower and flourish.

Farmer First adopted a holistic view of rural people's development challenges, introducing for the first time a concept – 'sustainable livelihoods' – that has become the *leitmotiv* of the development industry in the early part of the new century. The principles of the sustainable livelihoods approach were first put forward in an IDS Discussion Paper (1992) jointly authored by Robert Chambers and Gordon Conway, and centred on ideas of capability, equity and sustainability. The livelihoods approach to rural development has subsequently been adopted by the UK's DFID (Department for International Development), Sweden's SIDA (Swedish International Development Co-operation Agency), the United Nations Development Programme (UNDP) and several other bilateral and multilateral development agencies. Chambers' work on the hidden facets of poverty and deprivation, and the need to inform our understanding of these problems based on the experiences of those who are beset by them, inspired a large-scale body of work, leading up to the production of the World Bank's *World Development Report 2000/01: Attacking Poverty*, entitled 'Consultations with the Poor' and 'Voices of the Poor' where, for perhaps the first time, the Bretton Woods institutions' understanding of the poverty phenomenon became to some extent informed by the experiences and opinions of the poor themselves.

Robert Chambers became increasingly involved in rising to the methodological challenge that had arisen from the need to understand poverty and deprivation from the perspective of the poor and deprived. The first challenge was to find a cost-effective means of probing and penetrating beneath the apparent surface within target rural communities. Up until the late 1970s, constraints of time, resources and willingness had contributed to the prevalence of what he called 'quick and dirty' local survey methods, where outside experts or support teams engaged in superficial and uncritical rural appraisal as a precursor to project development, often using questionnaires (survey slavery) that reflected their preconceived ideas about rural areas and the challenges of development. Chambers helped pioneer the tool of Rapid Rural Appraisal (RRA) which was neat and nimble, but most importantly incorporated a range of devices to 'include the excluded' and to obtain a broad, flexible, insightful and locally contextualised understanding of poverty and the poor, or of other local development challenges. RRA came to constitute not just a new research methodology but also a new way of understanding development. It centred on developing and using a range of imaginative techniques – key informants, group appraisal, triangulation, progressive learning, identification and incorporation of indigenous knowledge. These both yielded an insightful understanding of local conditions as revealed by a representative cross-section of the communities themselves and freed up time for learning with and from local people, and for searching out the poor, the marginal and the excluded.

RRA steadily evolved into Participatory Rural Appraisal (PRA), and subsequently into a wider set of participatory tools that now constitutes the dominant development methodology and ideology. PRA was presented in *Whose Reality Counts? Putting the First Last* (1997) (the sequel to *Rural Development: Putting the Last First*) as a methodological solution to the challenge of building development efforts around local knowledge, experiences and capabilities. It is based on the observed ability of local communities across the developing (and developed) world to express and analyse their own complex realities in a way which has often been shown to be at variance with the perceived realities held by professionals operating from the top down. PRA centres on a raft of recipes for including the excluded and enabling local communities to express their development realities in ways which were suitable to their own cultures, literacy levels and so on. Much PRA, Participatory Action Research and Participatory Learning and Action in recent years has been driven by innovative development workers in and from the South. A prerequisite for practitioner/facilitators is critical self-awareness: a reflexive sense of the baggage of values and behaviour that they bring to the table of development courtesy of their background and identity. A key word in PRA is 'empowerment'. Robert Chambers sees participatory development as an 'exciting revolution in learning and action', and claims that the turn of the twenty-first century has been one of the most stimulating and challenging periods to be involved in the development profession.

Robert Chambers would, I'm sure, be the first to emphasise that his contribution to the reversal of development directionalities and the construction of appropriate methodological tools has been a collaborative and participatory endeavour, be that with colleagues and passers-through at the IDS, practitioners in the field and in big offices, or the poor and the peripheral themselves. Nonetheless, his thumbprint is everywhere one looks in the modern development field: bottom-up development, participatory development, sustainable livelihoods, the redefinition of poverty, the strengthening of civil society, and the entire ethos of appropriate development. In both the developing and development worlds, Robert Chambers has become one of the best-known, most influential and most widely followed persons in contemporary development over the last three decades. So many of us who, as outsiders, work on development issues genuinely but forlornly hope that our efforts will make a difference to the problems with which we are concerned; Robert Chambers is one among very few who can confidently claim to have made a profound difference to the ideology, paradigm and practice of development.

Major works

Chambers, R. (1981) 'Rural Poverty Unperceived: Problems and Remedies', *World Development* 9(1): 1–19.

—— (1983) *Rural Development: Putting the Last First*, Harlow: Longman (translated into Arabic, French, Indonesian, Italian, Japanese, Portuguese and Vietnamese).

—— (1989) *Farmer First: Farmer Innovation and Agricultural Research*, ed. with Arnold Pacey and Lori Ann Thrupp, London: Intermediate Technology Publications.

—— (1992) 'Sustainable Rural Livelihoods: Practical Concepts for the 21st Century', with Gordon Conway, *IDS Discussion Paper* 296.

—— (1993) *Challenging the Professions: Frontiers for Rural Development*, London: Intermediate Technology Publications.

—— (1994a) 'The Origins and Practice of Participatory Rural Appraisal', *World Development* 22(8): 953–69.

—— (1994b) 'Participatory Rural Appraisal (PRA): Analysis of Experience', *World Development* 22(9): 1253–68.

—— (1994c) 'Participatory Rural Appraisal (PRA): Challenges, Potentials and Paradigms', *World Development* 22(10): 1437–54.

—— (1997) *Whose Reality Counts? Putting the First Last*, London: Intermediate Technology Publications.

Michael Parnwell

HOLLIS B. CHENERY (1918–94)

Hollis Chenery's importance as a development economist reaches beyond his contributions to the academic debate, which were considerable. As Vice-President of the world's premier multilateral development institution, the World Bank, Chenery's ideas and work widely influenced development thinking and practice during the 1970s and 1980s.

Hollis Chenery was born into a wealthy family in Richmond, Virginia, USA, in 1918. His father was an oil and gas magnate. He grew up in Virginia as well as Pelham Manor in New York state. His upbringing in affluent suburban settings there would leave a permanent mark on him. A private passion of his was racehorses and one of those he owned would once win the Triple Crown.

Before World War II, Hollis Chenery attended the universities of Arizona and Oklahoma, earning undergraduate degrees from both. During the war, Chenery served as an Army Air Officer in the US military. He then went on to earn two master's degrees from the California Institute of Technology and the University of Virginia respectively, before gaining a doctorate in economics from Harvard University in 1950.

From the outset, Hollis Chenery's career combined academia with positions in international development. After World War II, Chenery served as a Marshall Plan economist in Europe, throwing himself into the European reconstruction programme. This job and being part of the post-war reconstruction process greatly influenced his thinking regarding industrial development, economic growth and societal change. It was decisive in focusing Chenery's considerable intellectual energies on development economics, the field which he would make his life's work.

Chenery's initial academic position after returning to the United States and receiving his PhD was as Professor of Economics at Stanford, a post which he held from 1952 until his appointment as a Guggenheim Fellow in 1961. That same year he joined the newly-established United States Agency for International Development (USAID), which was created by decree in terms of President John F. Kennedy's Foreign Assistance Act. Chenery quickly rose through the Agency's ranks to become an Assistant Administrator. Yet, he stayed with USAID only until 1965, when he returned to academia. He was appointed Professor of Economics at the university where he had earned his doctorate fifteen years earlier. Harvard would remain Chenery's academic home for the rest of his life, and one to which he would return after his tenure at the World Bank.

Chenery's academic work prior to joining the international development institutions was focused on empirical studies regarding factors of growth and development. His work was concerned with the role of investments and industrialisation in the development process, and the design of development projects and programmes (e.g., 1953, 1955, 1958, 1959). Chenery's book with Paul Clark on *Interindustry Economics* (1959) combined the qualities of a textbook with a systematic reader for professionals. His 1961 'Comparative Advantage and Development Policy' was received as a significant examination of the debate on resource allocation in less-developed economies and the classical principle of comparative advantage. The paper considered development policy from an economic theory standpoint, yet based on an analysis of actual government policies. He concluded that much of the confusion resulted from the failure to distinguish theoretical and operations research.

Hollis Chenery joined the World Bank in Washington, DC, in 1970 to serve as Economic Adviser to the Bank's then President, Robert McNamara, the former US Secretary of State. Two years later, McNamara handpicked Chenery as Vice-President for Development Policy at the Bank, a position he would hold for more than a decade until 1983. In this position, Chenery played an important role, influencing both the World Bank's internal operations as well as its lending policy towards developing countries.

Hollis Chenery was one of the first development economists to suggest that there was no linear path to development, thus rejecting the views of earlier development theorists such as economic historians **Alexander Gerschenkron** and **Walt W. Rostow**. Their 'stage theory' had stipulated that all countries passed through the same historical stages of economic development that the now industrialised had done earlier. According to this view, the developing countries were only at an earlier stage of development, i.e. primitive versions of developed countries. Furthermore, this view tended to equate development simply with industrialisation and economic growth, which would lead to a trickle-down effect increasing the overall wealth of the population. Based on his broad empirical work, Chenery recognised that there were differences in the experiences of individual countries, while he at the same time acknowledged that there were similar patterns that could be identified in different countries' paths to development. He understood that it was not inevitable that countries would all pass through the same linear phases. For that reason, he saw it as a major task for development economists to suggest ways in which developing countries could leap-frog over certain stages to catch up with industrialised countries. These shortcuts would be based on a systematic learning from the development histories and experiences of developed and more advanced underdeveloped countries.

Hollis Chenery brought these more nuanced views of the development process with him to the World Bank. His worldview coincided well with that of Robert McNamara, reflecting the liberal developmentalism of the era. This thinking was conditioned by the idea that poverty was the key factor in the spread of communism and it was, therefore, essential to combat poverty in Asia, Latin America and Africa to prevent them from falling into the communist bloc. Poverty reduction and distribution of the fruits of economic development, not growth at an aggregate level, thus became important to the World Bank. An influential book on *Redistribution with Growth* edited by Chenery with others (1974) epitomised this change in the Bank policy.

As the Vice-President for Development Policy in the mid-1970s, Chenery initiated a study on the role of the Bank in promoting economic development since 1950. The study, which was carried out by an outside consultant, David Morawetz, found that economic growth around the world had been swift and spectacular. However, its results were generally poorly distributed. At the same time, time-series data showed that economic growth was essential for poverty reduction and that poverty levels only exceptionally rose with growth (Ahluwalia, Carter and Chenery 1979). As a consequence, governments and the World Bank as an international financial institution needed to refocus their efforts on poverty eradication, even at the cost of increasing indebtedness. During McNamara's tenure, supported by the empirical research and theoretical thinking under Chenery's leadership, the World Bank's lending was expanded considerably and an explicit poverty alleviation focus was brought to its operations. Chenery chaired an in-house task force established in 1981 that articulated the Bank's poverty focus.

When the first oil crisis hit from 1973 onwards, Chenery was concerned with its impacts on the world economy. In the spirit of the era, as exemplified by the studies of the Club of Rome and writings of Paul and Anne Ehrlich, Chenery had adopted a view that the rapid growth in the world economy was pushing against the productive capacity and the potential rate of expansion of world supplies of many raw materials. He wrote an article in the influential *Foreign Affairs* journal in which he argued that the 'symptoms of underlying stress have been manifested … in the form of raw-material shortages, a food and fertiliser crisis, a dramatic rise in petroleum prices, and finally, worldwide inflation and threats of impending financial disaster' (1975b). His worst fears did not materialise, although a second oil crisis in 1979–80, also triggered by political events in the Middle East, forced further adjustments in global energy policies. Chenery published a subsequent article in *Foreign Affairs* (1981), in which he argued, with hindsight over-optimistically, that the need for this adjustment was bringing forth an energy transition that would reduce the need for formal international agreements that had proven to be elusive.

Overall, Chenery had a dramatic impact on the World Bank's research programme and its influence on the Bank's policy-making. According to him, the World Bank's best research was in the field of empirically based comparative studies. He saw that the World Bank's research programme could optimally combine micro-oriented studies based on its own lending operations in various countries with more theoretical, speculative macroeconomic research. He pursued this policy successfully, getting the Bank's Board to accept the approach as a legitimate way of influencing policy. Chenery took it upon himself to build up the research capacities within the Bank by recruiting a number of highly qualified economists, many from academia. He bridged the gap between research and operations by establishing a rotational system for staff. The in-house research capacity was complemented by bringing in outside expertise through consultants and academics participating in research work. A Research Committee was established and research findings were widely disseminated internally through management seminars and internal discussions based on the cases of individual borrower countries. These all were tools intended to improve the Bank's analysis and its consequent lending policies.

During Hollis Chenery's time, the World Bank started producing a wide variety of policy papers that became authoritative statements on development issues. The Bank, jointly with the International Labour Organization (ILO), worked on the 'basic needs' concept and its implications for lending operations. Hollis Chenery initiated the work that would lead to the preparation of the first annual *World Development Report*, the Bank's flagship publication to date.

After Robert McNamara's retirement from the World Bank presidency in 1981, the new management under Alden Winship (Tom) Clausen changed course. This coincided with the emergence of neoliberal ideology as personified by the rise to power of Ronald Reagan and Margaret Thatcher in the USA and UK respectively. At the World Bank, Anne Krueger was hired to replace Hollis Chenery, with the result that neoliberal dogma replaced the poverty alleviation focus. However, while the official policy of the Bank changed, at the operational levels many staff economists still maintained their pragmatic approaches to development policy.

After leaving the Bank, Hollis Chenery returned to Harvard where he continued to put his empirical experiences with international development to good use in writing and teaching. He collaborated with T.N. Srinivasan in editing a *Handbook of Development Economics* (1988, 1989), which became the most comprehensive sourcebook on development economics available.

International development theory has undergone considerable changes since the 1970s and 1980s, when Hollis Chenery was an active participant in its making. Some of his work has inevitably been overtaken by events and advances in our understanding of the development process. Nevertheless, Chenery made a lasting impact on international development theory, policy-making and praxis. Despite significant ups and downs since his move from the World Bank, his legacy lives on and is today, perhaps paradoxically, more evident again in the institutional policies than immediately after his departure, because of the sharpened poverty focus of its current activities.

Major works

Adelman, I. and Chenery, H.B. (1966) 'Foreign Aid and Economic Development: The Case of Greece', *Review of Economics and Statistics* 48(1): 1–19.

Arrow, K.J., Chenery, H.B., Minhas, B.S. and Solow, R.M. (1961) 'Capital–Labor Substitution and Economic Efficiency', *Review of Economics and Statistics* 43(3): 225–50.

Chenery, H.B. (1952) 'Overcapacity and the Acceleration Principle', *Econometrica* 20(1): 1–28.

—— (1960) 'Patterns of Industrial Growth', *American Economic Review* 50(4): 624–54.

—— (1961) 'Comparative Advantage and Development Policy', *American Economic Review* 51(1): 18–51.
—— (ed.) (1971) *Studies in Development Planning*, Harvard Economic Studies, Cambridge, MA: Harvard University Press.
—— (1975a) 'A Structuralist Approach to Development Policy', *American Economic Review* 65(2): 310–16.
—— (ed.) (1979) *Structural Change and Development Policy*, A World Bank Research Publication, New York: Oxford University Press.
—— (1983) 'Interaction Between Theory and Observation in Development', *World Development* 11(10): 851–903.
Chenery, H.B. and Clark, P.G. (1959) *Interindustry Economics*, New York: John Wiley and Sons.
Chenery, H.B., Duloy, J.H. and Jolly, R. (eds) (1974) *Redistribution with Growth: Policies to Improve Income Distribution in Developing Countries in the Context of Economic Growth*, London: Oxford University Press.
Chenery, H.B. and Strout, A. (1966) 'Foreign Assistance and Economic Development', *American Economic Review* 56(4): 679–733.
Syrquin, M. and Chenery, H.B. (1989) *Patterns of Development, 1950–1983*. Discussion Paper, Washington, DC: World Bank.

Further reading

Ahluwalia, M., Carter, N. and Chenery, H.B. (1979) 'Growth and Poverty in Developing Countries', in H.B. Chenery (ed.), *Structural Change and Development Policy*, New York: Oxford University Press.
Asher, R. (1983) 'Chenery, Hollis B. (Tenure: 1970–1983)', World Bank Group Oral History Program Record Series: S4100, Washington, DC: World Bank.
Chenery, H.B. (1953) 'The Application of Investment Criteria', *Quarterly Journal of Economics* 67(1): 76–96.
—— (1955) 'The Role of Industrialization in Development Programs', *American Economic Review* 45(2): 40–57.
—— (1958) 'Development Policies and Programmes', *Economic Bulletin for Latin America* 3(1): 51–77.
—— (1959) 'The Interdependence of Investment Decisions', in M. Abramovitz *et al.* (eds), *The Allocation of Economic Resources: Essays in Honor of Bernard Francis Haley*, Stanford, CA: Stanford University Press.
—— (1975b) 'Restructuring the World Economy', *Foreign Affairs*, 53(2): 242–63.
—— (1981) 'Restructuring the World Economy: Round II', *Foreign Affairs*, 59(5): 1102–20.
—— (1982) 'Industrialization and Growth: The Experience of Large Countries', World Bank Staff Working Paper, No. 539, Washington, DC: World Bank.
Chenery, H.B. and Srinivasan, T.N. (eds) (1988 and 1989) *Handbook of Development Economics*, Vols I and II, Amsterdam: North Holland.
Easterly, W. (2001) *The Elusive Quest for Growth: Economists' Adventures and Misadventures in the Tropics*, Cambridge, MA: MIT Press.
Pace, E. (1994) 'Hollis B. Chenery Dies at 77; Economist for the World Bank', *New York Times*, 5 September.

Juha I. Uitto

DIANE ELSON (1946–)

Since the late 1970s, and throughout numerous institutional and disciplinary incarnations, Diane Elson has been at the forefront of gender and development (GAD) scholarship at a world scale. Under the broad umbrella of analysing socio-economic processes from a gender perspective, Elson's work has spanned a diverse range of themes. These include women's position in the international division of labour, gender dimensions of structural adjustment, the significance of gender inequality and unpaid labour within the household or 'reproductive economy' (sometimes called the care economy, or sphere of social reproduction) for the operation of national economies and international trade and finance, gender budget initiatives (GBIs), and gender and globalisation. As one of a still relatively small number of feminist economists able to draw from an impressive multidisciplinary expertise encompassing sociology and politics as well as economics, Elson's work has been distinguished, *inter alia*, by its ability to challenge and make accessible to non-economists the concepts underlying thinking and modelling in the predominantly male field of macro-economics. Macro-economic models often masquerade under the guise of gender neutrality, but in reality emanate from gender-blindness, in turn resulting in gender bias. As Elson has so eloquently argued, the need to 'talk to the boys' (Elson 1998a), in the form of making gender heard in an arena where there are few feminist voices conversant with the language and concepts deployed within international development economics has been an abiding principle in much of her work over three and a half decades. The call to 'engender' macro-economics and international economics has not only featured in Elson's numerous academic publications and policy reports, but in major curriculum innovations in the UK and overseas.

Elson was born on 20 April 1946. She grew up in the English Midlands, attending Nuneaton High School for Girls. She was the first member of her family to go to university, and graduated from Oxford with a BA (Hons) in Philosophy, Politics and Economics in 1968. Elson then worked for three years as a Research Assistant at the Institute of Commonwealth Studies at Queen Elizabeth House, Oxford and was a member of St Antony's College. In 1985 she was appointed as Lecturer in Development Economics at the University of Manchester (where she later attained her PhD by publication and became Professor of Development Studies and Research and Graduate Dean). In the meantime, Elson held a range of teaching and research positions at the Universities of Sussex, York and Oxford, as well as fitting in consultancy and lecturing at the Open University and Manchester with the birth and pre-school care of her son, Paul. After fifteen years of full-time employment at Manchester, Elson moved to a Chair in Sociology at the University of Essex, where she teaches predominantly on gender, the sociology of development and human rights.

Elson stands out as an agenda-setting individual in her own right, particularly in opening up and establishing new lines of enquiry and conceptual development, but she has also conscientiously and productively engaged in numerous collaborations with fellow academics and with writers and researchers from other disciplines and backgrounds, both in the UK and internationally. This has included prominent GAD scholars such as Ruth Pearson, Nilufer Cagatay, Caren Grown and Debbie Budlender, as well as more junior research teams. Some of Elson's collaborative ventures have resulted from her consultancy, advisory and research work for major development institutions such the Swedish International Development Co-operation Agency (SIDA), the United Nations Development Programme (UNDP), and UNIFEM (The United Nations Development Fund for Women), where between 1998 and 2000 she was a special advisor to the Executive Director, Noeleen Heyzer. Elson has also maintained substantial involvement in women's groups and networks, and in several forms of public service. The latter includes membership of the Management Committee of the UK Women's Budget Group, of the Board of Directors of the NGO 'Women Working Worldwide', and in UN teams and committees such as the UN Taskforce on the Millennium Development Goals. In the interests of advancing gender equality through changing the ways in which public revenue is raised and expended, Elson has also helped to co-found a global network of gender budget initiatives comprising civil society groups and government personnel.

Elson's longstanding commitment to movements and initiatives for change outside as well as within academia has strengthened the applied as well as the theoretical aspects of her work. Her writings are distinguished, for example, by a remarkable accessibility given the complexity of some of the issues she has dealt with, especially in the realm of macro-economics, trade and finance (Elson *et al.* 2000). In turn, a keen commitment to policy relevance, to judicious and critical interrogation of empirical data, and to linking the macro with the micro, has lent itself to an enviable capacity to bridge the often wide divide between analysis and action. This has helped to earn Elson credibility across several different sets of stakeholders in the development field as a whole, including policy makers, grassroots activists and countless PhD students who, under Elson's tutelage, have between them worked on gender issues in most parts of the world. Among GAD scholars and practitioners more generally, Elson is recognised as having been one of the movement's most valuable assets in respect of establishing gender as a legitimate and enduring issue, both conceptually and pragmatically.

Elson is the author and editor of numerous books, including the widely cited *Male Bias in the Development Process* (1991), as well as special journal editions, papers and policy documents. Many of her writings have rapidly assumed the status of 'classics' in the GAD field, as evidenced *inter alia* by their extensive reprinting and translation into other languages. In the light of the prolific nature of Elson's work, it is clearly hard to do justice to the vast repertoire of ideas which she has injected into different debates, and which indeed have spawned debate in their own right. One of her earliest and most significant contributions was a co-authored article (Elson and Pearson 1981), which not only provided one of the first comprehensively researched yet concise accounts of women's involvement in export-oriented industrialisation, but also introduced the idea which, contrary to conventional WID (Women in Development) orthodoxy that women had been 'left out' of the development process, drew attention to the need to focus on the frequently exploitative ways in which women were actually *integrated*. More specifically, Elson and Pearson identified the need to distinguish between three, non-mutually exclusive, tendencies in the relation between the disproportionate concentration of women in un- and semi-skilled jobs in export-oriented factories, and their subordination as a gender. These were, respectively: 'a tendency to *intensify* the existing form of gender subordination; a tendency to *decompose* existing forms of gender subordination; and a tendency to *recompose* new forms of gender subordination' (*ibid*.: 110, emphasis in original). Various possibilities for how women workers could challenge these different forms of subordination were drawn out. Even if women made few instrumental gains in their positions (for example, improvements in pay and in working conditions), evaluation of women's struggles as a gender should be made in relation to the way that 'the struggle itself develops capacities for self-determination' (*ibid*.: 105). In the rich and ever-growing literature on gender and export manufacturing to which Elson herself has continued to contribute (Elson 1996), the 1981 paper remains a central point of reference.

Emphasis on the potential for women's resistance, despite considerable barriers to their mobility and empowerment, also appears in Elson's pioneering work on gender and neoliberal restructuring, which became a major new genre of enquiry in the late 1980s and early 1990s. Highlighting how women frequently absorbed the costs of structural adjustment policies by increasing their loads of paid, and in particular, unpaid labour, Elson argued not only for the need for macro-reforms such as the democratisation of states and the restructuring of the international financial system, but emphasised how seeds for transformation also lay at the level of women's involvement in self-help groups at the community level (Elson 1992a). Yet, Elson's contribution to feminist critiques of structural adjustment went far beyond this insofar as she synthesised the observations drawn from localised case studies, linked the micro with the meso and macro, made the most convincing case for economies being gendered structures, and demonstrated the extreme gender bias inherent in structural adjustment programmes (along with macro-economic models more generally) (Elson 1993). More particularly, Elson drew attention to the fact that failure to incorporate gender in the macro-economic modelling of restructuring, marked above all by complete disregard for women's extensive unpaid labour in the 'reproductive economy', led to pernicious effects on not only women's own quality of life, but, given their role in (re)producing human resources (and *ipso facto*, the labour force), to economically inefficient outcomes (Elson 1995; Cagatay *et al.* 1995). Out of this came the conclusion that any macro-economic model which neglected gender and the reproductive economy was not only fundamentally flawed on conceptual grounds, but could exacerbate social inequality and render possibilities for sustainable economic growth untenable. As articulated in her widely cited *World Development* article:

> Though maintaining gender inequality may sometimes lead to higher profits in the short run, it also tends to generate negative feedbacks which hamper the process of restructuring to a development path which is sustainable in the long run. The outcome of macro-economic policies depends upon the gendered social matrix in which they are introduced. But that gendered social matrix could in principle be changed in ways that promote both more effective macro-economic policies and greater gender equality.
>
> (Elson 1995: 1865)

Her continuing enthusiasm for taking on big ventures, and for transforming policy, includes her lead authorship on the 2000 and 2002 UNIFEM reports on *The Progress of the World's Women* (Elson 2000; Elson with Keklik 2002), and her work on GBIs (Elson 1998b, 2002, 2004). The 2002 UNIFEM report, which is impressively detailed yet accessibly presented, concentrates on the third UN Millennium Development Goal (MDG), namely to 'promote gender equality and empower women'. Elson criticises the four indicators selected to represent the target (the ratio of boys to girls in primary, secondary and tertiary education, the ratio of literate women to men in the 15–24 age group, the share of women in non-agricultural employment, and the proportion of parliamentary seats held by women), for their inherent weaknesses as well as continued data deficiencies, and also argues that more targets should be included, such as gender equality in the labour market. This constitutes an important contribution to a larger and growing body of feminist critique of the gap between rhetoric and genuine commitment to addressing gender inequality in the MDGs.

With regard to her work on GBIs, Elson's contributions not only include key justifications for integrating a gender perspective into national budgetary processes, but also provide important guidelines for 'engendering' budgets, as well as documenting several case study examples from among the forty or so countries in which such initiatives have thus far taken place.

It should also be noted that Elson's writings have embraced more general development issues outside GAD, including development theory (Elson 1998b), the socialisation of markets (Elson 1992b), and the UN Global Compact (Elson 2003). Often richly informed by Elson's feminist and socialist political perspectives, such contributions have made her a major name in the development field in general, as well as within GAD circles.

It is within GAD, however, that Elson's pioneering and enduring contribution to scholarship is most apparent. No student of gender and development today completes a course at undergraduate or postgraduate level without some exposure to Elson. Given Elson's ongoing ventures into cutting-edge issues in the GAD field – currently comprising fiscal policy in relation to women's economic and social rights, and globalisation and gender equality – and with their hallmark rigour and vision, this situation is unlikely to change for a long time to come. Elson will also be present as one of GAD's outstanding scholars and ambassadors in any retrospective review of gender and development that may be compiled in the future.

Major works

Cagatay, N., Elson, D. and Grown, C. (eds) (1995) Special Issue 'Gender, Adjustment and Macro-economics', *World Development* 23(11): 1825–2017.

Elson, D. (ed.) (1991) *Male Bias in the Development Process*, Manchester: Manchester University Press, 1st edn, chaps 1, 7 and 8 written by Elson, pp. 1–28, 164–210; 2nd edn 1995, with two additional chapters written by Elson, pp. 211–79. Chapter 7 'Male Bias in Macro-economics: The Case of Structural Adjustment' reprinted in A.V. Dutt (ed.) (2002) *The Political Economy of Development*, Vol. 2, Cheltenham: Edward Elgar.

—— (ed.) (2000), *Progress of the World's Women 2000*, New York: UNIFEM, pp. 1–164. Also in French and Spanish translation.

Elson, D., Budlender, D., Hewitt, G. and Mukhopadhyay, T. (2002) *Gender Budgets Make Cents*, London: Commonwealth Secretariat.

Elson, D., Cagatay, N. and Grown, C. (eds) (2000) Special Issue 'Growth, Trade, Finance and Gender Inequality', *World Development* 28(7): 1145–1390.

Elson, D., Faune, M., Gideon, J., Gutierrez, M., Lopez de Mazier, A. and Sacayon, E. (1997) *Crecer con La Mujer*, San Jose, Costa Rica: Embajada Real de los Paises Bajos.

Elson, D. with Keklik, H. (2002) *Progress of the World's Women 2002*, New York: UNIFEM.

Further reading

Elson, D. (1992a) 'From Survival Strategies to Transformation Strategies: Women's Needs and Structural Adjustment', in Beneria, L. and Feldman, S. (eds), *Unequal Burden: Economic Crisis, Persistent Poverty and Women's Work*, Boulder, CO: Westview, pp. 26–48.

—— (1992b) 'The Economics of a Socialised Market', in R. Blackburn (ed.), *After the Fall – The Failure of Communism and the Future of Socialism,* London: Verso.

—— (1993) 'Gender-Aware Analysis and Development Economics', *Journal of International Development* 5(2): 237–47. Reprinted in Jameson, K.P. and Wilber, C.K. (eds) (1996) *The Political Economy of Development and Underdevelopment*, New York: McGraw-Hill, pp. 70–80 and in Benería, L. and Bisnath, S. (eds) (2001), *Gender and Development: Theoretical, Empirical and Practical Approaches*, Vol. 1, Cheltenham: Edward Elgar.

—— (1995) 'Gender Awareness in Modelling Structural Adjustment', *World Development* 23(11): 1851–68. Reprinted in Benería, L. and Bisnath, S. (eds) (2001), *Gender and Development: Theoretical, Empirical and Practical Approaches*, Vol. 2, Cheltenham: Edward Elgar.

—— (1996) 'Appraising Recent Developments in the World Market for Nimble Fingers: Accumulation, Regulation, Organisation', in Chhachhi, A. and Pittin, R. (eds), *Confronting State, Capital, and Patriarchy: Women Organising in the Process of Industrialisation*, London: Macmillan, pp. 35–5.

—— (1998a) 'Talking to the Boys: Gender and Economic Growth Models', in Jackson, C. and Pearson, R. (eds), *Feminist Visions of Development*, London: Routledge, pp. 155–70.

—— (1998b) 'Integrating Gender Issues into National Budgetary Policies and Procedures: Some Policy Options', *Journal of International Development* 10: 929–41.

—— (2000) 'Theories of Development', in Janice Peterson and Meg Lewis (eds), *Elgar Companion to Feminist Economics*, Aldershot: Edward Elgar. Reprinted in Benería, L. and Bisnath, S. (eds), *Gender and Development: Theoretical, Empirical and Practical Approaches*, Vol. 1, Cheltenham: Edward Elgar, 2001.

—— (2002) 'Gender Responsive Budget Initiatives: Key Dimensions and Practical Examples', in Judd, K. (ed.), *Gender Budget Initiatives*, New York: UNIFEM.

—— (2003) 'Human Rights and Corporate Profits: The Case of the UN Global Compact', in Benería, L. and Bisnath, S. (eds), *Global Tensions: Challenges and Opportunities in the World Economy*, London: Routledge.

—— (2004) 'Engendering Government Budgets in the Context of Globalisation(s)', *International Feminist Journal of Politics* 6(4): 623–42.

Elson, D. and Pearson, R. (1981) 'Nimble Fingers Make Cheap Workers: An Analysis of Women's Employment in Third World Export Manufacturing', *Feminist Review* Spring: 87–107.

Jackson, C. and Pearson, R. (eds) (1998) *Feminist Visions of Development*, London: Routledge.
Molyneux, M. and Razavi, S. (eds) (2002) *Gender Justice, Development and Rights*, Oxford: Oxford University Press.
Rai, S. (2002) *Gender and the Political Economy of Development*, Cambridge/Oxford: Polity in association with Blackwell.

Sylvia Chant

ANDRE GUNDER FRANK (1929–2005)

Indisputably one of the world's most prominent and productive radical political economists, Andre Gunder Frank's massive corpus of work[1] – forty-four books in 140 different editions, 169 chapters in 145 edited books covering thirteen languages, and 400 articles translated into twenty languages – has left an indelible mark on the critical study of development and the world system. Frank led an extraordinarily peripatetic and interesting life, brushing shoulders with everyone from Milton Friedman to Che Guevara. His name will always be associated with the body of radical scholarship that emerged from Latin America in the 1960s operating under the sign of dependency theory, but his most enduring contributions reside elsewhere (see below). A combative and iconoclastic intellectual, Frank's work has always been controversial and provocative, drawing as much fire as it has acclaim. It is a measure of his stature and influence that in a collection of essays published upon his retirement, his interlocutors included Immanuel Wallerstein, **Samir Amin**, and **Eric Wolf**. His stature as a theorist of the world system – certainly the equal of Fernand Braudel – is irrefutable (Frank *et al.* 1996).

Born in Berlin in 1929, Frank left Germany aged four. His parents fled to Switzerland as political exiles (his father was a novelist and pacifist) in the wake of Hitler's ascent. Frank would return to Germany forty years later in exile himself, fleeing the military coup in Chile. This early exile marked the beginning of a nomadic life which has no equal in the world of the academic activist: he became an internationalist *à la lettre*. By his own reckoning, Frank rarely stayed more than one to two years in any one place (Chile, England, the US and Holland are the exceptions) and since 1961 he had been pursuing what he calls his 'Oddissey around the world'.[2] His nomadism was to some extent a product of his own restless, inquisitive nature; but in equal measure he was in a sort of permanent exile, at once too radical, too ornery and too unconventional for most universities on both sides of the Atlantic. His life was intimately shaped by the Cold War and the rise of Third World development policy and practice.

As a young boy, he migrated across Switzerland's three principal language areas and this laid the foundation for what became his multilingual fluency (he has control over seven languages). In 1941 he moved with his parents to the USA, where he remained until he was thirty-one years old. As a teenager his life was equally mobile – California, Idaho, Michigan, New York. He finished school at Ann Arbor High in Michigan[3] and entered college at Swarthmore, where he became – by his own admission – a 'Keynesian economist'.

It is one of the enduring paradoxes of radical social science that Frank was educated in Economics at the University of Chicago, where he studied with Milton Friedman. He was awarded the PhD in 1957 for a dissertation on Soviet agriculture in the Ukraine. Not surprisingly, his career at Chicago was unorthodox. Bert Hoselitz in Anthropology introduced him to development and modernisation theory on which he was to unleash his ferocious critique. **Walt Rostow**, whom he later met at MIT in 1958, was, of course, his *bête noire*. Frank wrote little on his dissertation during the period 1957 to 1962 when he taught at Michigan State University, Wayne State and the University of Iowa. He visited the Soviet Union in 1960 but as he put it 'all official doors in Moscow and Kiev were closed to me'. He later returned to the question of socialism but this time to its vulnerabilities in relation to the world system and the unevenness of its cyclical development. Frank was sometimes credited with having predicted the collapse of the Soviet Union in 1989.

In 1961 Frank resigned from his position at Michigan State, disillusioned by both the impress of the Cold War on academia and by the realisation that most social science research was 'part of the problem'. His radicalisation had begun earlier, largely through self-discovery and what he calls 'years of wandering through the woods'. His resignation marked the beginning of a world-wide trek across many countries, universities and disciplines. In 1962 he taught at the University of Brasilia (unable to teach at Leipzig or Havana as he intended), and thereafter successively at UNAM in Mexico City, George William University in Montreal, and the International Labor Organisation (ILO) in Chile (from where he was fired). In 1968, at the instigation of Salvador Allende, he was appointed to the Faculties of Sociology and Economics in Santiago. Any return to the US was foreclosed by a 1965 decision to deny him entry on the grounds of his writing for *Monthly Review* and 'his identification with the Chinese Communist position' as the Immigration and Naturalisation Department put it. His Chilean sojourn was foundational for his work on dependency theory but it also marked the beginning of a new concern with the need for global resistance to the crisis of capitalist accumulation that in his view had begun in 1970. Frank's Latin American sojourn was brought to a dramatic halt, however, with the overthrow of the Allende government in 1972 and his flight into exile. For three unhappy years he moved between Berlin, Munich and Frankfurt, before being appointed Professor of Development Studies at the University of East Anglia in 1978. From there he moved to Amsterdam in 1983. Over this period Frank claimed to have applied unsuccessfully for eighty positions in the USA.

Frank taught in at least sixteen universities and across ten different disciplines. In 1994 he went into mandatory retirement from the Faculty of Economics at the University of Amsterdam. At the time of his death, he was a member of the Graduate Faculty in Sociology at the University of Toronto but between 2000 and 2004 he held visiting appointments at the Universities of Miami, Nebraska, Northeastern and Calabria. For almost half a century – a stretch of time that included the death of his first wife, Marta Fuentes, in 1993, the divorce from his second, Nancy Howell, and four major medical operations – his productivity and intellectual vitality did not wane. He is survived by his two sons and third wife, Alison Candela.

Frank's large body of work was all directly shaped by **Marx** (though he has rarely engaged with the larger body of modern Marxist theory from Luxemburg to Lenin to Trotsky to Gramsci to Horkheimer). The centrality of accumulation, and particularly accumulation on a world scale, encompasses the five major themes in his corpus. The first addresses the socialist bloc, beginning with his earliest work on socialist productivity in the Soviet Union and the question of political economic organisation in relation to economic flexibility and rigidity. He returned to the socialist economies in the 1970s to explore how they were being transformed by the world economic crisis outside of the socialist bloc (Frank 1980, 1983). This led him to explore both the internal contradictions within eastern Europe generated by the world crisis and how the former socialist economies are being transformed under the influence of the IMF, and the consequences of the destruction of COMECON (Council for Mutual Economist Assistance) on the global division of labour. A second theme was social and anti-system movements, a project initiated with Marta Fuentes and developed further in a collective project (1990) with **Samir Amin**, Immanuel Wallerstein and Giovanni Arrighi. The argument turned on the decline of Left parties and organisations and the flowering of social movements of all sorts which carried the hopes of social transformation without the capture of state power. These movements were placed on a larger canvas of world historical mobilisation, and confronted the uncomfortable fact that much of the flowering of civil society could be markedly 'uncivil' (i.e. unprogressive).

The third theme – world system development and underdevelopment – began with his move to Latin America. He first deployed the term world system and dependence in the mid-1960s and these ideas shaped his understanding of Third World underdevelopment. Using case studies of Mexico, Chile and Brazil, Frank laid out a historical account of the creation of complex dependent relations linking the metropole and the periphery in the wake of the fifteenth-century conquest. Rejecting the idea that Latin America was or had been feudal, Frank explored the processes of incorporation into the capitalist world market and the structures of dependence (from inter-state to intra-village) so created. A number of other key theorists – especially **Fernando Cardoso** – developed these ideas through detailed case studies, and the ideas were picked up and developed in other ways by the Economic Commission for Latin America (ECLA) and UNCTAD (United Nations Conference on Trade and Development), and came to have enormous influence in academic and policy circles. Dependency emerged from both the Marxist and structuralist antecedents (Love 1990), especially the trade-based theorists like **Raúl Prebisch** at ECLA, but the Marxist-inspired work of Frank (1967, 1969) and Cardoso and Faletto (1969) provided its foundations. Frank's work, in particular, unleashed a firestorm of debate from both Right and Left. His work was a direct challenge to the Latin American Communist Parties but also, of course, to the modernisation theorists. Dependency was also attacked from within Marxian theoretical circles for its 'circulationist' (i.e. exchange-based) view of capitalism, its functionalism, and its failure to engage with the structural Marxism of Althusser and his colleagues in France (see Brenner 1977; Laclau 1971; Bernstein and Nicholas 1983). Whatever one thinks of this 1960s dependency theory (and its promotion of autarchy), Frank had identified the existence of a system of world accumulation marked by the rupture of the fourteenth century and linked what he called the development of underdevelopment to its rhythms, crises and conjunctures. Here he had independently arrived at a similar point to that of **Samir Amin** (1972) and Immanuel Wallerstein (1974), who were to lay out their respective visions of the world system a little later.

By the time dependency had reached the elevated levels of policy, Frank was already troubled and was moving toward his fourth theme, contemporary international political economy (1978). In the early 1970s, Frank predicted a worldwide economic crisis that would both integrate the socialist bloc into its capitalist orbit and impose export intensification on the Third World through force rather than consent; he also anticipated the 'oil shock' as the harbinger of a series of successive and ever deeper recessions within a world economic crisis (see 1980, 1982, 1983). Here much of his work addressed such questions as the regionalisation of the world economy, the consequences of Thatcherism and Reaganism for the 1979–82 world recession, and successfully predicted some of the world-historical changes in the late twentieth-century political economy.

Having spent more than two decades exploring (and forecasting) contemporary political economy on the world scale, Frank returned, in his fifth theme, to world history. His earlier concern had been with 500 years of capitalist world development but this time he sought to recast this period by breaking irrevocably with its Eurocentrism, and by pushing backwards the origins of the world system 5000 years (and in so doing came to abandon completely the category of capitalism itself and break irrevocably from his old adversaries Wallerstein, Amin and Braudel). His book *ReOrient* (1998) starts from the pathbreaking work of Janet Abu-Lughod (1989) on the thirteenth-century world system, and of **Jim Blaut** (1992) on the flaws of Eurocentric history (all of Weberian and Marxist history is deemed to be Eurocentric by Frank), and endeavours to reinterpret the modern age and the rise of Europe. *ReOrient* makes two broad arguments. One demolishes the pillars of Eurocentrism by showing that Asia was outpacing Europe until the mid-1700s and its decline came about because of internal crisis, not the rise of Europe. The second turns on the conjunctural circumstances that permitted the rise of Europe around 1800. His major thesis is that 'there was a single global world economy with a worldwide division of labor and multilateral trade from 1500 onward' (1998: 52). But the core of the system was Asia; it dominated the intense and tight trade linkages in such a way that the whole was not a 'European world system'. Inevitably this book has been received in highly polarised ways (for example, the responses to it by Wallerstein 1999, Amin 1999 and Arrighi 1999) but the broad lineaments of the argument are consistent with some of the more important and original work emerging from the study of China (Pomerantz 2000).

Andre Gunder Frank remained as productive as ever until shortly before his death from cancer. His return to world history launched his final project, which was a tripartite analysis of first what he called the 'new world [dis]order', second the 'reOrienting of the Nineteenth century' (explaining why the relative powers of the Orient and the West were inverted, building upon Pomerantz's *The Great Divergence* (2000)), and finally 'Bronze Age World System Cycles' in which he attempted to chart global accumulation (as he understood it) into the fourth and fifth millennia BP. Frank ended his writing career surrounded by the same sort of debate, the same sort of iconoclastic style, and the same controversy that marked his entry into the staid world of the academy four decades earlier.

Notes

1 His bibliography covering the period 1955–95 contains 880 publications in twenty-seven languages; by 2003 the total figure exceeded 1000 publications. The International Institute of Social History in Amsterdam received Frank's papers in 1995; they cover 35 metres of shelf space (the inventory is available online at: http://rrojasdatabank.info/afgfrank/IISG_Pubs.html). His work has been cited in journals on average over the period 1975–present about 200 times per annum.

2 All quotations by Frank, unless otherwise indicated, are taken from his 'Cold War and Me' essay and his short political memoir published on his website.

3 He changed his name from Andrew (anglicised by his parents from Andreas) to Andre, adopting the appellation Gunder in high school (a nickname derived from a Swedish world record distance runner).

Major works

A vast amount of work in many languages is available at Frank's website: http://rrojasdatabank.info/agfrank/personal.html.

Frank, A.G. (1967) *Capitalism and Underdevelopment in Latin America*, New York: Monthly Review.

—— (1969) *Latin America: Underdevelopment or Revolution*, New York: Monthly Review.

—— (1978) *World Accumulation: 1492–1789*, New York: Monthly Review.

—— (1980) *Crisis in the World Economy*, New York: Holmes.

—— (1982) *Reflections on the World Crisis*, with S. Amin, G. Arrighi and I. Wallerstein, New York: Monthly Review.

—— (1983) *The European Challenge*, Nottingham: Spokesman.

—— (1990) *Transforming the Revolution*, with S. Amin., G. Arrighi and I. Wallerstein, New York: Monthly Review.

—— (1998) *ReOrient: Global Economy in the Asian Age*, Berkeley, CA: University of California Press.

Frank, A.G., Chew, S.C. *et al.* (1996) *The Underdevelopment of Development: Essays in Honor of Andre Gunder Frank*, Thousand Oaks, CA: Sage.

Further reading

Abu-Lughod, J.L. (1989) *Before European Hegemony*, New York: Oxford University Press.

Amin, S. (1972) *Accumulation on a World Scale*, New York: Monthly Review.

—— (1999) 'REORIENTALISM? – History Conceived as an Eternal Cycle', *Review* 22(3): 36.

Arrighi, G. (1999) 'REORIENTALISM? – The World According to Andre Gunder Frank', *Review* 22(3): 28.

Bernstein, H. and Nicholas, H. (1983) 'Pessimism of the Intellect, Pessimism of the Will – a Response to Andre Gunder Frank', *Development and Change* 14(4): 609–24.

Blaut, J.M. (1992) *1492: The Debate on Colonialism, Eurocentrism, and History*, Trenton, NJ: Africa World Press.

Brenner, R. (1977) 'The Origins of Capitalist Development', *New Left Review*, 104: 25–92.

Cardoso, H. and Faletto, E. (1969) *Dependencia y Desarallo en America Latina*, Mexico: Siglo XXI Editores.

Laclau, E. (1971) 'Feudalism and Capitalism in Latin America', *New Left Review* 67: 19–38.

Love, J.L. (1990) 'The Origins of Dependency Analysis', *Journal of Latin American Studies* 22(1): 143–68.

Pomerantz, C. (2000) *The Great Divergence*, Princeton, NJ: Princeton University Press.

Wallerstein, I. (1974) *The Modern World System*, Vol. 1, New York: Academic Press.

—— (1999) 'REORIENTALISM? – Frank Proves the European Miracle', *Review* 22(3): 18.

Michael Watts

PAOLO FREIRE (1921–97)

In his struggle for liberation of the poor and voiceless, Paolo Freire was something of a refugee searching for a state that would accept his basic thoughts. His life journey started in Recife, northeastern Brazil, where he was born in 1921. In spite of his middle-class background, the prevailing economic hardships made him suffer from hunger and poverty. This personal experience made him realise the close relationship between social class and knowledge. Freire started to wonder about the apparent indifference by the poor towards their obvious oppression. The Latin American people had basically subsumed to what Freire called a culture of silence, quietly accepting the superiority of the elite over the people. To be able to strike back, the people had to regain their language.

As his family's economic conditions improved, Freire was eventually able to enter the University of Recife to study law, which he initially considered as useful in his struggle for the poor. However, at the time he was also influenced by philosophy and language. His own way of thinking developed through reading Catholic ideologies, as well as **Karl Marx** and French existentialists such as Sartre. Parallel to legal studies, his interest in educational matters grew, and after completing his degree he turned to social work and education. Within the field of education he was also greatly influenced by his wife since 1944, Elza Maia Costa de Oliveira, who was a school teacher.

Working on social welfare he came into close contact with the plight of the urban poor. It was at this early stage that the basic concepts for his dialogue method in adult education started to take shape. He was eventually awarded a PhD degree in 1959 for his subsequent work in education at the

University of Recife. From his base at that university, he was able to initiate a very successful literacy programme that was diffused to the whole of Brazil. In the midst of a radical excitement, the Freirean foundation of building his pedagogy on resistance was in demand across a wide sector of the population. In today's terms, we could say that the literacy courses offered were an instrument for empowerment. The open defiance created was not desired by the new Brazilian regime then emerging from the military coup of 1964. To Freire, the coup was caused by inconsistencies within the Left in its claim for power that was not really available to them. That scared the Right into building a force that eventually led to this military take-over (Freire 1998). After spending a period in prison, Freire embarked on his long journey by being sent into exile in Chile via a short spell in Bolivia.

The adult education programme he was involved in at that time was recognised internationally by UNESCO among others. After some five years in Chile, he was invited to the USA and Harvard University as a Visiting Professor. What he now experienced provided a new input to his thinking in education, and its part in the process of social change. It was in the USA that Freire realised that exclusion was not only part of what was called the Third World, but was very much part of life in the North. Consequently, liberation was not only a struggle confined to the people of Latin America, Africa and Asia, but the USA and Europe as well. It is with these new experiences in mind that Freire (1970) finalised his most influential book, *The Pedagogy of the Oppressed.* As with his other books, this one can be seen as building directly on his own practice.

In 1970 Freire continued his journey to carry his mission further, as he ended up at the World Council of Churches in Geneva. During this period he was given an opportunity for lecturing and promoting his concepts on education in Africa and Asia. Undoubtedly Freire's intellectual work proved a major encouragement in attempts to establish a new kind of educational thinking in many of these countries. However, many regimes found the practice of this education for liberation difficult to accept. Ivan Illich (1971), his close friend and collaborator, commented on this by saying that Geneva was an exile for Freire, while his friends and followers had been imprisoned or exposed to torture. It is difficult for any government to incorporate the Freire agenda but it is a very powerful alternative for the poor in society to decide on their own future.

In 1979 Freire was invited back to his native Brazil to take up a position at the University of São Paolo. Some years later he was even appointed to the position of Minister of Education for São Paolo state, which gave him authority to direct school reforms in a major part of the country. He passed away in 1997, leaving a legacy of an educational theory but an even more powerful practice. To Freire, education could never be neutral; it was always to constitute part of a political agenda.

Freire's significance is that his work has provided essential inputs to the theoretical debate on education, but at the same time the practice he has generated is even more important. The protest against the culture of silence is summarised in a critique against the banking methodology, as a conventional custom in school. According to the Banking Model of Education, teachers have all the necessary knowledge and make deposits as to a bank. The students are the depositories that receive and memorise the information given to them passively. As in a bank, information stored in this way can be withdrawn as required at any time. This pedagogy is a reflection of the oppressive society against which Freire reacted at an early stage of life. Some of the obvious contradictions in this methodology were brought out by Freire:

- the teacher teaches – the students are taught
- the teacher knows all – the students know nothing
- the teacher thinks – the students are thought about
- the teacher talks – the students listen
- the teacher disciplines – the students are disciplined
- the teacher chooses and enforces the choice – the students comply
- the teacher acts – the students have an illusion of acting
- the teacher chooses the programme content – the students adapt to it
- the teacher is the subject – the pupils are mere objects (adapted from Freire 1970).

By this methodology, students store the knowledge that in actual fact oppresses them – making them accept things as they are. Part of the teaching by means of the banking method is to conceal certain facts that could demystify our entire existence. Therefore, they will not develop a critical consciousness that would make them able to intervene in world affairs. Banking education does not stimulate the critical thinking needed to challenge their oppressors.

Contrary to Illich, this does not lead Freire to the conclusion that the way out is *deschooling*, but rather to a way to change education. His proposed alternative is an educational model that is *problem-posing* that would replace the banking education. A key element in this is a changed relationship between the teacher and the student, in which they are involved in a dialectical *dialogue*. In this new pedagogy, teachers and students are turned into both teachers and learners at the same time, now all being subjects exchanging and recreating knowledge. In a dialogue, a question or a concept can be raised by the teacher, which in return will be answered by another question or idea. From this a synthesis will be found that, in itself, will pose a new question or thesis. Both teachers and students will learn, teach each other, as well as critically find a new way of thinking.

From a practical point of view, the pedagogy was built on the reality of the oppressed themselves, often illustrated by using a set of pictures to spark a discussion. This was combined with selecting words that could be divided into some commonly used syllables. From this, new words were constructed, gradually going from the simple to the more complex. It must be noted here that this was easy in the case of Portuguese and Spanish, but is difficult to adapt to other languages. Finally, the words were structured in such a way that it was possible to contextualise them in terms of the social and political realities. In Freire (1973), some graphic expressions of his practical approach to education are given.

The purpose of the dialogue education was to achieve a kind of political literacy – a process of *conscientisation.* With a critical pedagogy, educators and learners were to be involved jointly in a practice to demystify reality. When oppressed persons begin looking critically at the world through a joint dialogue, it constitutes a challenge to the culture of silence. The oppressed person will no longer be an object accepting the *status quo* of society, but becomes a subject reacting more actively by questioning surrounding social forces. Overcoming a state of dependency is initiated by realising that dependency exits.

Freire was a contemporary of some of the early *dependistas* in Latin America, and they were inspired by a similar intellectual tradition. In both cases, a starting point could be found in **Marx**ism or neo-Marxism. We can also see how Freire uses the example of Che Guevara, as a revolutionary who practised a constant dialogue with the people to achieve an authentic strategy for liberation action (Freire 1998). On the other hand, he is highly critical of most existing socialist societies, except Cuba and China, because they were unable to go beyond bourgeois education to create a critical mass of the new breed of people.

In Latin America the closed society, based on a culture of silence, took shape with the conquest by the Spanish and Portuguese. Since that time, only the centre of decision-making has changed – Spain, Portugal, England and the USA. To Freire, the economic dependence is obvious, but he also focuses on the educational system that serves the *status quo*. This position is not created in a specialised laboratory and brought down to Latin America. It is part of the whole relationship between the dominated in the Third World and the dominant forces of the affluent North. To aid understanding, it is necessary to analyse connections between the culture of silence and the culture that has a voice (Freire 1972).

It has been natural that much of the practical influence of the Freire pedagogy has been concentrated in Latin America. Apart from his native Brazil, we can note that Chile under Allende's socialist government (1970–73) was inspired by Freire's thinking. Another important example is the Sandinista period in Nicaragua (1979–90), which signifies one of the sincere attempts to create a revolutionary transformation, not least with an educational system appropriate to that agenda. In Africa his ideas were first taken up by some of the former Portuguese colonies, such as Angola and Guinea Bissau, both of which had a similarly progressive approach to their educational practice at an early stage. The practical experiences in Guinea Bissau have been elaborated on in Freire (1978).

One of the outstanding proponents of an alternative development philosophy for Africa was **Julius Nyerere**, in his capacity as president of Tanzania. Early on, the country was praised for its achievements in adult education, which were largely a product of the Freirean concept. Even if there were some drawbacks as Tanzania was forcefully brought into mainstream neoliberal global politics, there was a return to a new progressive literacy agenda in the late 1990s – the Integrated Community Based Adult Education (ICBAE). In addition, these kinds of programmes – the Freirean approach to education – have been adopted by various grassroots organisations and liberation movements. Among others, the Oromo Liberation Front (OLF) in Ethiopia has used this pedagogy to increase awareness among its cadres. In the North, we find that Freire has been of influence to the Black Panther Movement and feminist groups in the USA, among others. This illustrates the usefulness of these pedagogical concepts to the industrialised world as well, and to the oppressed people found in the midst of plenty.

Major works

Freire, P. (1970) *Pedagogy of the Oppressed*, New York: Seabury.
—— (1972) *Cultural Action for Freedom*, Harmondsworth: Penguin Books.
—— (1973) *Education for Critical Consciousness*, New York: Seabury.
—— (1978) *Pedagogy in Progress: The Letters to Guinea Bissau*, New York: Seabury.
—— (1985) *The Politics of Education: Culture, Power and Liberation*, London: Macmillan.
—— (1998) *Pedagogy of the Heart*, New York: Continuum.

Further reading

Illich, I. (1971) *Celebration of Awareness: A Call for Institutional Revolution*, London: Marion Boyars.

Anders Närman

JOHN FRIEDMANN (1926–)

John Friedmann is a key development thinker notable for both the quality and quantity of his contributions and for a remarkable career in which he continues to provide insights on the intellectual frontiers of development. Friedmann's line of original and formative thought over half a century is distinctive in its breadth, depth and originality. His work helped define the new field of 'regional development' and in addition he both originated and facilitated key development concepts such as core–periphery, urban fields, agropolitan development, world cities, hyperurbanisation, polarised development, empowerment and civil society – these will be explained below.

Distinctively, he has always been able to bring an interdisciplinary breadth and innovative perspective to topics at the forefront of development thinking. He understands power, space, sociology and economics. He continually derives innovative ways of combining these rich intellectual concepts to elucidate new and important meanings in our understanding of development.

Wander through the administrative halls of today's academia and the word 'interdisciplinary' brings forth positive affirmation in every corridor. It is noteworthy that John Friedmann was trained in a unique interdisciplinary tradition half a century ago and that much of his life's work can honour that tradition.

John Friedmann was born in Vienna, Austria in 1926. His family fled the Nazis in 1940 and, after a somewhat peripatetic existence (during which he became a US citizen in 1944), he earned his MA and eventually PhD from the University of Chicago in 1955. It is notable that his PhD was undertaken in 'a short-lived Program for Education and Research in Planning' under the direction of Harvey Perloff (Friedmann 2002: 119). Uniquely, this programme was interdisciplinary and his PhD was conferred in the fields of 'planning, economics and geography'.

It is important to realise that John Friedmann and his mentors were among the first scholars formally to break disciplinary boundaries. Normally, planners were trained within schools of architecture. Instead, John Friedmann received significant training in the social sciences and began to think of planning as an applied social science. This approach resounds in his life's work.

In a personal retrospective (*ibid.*: 2002), Friedmann acknowledges several important early influences. His father, Robert Friedmann, was a historian and philosopher who introduced him to the works of American urbanists Reinhold Niebuhr and Lewis Mumford, and the political theorist Hannah Arendt, and encouraged him to wander through the intellectual minefields of the philosophy of science. He also engaged with the Jewish philosopher, Martin Buber. 'I inherited from this a strong sense of moral purpose, a philosophical disposition, and (ultimately) a strong sense of history that questioned the possibilities of reason as an active force in history' (Friedmann 2002: 120). Linking the large concepts of society, power and space established Friedmann as a 'new thinker' in development.

Friedmann's unique degree at the University of Chicago carried a fine pedigree – the new degree programme having been started by the former Governor of Puerto Rico, Rexford Tugwell. Tugwell brought in many scholars, including planner Harvey Perloff, who was to become Friedmann's Advisor and eventual colleague at the University of California at Los Angeles (UCLA). John Friedmann also worked with urban designer Melville Branch, economic planner Julius Margolis, technology specialist Richard Meier and housing expert Martin Meyerson during his graduate years. Friedmann was also a witness to the inception of neoliberalism (a firm conservative belief in the market system) by economists Milton Friedman and Theodore W. Schultz that led the University of Chicago's Department of Economics (and the country and the globe) on a political economic course that today dominates global policy.

Regional development was a very new concept at the time, and Friedmann's dissertation study of the Tennessee Valley Authority was a pioneer effort to indicate how decisions could be made in 'regional' ways that included multiple states, counties and local authorities. Friedmann showed that 'watershed' boundaries were not necessarily appropriate to a dynamic post-war economy. The effect on cities was a key, but so, too, was the persistence of rural poverty.

Following the award of his PhD, Friedmann spent six years in Brazil and Korea working with US development agencies. He taught, he wrote, he absorbed, he planned (alas, his plans did not gain support from the United States Agency for International Development – USAID). In 1960, the Massachusetts Institute of Technology (MIT) – a premier institution in the planning field – recruited Friedmann to its faculty. He quickly became engaged in a joint research programme in Venezuela focused on regional development.

In 1964, Friedmann left MIT to be the Director of the Ford Foundation programme in Chile. As for most planners, the trade-off between taking action and simply thinking is a true challenge, although they can be happily and appropriately interspersed. Friedmann accomplished this. During this period he identified the concept of *hyperurbanisation*, which he defined as a doubling rate in the size of urban population in less than twenty years. He accurately noted that this process would lead to a new situation of political awareness that would substantially change the *status quo*. The political history of Chile proves his case.

In 1965, Friedmann and John Miller (his graduate student) published an article on 'Urban fields' that extended Jean Gottmann's (1961) analysis out of a single case and showed how urban areas across the United States were coalescing in a variety of important ways in economic and spatial dimensions. They predicted that 85–90 per cent of the American populace would soon be living in an 'urban field'.

The *core–periphery concept* has its strongest origin, especially at the regional scale, in Friedmann's (1966) work during this period. He showed clearly that core–periphery relationships worked strongly on the regional scale, and brought space, power and society into unequal relationships. This concept eventually became a critical component for world systems theory as formulated by Immanuel Wallerstein (1974). Friedmann developed the concept of nearly immutable power relationships that transcend scales between rich and poor and translate into economic/development relationships. Indeed, his work was somewhat scale-related to regional levels. Nonetheless, these relationships soon proved to be scale-transcending and proved to be the basis of a large corpus of local–global theory on scale relationships.

In 1969 John Friedmann accepted an academic position in planning at UCLA. He clearly helped build UCLA into one of the best planning programmes in the country. Graduate students were mesmerised by the onslaught of knowledge he provided, both in his classes and in the programme. Somehow he managed to have the likes of David Harvey, **Samir Amin**, Janet Abu-Lughod, Manuel Castells and other famous figures visit for a whole week. They would give a colloquium, but graduate students would also meet with them after reading their works, and they would participate in scheduled seminars. Many of the graduate students interviewed admitted they felt they were at the centre of the intellectual universe. Through this process, Friedmann not only gave many students a special education but also provided a model for how best to teach graduate students.

During this period, John Friedmann engaged the popular concept of *growth centres*. His work on *polarised development* (1972) called into question the spatial dynamics and possible patterns of growth centre implementation and harkened back to his earlier arguments on the core–periphery concept. This work was further elucidated in *Urbanisation, Planning and National Development* (1973) which addressed questions of the role of the nation-state in urban and regional planning.

A *pragmatist* is someone who has an approach to the world that asks two questions: (1) is there a problem? and (2) can we help solve it? As a probable pragmatist, Friedmann wrote *Retracking America* (1973). In this book he introduced the 'communicative turn' into planning and emphasised the possibilities of transactive planning. One real question is whether Friedmann is a simple pragmatist or a 'radical' as he claims himself to be. We will return to this.

John Friedmann has always been engaged with urban questions that involve space, society and economy. In *Territory and Function* (Friedmann and Weaver 1979), the concept of *agropolitan development* acknowledged *the* rural–urban linkage that was of vital importance in the economics of the developing world and spoke to the importance of the agricultural hinterland that supported a city's economic existence and viability. In this concept, Friedmann refocused planning on rural development as the engine that drives urban development in small- and medium-scale urban centres. Again, he re-emphasised the region as a critical factor in development.

As John Friedmann himself notes, ' "World City" was an idea waiting to be born' (2002: 145). As a major development concept, *World City* received an initial formalisation by Friedmann and Wolff (1982) and was further elaborated as Friedmann's (1986) 'The World City Hypothesis'. Simply, the 'World City' exploded the nation-state-based urban hierarchy to reflect major city roles in a new globalising economy. This is a powerful extension of urban thought, reflected, for example, in a major evaluative and prospective conference on the hypothesis and its application in 1993, at which he was guest and keynote speaker (Friedmann 1995; Knox and Taylor 1995).

Empowerment (1992) encapsulated Friedmann's approach to development. As the world changed with the collapse of regimes in the Soviet Union and Eastern Europe, Friedmann argued for an alternative development theory. He explicitly inserted the human development link to economic development, asking important questions that once again link political power and economic development. Disempowerment is seen as coming from poverty. This disempowerment has social, political and psychological components. This is very much in line with **Amartya Sen**'s (1999) later work on *Development as Freedom* that provided a 'morally informed framework' for embedding policies that are pro-poor in the structure of mainstream development planning. Friedmann's empowerment work benefits from a 'creative tension' between mainstream and poverty initiatives that outlined viable futures for the poor of the world.

John Friedmann (2000) even engaged the *sustainable city* initiative with his work on Mexico City. This is just another instance in which he is not following, but leading the debate on important urban and regional issues.

His current research focuses on the power of the household in development, the significance of livelihoods research, and also inspects the increasing importance of civil society. *The Prospect of Cities* (2002) continues to raise important questions, while his latest volume (Friedmann 2005) surveys the unprecedented urban transition in China.

John Friedmann not only continues to make major contributions to the field of development, but he does so with a remarkable persistence, resilience and a wise person's sense of what is needed. His book on philosophy, *The Good Society* (1979), should be required reading by all who have some allegiance to the pragmatic tradition.

John Friedmann made the 'regional' important to development. His interdisciplinary focus provided numerous signal intellectual contributions that ranged across disciplines. He is clearly a 'key thinker' in development, but he is best recognised for his role as an ethical and moral leader in development.

Major works

Friedmann, J. (1966) *Regional Development Policy: A Case Study of Venezuela*, Cambridge, MA: MIT Press.

—— (1972)'A General Theory of Polarised Development', in N.M. Hansen (ed.), *Growth Centers in Regional Economic Development*, New York: Free Press, pp. 82–107.

—— (1973) *Retracking America: A Theory of Transactive Planning*, Garden City, NY: Doubleday/Anchor.

—— (1973) *Urbanisation, Planning and National Development*, London: Sage.

—— (1979) *The Good Society: A Personal Account of its Struggle with the World of Social Planning and a Dialectical Inquiry into the Roots of Radical Practice*, Cambridge, MA: MIT Press.

—— (1986) 'The World City Hypothesis', *Development and Change* 17(1): 69–84.

—— (1992) *Empowerment: The Politics of Alternative Development*, Oxford: Blackwell.

—— (1995) 'Where We Stand: A Decade of World City Research', in Knox, P.L. and Taylor, P.J. (eds), *World Cities in a World-System*, Cambridge: Cambridge University Press.

—— (2000) *Human Settlements and Planning for Ecological Sustainability: The Case of Mexico City*, Cambridge, MA: MIT Press.

—— (2002) *The Prospect of Cities*, Minneapolis: University of Minnesota Press.

—— (2005) *China's Urban Transition*, Minneapolis: Minnesota University Press.

Friedmann, J. and Miller, J. (1965) 'The Urban Field', *Journal of the American Institute of Planners* 31(4): 312–19.

Friedmann, J. and Weaver, C. (1979) *Territory and Function: The Evolution of Regional Planning*, Berkeley, CA: University of California Press.

Friedmann, J. and Wolff, G. (1982) 'World City Formation: An Agenda for Research and Action', *International Journal for Urban and Regional Research* 6(3): 309–44.

Further reading

Gottmann, J. (1961) *Megalopolis: The Urbanized Northeastern Seaboard of the United States*, New York: The Twentieth Century Fund.
Knox, P.L. and Taylor, P.J. (eds) (1995) *World Cities in a World-System*, Cambridge: Cambridge University Press.
Sen, A. (1999) *Development as Freedom*, New York: Knopf.
Wallerstein, I. (1974) *Capitalist Agriculture and the Origins of the European World Economy in the Sixteenth Century. Vol. 1 of the Modern World System*, New York: Academic Press.

Gary Gaile

MOHANDAS (MAHATMA) GANDHI (1869–1948)

> If you cannot change yourself, how can you change the World?
> Gandhi

Mohandas Karamchand Gandhi, known as '*Mahatma*', meaning 'great-souled' as people called him, was born on 2 October 1869 in Western India and went to England in 1888 to study law. He returned to India in 1891, and after practising law for some time he moved to South Africa in 1893, where he was a determined opponent of the 'pass laws' and other kinds of racial discrimination. There he organised '*Satyagraha*' (non-violent action, or 'passive resistance') in 1906, 1908 and in 1913. With a view to serving his country of birth, he returned to India in 1914 and soon became a leading figure in the cause of Indian nationalism and development movements. In 1915 Gandhi established the Satyagraha Ashram at Ahmedabad, and in 1917 moved it to the Sabarmati River. On 30 January 1948 Gandhi was shot dead by a Hindu fanatic, while holding a prayer meeting in Delhi.

In the late 1940s, the term 'development' was not in currency as it is today. Gandhi had, therefore, used the term 'progress' for development, more with respect to ethics and cosmic integrity. He said, 'by economic progress we mean material advancement without limit, and by real progress we mean moral progress, which again, is the same thing as progress of the permanent elements in us' (*The Collected Works of Mahatma Gandhi* (*WMG*), vol. 87: 249). He never made an academic contribution to 'development'; instead he pleaded for an organised effort to change the ruling paradigm and move towards a superhuman stage. The Gandhian idea of development is based on the foundational ethics of *ahimsa* (non-violence, which he interpreted as 'firmness in the truth'). Perhaps Gandhi's most important influence has been on the black civil rights movement in the USA led by Martin Luther King.

In the cosmic-moral organisation, faith is the path of spirituality, and spirituality in its true sense is the motive force behind development. Gandhi suggested 'Seven Social Sins' to be avoided. Of course, these are ideals, but they are more relevant in the present era of desperation and could easily be accepted. According to Gandhi, the 'Seven Social Sins' are:

(i) *Consumption without conscience.* The means and symbols of consumption are the 3Ps, namely property, power and prestige. Without conscience or moral responsibility, consumption always turns to evil, in other words, eating food is consumption, but without conscience it leads to sickness. Consumerism is one of the basic root causes of social conflict.

(ii) *Knowledge without character.* Knowledge is power, but without morality it is hypocrisy. With a lack of morality knowledge becomes a heavy burden on humanity.

(iii) *Wealth without labour.* In the modern era, wealth is the highest symbol of projecting the level of personality and status. This promotes the tendency of the 'rich becoming richer and the poor becoming poorer'. Such rich people never realise their social responsibility; this leads to social crisis.

(iv) *Business without morality.* According to the philosophy of welfare economics, the basis of economic success is morality. Gandhi said, 'Economics that hurt the moral well-being of an individual or a nation are immoral and, therefore, sinful' (*WMG*, vol. 13: 317, also *cf. WMG*, vol. 75: 158).

(v) *Religious duty without sacrifice.* Compassion, service, sacrifice and altruism are the basic rules of *dharma* (righteous duty). Lacking the sense of sacrifice promotes individualism, egocentricity and selfishness. Serving the poor is service to God.

(vi) *Science without human sense.* Science has the power of both creation and destruction; its nature depends on the way in which it is used by mankind. Misconception of science as a tool and technique for the use of resources has promoted the philosophy of consumerism, which finally resulted in environmental disorder, pollution and loss of peaceful life. Science is the means, but the end is human kindness!

(vii) *Politics without principles.* Politics without principles is the cause of the global crisis and destruction of social harmony. Gandhi emphasised the need for spirituality as the first step of politics and governance.

Whenever the ethical code of a society is lost, civilisation will end. In such a situation, a cultural tradition is unable to find a balance between the needs and expectations of society, and that is how it fails to illuminate a path of revolution. In this way, the road to progress and development comes to a dead-end. Gandhi proposed a philosophy of revival and peace which he called *Sarvodaya* ('well-being for all'). Enhancing personal welfare (*sva-*) to the level of communal well-being (*sarva-*) is the cultural code of *Sarvodaya*. History has proved that the alternative to war and conflict is *ahimsa* ('non-violence'), i.e. peaceful agitation, creating mass awareness of cultural unity.

There are two ideas inherent in the philosophy of *Sarvodaya*:

(i) *Democracy is a life style.* Democracy is not only a way of governance, it is also a way of life. Gandhi warned that a process of dialogue and criticism must always be maintained between the public and ruling powers. The continuing trend towards loss of public awareness is a sign of the decline of democracy.

(ii) *Machines have a cultural value.* The use of *machines* (a product of science) also has a cultural value. The application of science, or its tool, the machine, depends on human intention. Science is like a machine, the effects of whose use depend upon the attitude and motives of the person who has control over it. The path on which science is presently advancing will lead to a great dissolution. Cultural values are imposed upon machines by human beings (*cf. WMG*, vol. 25: 251–2).

Gandhi's view of non-violence, *ahimsa*, vegetarianism and *karma* (right action) is based on the idea of the total spiritual interconnectedness and divinity of life as a whole. All natural phenomena are, therefore, divine, sacred and of equal value. As human beings, we have to take the main responsibilities towards nature through a moral-ethical-religious approach. His theory of *ahimsa* was not strict like a sectarian rule; he said: 'whoever believes in *ahimsa* will engage himself in occupations that involve the least possible violence'.

Committed to non-violence *(ahimsa)* and self-realisation *(svachetana)*, Gandhi wanted to solve India's problems from the perspective of individual conversion and with the ideology that 'every man has an equal right for the necessities of life even as birds and beasts'. According to him, the 'right thing' is a moral order *(dharma)* operated by right action *(karma)*. Gandhi's emphasis upon self-realisation and rules of conduct and virtues is essential for spiritual life and also for the maintenance of the social order. Gandhi warned politicians about the social and political evils of which they become part!

Gandhi said, 'every human being has a right to live and, therefore, to find the wherewithal to feed himself and where necessary to clothe and house himself' (*WMG*, vol. 38: 197). He said, 'there's enough in the world to meet the needs of everyone but there's not enough to meet the greed of everyone'. If every little village became self-sufficient then one would have an ideal setting for life. Therefore he did not advocate macro-economic policy. All villages should become self-sufficient. It must work from the bottom upwards and not the top down.

As we work in different walks of life, we should continue to be inspired by his words that we must 'recall the face of the poorest and the weakest man (and woman) whom you have seen and ask yourself if the step you contemplate is going to be of any use to him (and her)' (*WMG*, vol. 89: 125).

Gandhi practised what he preached. If he was concerned about the 'poorest of the poor' he adopted a life style which reflected his constituency. And if he preached 'cleanliness' and the uplifting of Harijans ('untouchables'), he himself undertook the scavengers' work, and emphasised the importance of inner and outer cleanliness. To follow Gandhi is difficult and yet millions did. He emphasised throughout that we must change ourselves before we can change others and that our real enemies are within.

The essence of natural cures is that we learn the principles of hygiene and sanitation and abide by those laws as well as the laws relating to proper nutrition. If rural reconstruction does not include rural sanitation, our villages will remain the dung-heaps that they are today.

Gandhi's twenty-one years of experience in South Africa transformed his views on life and human existence, which he again experimented with in India. He started to look at the world from a poverty-trapped peasant's perspective, rather than from a middle-class bourgeois perspective. This led him to the creation of three principles of sustainable development: *Sarvodaya*, *Swadeshi* and *Satyagraha*. The other very important feature of sustainable development which he propagated was the whole question of local economy where everybody in the area would be self-sufficient. They would be employed and could sustain themselves and their families with dignity and work.

For Gandhi the *swadeshi* spirit extended to all the elements composing the *desh* (community) and implied a love of not only the traditional way of life but also the natural environment and especially the people sharing it. Gandhi used the term *swaraj* to describe a society run in the *swadeshi* spirit. It meant self-rule or autonomy and implied not only formal independence but also cultural and moral autonomy.

The removal of untouchability, an end to Hindu–Muslim enmity, economic self-sufficiency, and making non-violence effective were fundamental elements of his vision. Gandhi offered a viable alternative for self-rule and self-government. He focused on the welfare of all – *sarvodaya* – and the method he followed was that of *satyagraha*.

Sarvodaya ('the uplifting of all') was a philosophical position that Gandhi maintained. Society must strive for the economic, social, spiritual and physical well-being of all, not just the majority. He favoured a holistic approach to well-being, and a total approach to the community. For him the well-being of every individual was an important concern.

He advocated that the locus of power must be situated in the village or neighbourhood unit. He believed that there should be an equitable distribution of resources and that communities must become self-sustaining through reliance on local products instead of large-scale imports from outside. Gandhi was opposed to large-scale industrialisation, and favoured small local industries that promote local self-sufficiency, which he called *Swadeshi*. In current terms, it means buy local, be proud of local, support local, uphold and live local. Economic equality should never mean possession of an equal amount of worldly goods by everyone. It does mean, however, that everyone should have a proper house, an adequate and balanced diet, and sufficient cloth(ing). It also means that the cruel inequality that obtains today will be removed by purely non-violent means.

Finally, Gandhi's best-known theory of *satyagraha*, 'truth force' or non-violent direct action, is actually a way of life, not just an absence of violence. It also entailed respect for all beings regardless of religious beliefs, caste, race or creed, and a devotion to the values of truth, love and responsibility. Mass awakening exemplified by the *satyagraha* movement, is 'the perfect example of how one could confront an unjust and uncaring, though extremely superior power' (*WMG*, vol. 9: 118, and *cf.* Gandhi 1990: 21). The achievement of political and moral ends through *ahimsa* is what Gandhi called *satyagraha*. This notion of non-violent action is the crucial part of Gandhi's political theory. *Satyagraha*, in fact, is a theory of action. It calls for courage, strength of character and positive commitment to a righteous cause. In some circumstances, e.g. inhuman acts like molestation or killing, it might be better to choose violence than craven submission to injustice. Gandhi thought that 'total non-violence' might be feasible only when mankind has acquired superhuman qualities.

The spectres of global warming, lack of water through deforestation, and continued depletion of natural resources and diversity on Earth, are some of the results of 'unsustainable development' and economic growth. Should we not follow the path of Gandhi? He corrected himself over and over in his striving for moral-spiritual perfection while doing his best to reform existing institutions and social practices. He wrote in 1932, 'I do not accept defeat but hope, with God's grace, to melt the stoniest heart and, therefore, continually strive to perfect myself' (*WMG*, vol. 50: 451).

We are slowly realising that reducing poverty or moving towards sustainable development is not just an economic or a technical problem or one of acquiring greater financial inputs. All these are important but achieving these goals also needs an inner awakening, an inner transformation of man – a path already paved by Gandhi. If we ignore Gandhi 'we ignore him at our own peril', Martin Luther King once said.

Major works

Gandhi, Mohandas Karamchand (1927) *An Autobiography or the Story of My Experiments with Truth*, Ahmedabad: Navajivan.

—— (1928) *Satyagraha in South Africa*, Ahmedabad: Navajivan.

—— (1958–94) *The Collected Works of Mahatma Gandhi*, 100 vols, New Delhi: Publication Division, Ministry of Information and Broadcasting, Govt of India.

—— (1990, 1997) *Hind Swaraj and Other Writings*, ed. A.J. Parel, Cambridge: Cambridge University Press.

Further reading

Adams, I. and Dyson, R.W. (eds) (2003) 'Mohandas Gandhi', in *Fifty Major Political Thinkers*, London and New York: Routledge, pp. 196–9.

Bondurant, J.V. (1965) *Conquest of Violence: The Gandhian Philosophy of Conflict*, Berkeley, CA: University of California Press.

Hick, J. and Hemple, L. (1989) *Gandhi's Significance for Today*, New York: St. Martin's Press.

Parekh, B. (1997) *Gandhi*, Oxford: Oxford University Press.

Parel, A.J. (2000, 2002) *Gandhi, Freedom, and Self-Rule*, Lanham, MD: Lexington Books; Indian edition, New Delhi: Vistaar Pubs.

Pyarelal (1986) *Mahatma Gandhi: The Birth of Satyagraha*, Ahmedabad: Navajivan.

Singh, R.P.B. and Singh, R.S. (2002) 'Development in India: Scenario and Ideology', in Anders Närman and K. Karunanayake (eds), *Towards a New Regional and Local Development Research Agenda*, Dept of Geography, Göteborg University, Sweden, Series B, No. 100 and Centre for Development Studies, University of Kelaniya, Sri Lanka, No. 1, pp. 65–78.

Swan, M. (1985) *Gandhi: The South African Experience*, Johannesburg: Ravan.

Tendulkar, D.G. (1951–4) *Mahatma: The Life of M.K. Gandhi*, 8 vols, New Delhi: Ministry of Information and Broadcasting.

Terchek, R.J. (1998) *Gandhi: Struggling for Autonomy*, Lanham, MD and New York: Rowman & Littlefield.

Rana P.B. Singh

SUSAN GEORGE (1934–)

Susan George is unique among her contemporaries within the world of development thinking in having consistently identified the most fundamentally important global issues of our time before they were universally recognised. As a progressive scholar-activist, her highly innovative work has ranged from a focus on world hunger, poverty, debt, and North–South relations, to a series of critiques of the world capitalist system and the major players within it, particularly the World Trade Organisation, World Bank and other international financial institutions. A deep-seated concern throughout all her work has been with global injustice. Inherent in this has been a practical preoccupation with campaigning for change in the world at large, reflected particularly in her more recent work on the corporate-led neoliberal globalisation and the global justice movement. Indeed, it is probably safe to say that Susan George has had an unrivalled prominence as a development thinker over the past four decades. This reflects her ability to engage not just with academics, but with activists, journalists, as well as untold thousands of school children and university students whose lives have been changed by her work. Her influence is not just restricted to the English-speaking world as her work has been widely translated and published in French, Spanish, Italian, German, Portuguese, four Scandinavian languages, as well as Thai, Korean, Bengali, Estonian and Japanese. While she communicates her research through books and journals, she also writes for, or is interviewed frequently for, newspapers and other media around the world, as well as speaking at scores of influential conferences, notably successive World Social Fora and European Social Fora. Also significant is that George is listened to by a range of different constituencies; unlike so many scholars, her work is read by world leaders, politicians, great thinkers of our time, as well as the general public and those specifically interested in development issues.

Born in the USA, Susan George has lived all her adult life in France, gaining French citizenship in 1994. While her early academic career was in the USA, where she studied for her BA degree in French/Government from Smith College, Northampton, Massachusetts in 1956, she later attained her doctorate from the Sorbonne in Paris in 1978 with a thesis focusing on the transfer of the food system from the USA to the rest of the world. This was later published (George 1981). She is currently the Associate Director of the Transnational Institute in Amsterdam where she has been based non-residentially since its foundation in 1973; this is a global network of scholar-activists who work together to analyse, challenge and provide alternatives to global inequality, poverty and injustice. She is also the Vice-President for ATTAC France (Association to Tax Financial Transactions to Aid Citizens) which was founded in France in 1998 and which currently operates in fifty-one countries. ATTAC is an 'action-oriented education movement' that aims to tax both financial markets and transnational corporations with a view to redistributing income across the world. George has also served on the board of Greenpeace International (1990–5) as well as Greenpeace France.

Not surprisingly, and by her own admission, Susan George is not a scholar in the conventional sense of being an economist or development theorist. Rather than hindering her career, however, this has been her greatest strength. Free of the strictures of academia where, in her view, orthodoxy is revered, George has been able to challenge the injustices of the world system with a passion and foresight often missing from academic writing. This talent for adopting a powerful, yet unconventional approach is reflected in her first book, *How the Other Half Dies* (1976). Based on her work with a team set up to write a report for the World Food Conference held at the Food and Agriculture Organisation (FAO) in Rome in 1974, the book provides a critique of the standard accounts of why world hunger prevailed at the time. Focusing on how the reasons for hunger are part of a wider capitalist system based on unequal power relations between North and South as well as between the poor and the elites within countries, she criticises a host of phenomena hitherto viewed as benign. These include the functioning of agribusiness, the Green Revolution, and the corrupt operation of food aid. However, her most path-breaking point in the book is to highlight that famine, poverty and hunger have little to do with natural disasters, but rather with global politics and exploitation on the part of the world's powerful.

As well as the content of the book, the context of its publication and aftermath are also significant. In her latest book, *Another World is Possible if ...* (2004), George explicitly acknowledges that she was writing against the grain of contemporary opinion at the time, and that while it was acclaimed by the general public, it did not receive a warm reception in academic circles. For this reason, she began to study for her doctorate to counteract the 'pompous, usually male, professors proclaiming that "this woman has no credentials for saying all these terrible things she's saying"' (*ibid.*: 205).

With these credentials duly earned, George continued to work on issues that others were either afraid or unable to identify as central to world politics. With her flair for presenting the facts in bold terms, and for using simple, yet sophisticated arguments, she carried on her pioneering work exposing the powerful actors benefiting from the inequalities of the global system, as well as the exploitative nature of the system itself. Again ahead of her time, Susan George began to work on the issue of debt through the publication of *A Fate Worse than Debt* (1987). This highlighted the debilitating effects of IMF and World Bank lending, together with loans from private banks or so-called 'money mongers', to the South. Caught in a vicious circle of debt and loan servicing, Southern countries had no choice but to institute the conditionalities imposed on them, often in the form of

Structural Adjustment Programmes. George also noted how the overall flow of money ended up in favour of the North in that the Southern countries remitted more in debt repayments and dividends to the North than the latter transferred in aid. Again, George pioneered work that highlighted the deleterious effects of these policies on the poor and marginalised of the South.

George and her associates at the Transnational Institute recognised too that debt also affected ordinary citizens of the North. In *The Debt Boomerang* (1992), George identified six so-called 'debt connections' which affected the North: the environment; the drugs industry; Northerners providing huge subsidies to private banks; the loss of jobs and markets; immigration; and conflict and war. While few of these issues had been highlighted previously, the consideration of the ecological damage of debt, also discussed in *A Fate Worse than Debt*, was especially important. Indeed, George maintains to the present day that 'capitalism and environmental sustainability ... are logically and conceptually incompatible' (*ibid.*: 29), with transnational corporations being responsible for much environmental destruction in the world, bolstered by the policies of the international financial institutions. Indeed, she views the ecological future of the planet in truly apocalyptic terms; only systematic public spending on the part of all countries of the world can halt the world's destruction.

The role of the international financial institutions and the potential evil of the capitalist system if not controlled are other enduring themes of George's work. While these issues were identified in her early publications, she focused more specifically on them in later writing, first, in *Faith and Credit* with Fabrizio Sabelli (1994), and later, in *The Lugano Report* (George 1999). The former focuses on the World Bank, interpreted as a religious institution with the ultimate goal of being 'the visible hand of the programme of unrestrained, free market capitalism' (1994: 248). *The Lugano Report* is a partly fictitious account purportedly based on the views of a 'working party' commissioned to examine the future of capitalism, focusing on the threats and obstacles to preserving it. At times rather fanciful and certainly very entertaining, it also illustrates why Susan George's work has such large public appeal; she is able to write in innovative ways and does not flinch from using iconoclastic devices to make her point.

The Lugano Report also follows the other thread in her writing, which is to highlight potentially apocalyptic scenarios (see above). In this case, she suggests that by the year 2020 capitalism cannot be preserved without the elimination of two-thirds of humanity, made possible through restricting reproductive freedom, war, famine, limiting access to land and water, and ultimately allowing diseases of poverty to flourish, not least AIDS.

However, despite the pessimism that sometimes afflicts George's work, she is also careful to offer strategies for change. In a *Fate Worse than Debt*, for instance, she outlined her '3-D solution of debt, development and democracy' otherwise known as 'creative reimbursement' where, among other things, debtor countries can pay back their loans in local currencies over a long period of time (1988: 243–62). However, George's ideas about change for the future are most concisely outlined in *Another World is Possible if* ... As well as providing an insightful analysis of neoliberal globalisation, this work also acts as a handbook for those wishing to contribute in some way to challenging global inequalities. While she supports and proposes such initiatives as taxing financial markets, including the Tobin Tax (see **James Tobin**) (as in her work with ATTAC), she also provides some clear practical guidelines for those wanting to become involved in the global justice movement such as her 'seven commandments' that include how to plan meetings and produce leaflets (2004: 164–6).

Ultimately, however, it is important to stress that while Susan George's work has been invaluable in highlighting the potentially catastrophic path that capitalist development might lead us down, and to press home the inordinate injustices that plague the world, she remains optimistic. Not only does she have faith in people working and organising at the grassroots in particular, she also truly believes that hope is the key, however tenuous. To end with her own words: 'Despite everything, however, I don't just go about mouthing the slogan "Another world is possible". I actually believe it, while recognising that it's a very long shot and a fragile hope' (*ibid.*: 116).

Major works

George, S. (1976) *How the Other Half Dies: The Real Reasons for World Hunger*, London: Penguin.

—— (1981) *Les Stratèges de la Faim*, Geneva: Grounauer with the Institut Universitaire d'Études de Développement, University of Geneva.

—— (1987) *A Fate Worse than Debt*, London: Penguin.

—— (1990) *Ill Fares the Land*, London: Penguin.

—— (1992) *The Debt Boomerang: How Third World Debt Harms Us All*, London: Pluto Press with the Transnational Institute.

—— (1999) *The Lugano Report: On Preserving Capitalism in the Twenty-First Century*, London: Pluto Press.

—— (2004) *Another World is Possible if* ... , London: Verso.

George, S. and Sabelli, F. (1994) *Faith and Credit: The World Bank's Secular Empire*, London: Penguin.

Further reading

George, S. (2002) 'ATTAC: A Citizen's Movement for Global Justice Responds to September 11th', *Development* 45(2): 97–8.
www.tni.org/george.

Cathy McIlwaine

ALEXANDER GERSCHENKRON (1904–78)

Alexander Gerschenkron was born in 1904 in Odessa, Russia. His family moved to Vienna after the Russian Revolution and here he received his doctorate, at the University of Vienna in 1928, later becoming an academic at the Austrian Institute of Business Cycle Research (then headed by Friedrich Hayek). He left for the USA in 1938 on the day of the German invasion of Austria, and after working in Berkeley and Washington, became a Professor at Harvard University in 1946, where for many years he directed economic research in the Russian Research Centre (Blaug 1998: 82). He retired in 1974 and died in 1978. His thinking was heavily influenced by his association with, and analysis of, the economies of Russia and Eastern Europe, and he stressed the need for broad geographical frameworks in the study of economic history and development. 'Insularity is a limitation on comprehension,' he argued in his seminal essay (Gerschenkron 1962: 6). At Harvard he was joined by other European émigré scholars such as Bert Hoselitz and **Albert Hirschman**, whose theories of unbalanced growth have strong connections with Gerschenkron's ideas.

It is over half a century (1951) since Alexander Gerschenkron wrote his seminal essay, 'Economic Backwardness in Historical Perspective'. The essay was published eleven years later in a slightly revised form as the first chapter of a book of his essays with the same title (Gerschenkron 1962). The essay addressed the **Marx**ian generalisation that the history of advanced industrial economies traced out the pathway to development for more backward countries. Gerschenkron (1962: 7) thought that this was a half-truth because in many crucial respects 'the development of a backward country may, by the very virtue of its backwardness, tend to differ fundamentally from that of an advanced country'.

Gerschenkron saw a tension between the actual state of economic activities in a backward country on the one hand and the great promise inherent in industrial development on the other. He asked the question: what does it take for a 'latecomer' country to industrialise and become like the early industrialisers such as Britain and Germany? He took the 'latecomer' economies of late nineteenth-century central and eastern Europe as his source and discussed their links and involvement with an increasingly international economic system. His answer was essentially in the form of a metaphor. Latecomer countries would need 'to leap the gap of knowledge and practice separating the backward economy from the advanced' (Landes 1999). Economic history revealed that relative backwardness could be overcome by novel strategies and pathways to industrialism and accelerated growth. In his analysis, history could ride roughshod over any neat recipes of prerequisites as outlined by **Walt Rostow** (1960). The rivalry between Gerschenkron's ideas and those of Rostow since the 1960s has been analysed by a number of authors (Trebilcock 1981; Gwynne 1990; Gootenberg 2001).

Gerschenkron's main concept revolved around the complex nature of, and opportunities provided by, economic backwardness and deprivation. The model was concerned with the beginnings of industrial growth and assumed under-endowment as the starting point. Gerschenkron argued that as other countries advance and backwardness deepens, the underprivileged society will become increasingly sensitive to the contrast. Intellectuals and politicians could transfer this into national ideologies 'igniting the imaginations' of people and instilling faith that the 'golden age lies not behind but ahead' (Gerschenkron 1962: 24). As social tension increases, a vast effort is made to bridge the gap and a lunge for the benefits of industrial growth occurs. Dore (1990: 359) followed Gerschenkron in arguing that the sense of backwardness 'provided a charter for state action to mobilise resources and to take initiatives and risks'. From Dore's comparative work on Latin America and East Asia, he concluded that the urgency of the desire to catch up had been stronger among the political leaders and intellectuals of Asian countries after the Second World War due to cultural as much as economic reasons.

According to Gerschenkron, rapid industrial growth was deemed possible as backwardness could convey a number of economic advantages as well. The more backward the country, the more sophisticated will be the industrial equipment, technology and plant it can select for its manufacturing debut. The country can import technologically advanced machines with the theoretical advantage of enjoying the most significant economies of scale available and hence lower per unit costs. In theory, the costly stages of technological development may be skipped and new industrial systems installed.

This also provided the theoretical justification for a capital-intensive industrial pathway in contrast to a more labour-intensive one. Gerschenkron (1962: 9) emphasised the problems inherent in creating a large and efficient industrial labour force. These ideas were influential in the capital-intensive policies linked to import-substitution industrialisation

(ISI) in Latin America. Bennett and Sharpe (1985: 41) pointed out that countries in Latin America entered into ISI not with Gerschenkronian producer goods but with previously imported goods. Hence they coined the phrase 'late late industrializers' for countries that had even further to catch up and had to develop products and processes that were even more costly, risky and technologically sophisticated than the 'late industrialisers' of Russia and Eastern Europe.

Gerschenkron's comparative approach can be seen fifty years later in the study of 'duelling peripheries' of the global economy. How apt and useful are the development and historical comparisons between Eastern Europe and Russia, East Asia and Latin America (Gwynne, Klak and Shaw 2003)? One example of comparative work between Eastern Europe and Latin America is Love (1996), who argues for a link between Romanian development debates (in their latecomer stage, 1880–1940) and the later Brazilian push towards rapid industrialisation. Gerschenkron had observed that some industrial latecomers during both the nineteenth and twentieth centuries grew at a faster rate than their predecessors. Gerschenkron relied on the work of his colleague at Harvard, Henry Rosovsky (1966), for material on the Japanese comparison. Since his essay, the rapid growth of South Korean and Taiwanese industry in the 1960s and 1970s (Wade 1990) and of China since the mid-1980s, has added more contemporary weight to this thesis that benefits can be conferred on industrial latecomers.

Another key concept focused on the role of institutions in this economic transformation. In situations of relative backwardness, the scales of institutions, firms and plants loom large. Gerschenkron saw the importance of large-scale 'universal' banks dedicated to long-term investment in heavy industry. For example, he argued that these banks (in which commercial and investment banking were combined) had been the key institution in Germany's rapid industrial growth in the late nineteenth century. Forsyth and Verdier (2003) have recently shown that financial systems were regionally more complex in European countries than in the interpretation put forward by Gerschenkron but they still used his framework for comparative purposes.

However, for more backward countries (and particularly his case study of Russia, where 'fraudulent bankruptcy had been almost elevated to the rank of a general business practice' – Gerschenkron 1962: 20) the creation of universal banking institutions was inconceivable. Thus Gerschenkron argued that under these conditions the role of the state became fundamental for late industrialisation. He did not idealise this pivotal role of the state and noted the inefficiencies, incompetence and corruption that had occurred in the Russian experience. However, in spite of this, the role of the state became pivotal to generate the necessary finance for and growth in large-scale heavy industry.

Indeed, as the input needed to launch industrialisation increased with backwardness, it was argued that the process required a correspondingly larger state response, particularly in terms of finance. This could be linked to the wider concept of 'substitutes'. If a key element for growth was lacking, such as access to finance, European societies 'proved adept at developing new institutions, such as state industrial banks' (Gootenberg 2001: 57). Thus, the greater the country's initial deprivation, the more coercive and comprehensive the state's action would have to be. Wade's (1990) thesis that East Asian industrialisation can only be understood through the state 'governing the market' is distinctly Gerschenkronian in this context.

Gerschenkron's theories did have an impact on the policies for promoting industrialisation in developing countries in the 1960s and 1970s, along with those of the Latin American structuralists (such as **Raúl Prebisch**) and students and former colleagues of Gerschenkron, such as **Albert Hirschman.** Gerschenkron's perspective had the distinct advantage of recognising that the position of developing countries in relation to the world market was fundamentally different to that of the core economies (such as the UK and USA) on the eve of their industrialisation.

The subsequent industrial experience of developing countries in Asia, Africa and Latin America did reveal some practical problems with the applicability of the model. First, there was the problem of the assumed benefits to the backward country of being able to import the latest technology. The burden of development has grown over time as the initial costs of technology and plant have increased due to rising capital intensity and the scale of modern production. Second, because these new technologies are capital-intensive, it seems irrational for them to be adopted by developing countries in which cheap labour is normally plentiful (Landes 1999).

It is interesting to ponder why the work of Alexander Gerschenkron has had such little resonance within development geography while attaining much greater significance within the realms of economic and political sociology. Going back to Keeble's (1967) early critique of models of economic development, ample room was given to a discussion of **Rostow**'s theories but none to those of Gerschenkron. Rostow was criticised for his unilinear version of development but Gerschenkron's more open theory of differentiated industrial transformations was basically ignored.

Such a pattern continues to this day. The Fourth Edition of *The Dictionary of Human Geography* (Johnston *et al.* 2000) can be seen as a guideline to the value of which concepts are important for contemporary geography. Gerschenkron gets but a passing mention in Michael Watts's treatment of growth theory. While Rostow has a section to himself under 'stages of economic growth', there is no reference to economic backwardness, late industrialisation or 'latecomer models' in the dictionary.

One possible explanation for the lack of attention to Gerschenkron within human geography in the 1960s and 1970s was that the discipline moved strongly towards more nomothetic approaches. Meanwhile, Gerschenkron was essentially developing historically-based ideas and models from the strongly idiographic vision of the economic historian. As geographers increasingly turned their back on the idiographic framework and the detailed and nuanced studies of different cultural regions, Gerschenkronian analysis may have appeared out of step.

For today's development specialists, Gerschenkron could be seen as a precursor of contingent capitalism or the idea of different pathways towards capitalism. The direction of capitalism in any one country can be seen as affected by the nature of 'relevant' institutions and ideologies that will support and assist industrial and economic growth.

Since the fall of the Berlin Wall and the end of communism, academics have searched for new interpretations of development and policy initiatives in what could be called the world's peripheral economies (Gwynne, Klak and Shaw 2003). It might be useful to go back to Gerschenkron's nuanced view on economic backwardness and the idea of many pathways out of such a condition. He privileged historical discontinuity over linearly conceived continuities and traditions. Furthermore, he was also explicitly relational in the sense that later industrialisers should be seen as different precisely because they were affected, often through demonstration effects, by the first-comers (Gootenberg 2001: 57).

Major works

Two main collections contain Gerschenkron's major essays:

Gerschenkron, A. (1962) *Economic Backwardness in Historical Perspective: A Book of Essays*, Cambridge, MA: Harvard University Press.

—— (1968) *Continuity in History and Other Essays*, Cambridge, MA: Harvard University Press.

Further reading

Bennett, D.C. and Sharpe, K.E. (1985) *Transnational Corporations versus the State: The Political Economy of the Mexican Auto Industry*, Princeton, NJ: Princeton University Press.

Blaug, M. (1998) *Great Economists since Keynes*, Cheltenham: Edward Elgar.

Dore, R. (1990) 'Reflections on Culture and Social Change', in Gereffi, G. and Wyman, D.L. (eds), *Manufacturing Miracles: Paths of Industrialisation in Latin America and East Asia*, Princeton, NJ: Princeton University Press.

Forsyth, D.J. and Verdier, D. (eds) (2003) *The Origins of National Financial Systems: Alexander Gerschenkron Reconsidered*, London: Routledge.

Gootenberg, P. (2001) '*Hijos* of Dr. Gerschenkron: "Latecomer" Conceptions in Latin American Economic History', in Centeno, M.A. and López-Alves, F. (eds), *The Other Mirror: Grand Theory through the Lens of Latin America*, Princeton, NJ: Princeton University Press, pp. 55–80 (excellent essay on the contemporary applicability of Gerschenkron's theories in Latin America).
Gwynne, R.N. (1990) *New Horizons? Third World Industrialisation in an International Framework*, Harlow: Longman (analyses Gerschenkron and his policy links to Third World industrialisation).
Gwynne, R.N., Klak, T. and Shaw, D.J.B. (2003) *Alternative Capitalisms: Geographies of Emerging Regions*, London: Arnold.
Johnston, R.J., Gregory, D., Pratt, G. and Watts, M. (2000) *The Dictionary of Human Geography*, Oxford: Blackwell, 4th edn.
Keeble, D.E. (1967) 'Models of Economic Development', in Chorley, R.J. and Haggett, P. (eds), *Socio-Economic Models in Geography*, London: Methuen, pp. 243–302.
Landes, D. (1999) *The Wealth and Poverty of Nations*, London: Abacus.
Love, J. (1996) *Crafting the Third World: Theorizing Underdevelopment in Rumania and Brazil*, Stanford, CA: Stanford University Press.
Rosovsky, H. (1966) 'Japan's Transition to Modern Economic Growth, 1868–1885', in Rosovsky, H. (ed.), *Industrialization in Two Systems: Essays in Honor of Alexander Gerschenkron by a Group of his Students*, New York: John Wiley, pp. 91–139.
Rostow, W.W. (1960) *The Stages of Economic Growth: A Non-Communist Manifesto*, Cambridge: Cambridge University Press.
Trebilcock, C. (1981) *The Industrialisation of the Continental Powers, 1789–1914*, London: Longman (strongly supports Gerschenkron's ideas in terms of industrialisation in eastern and southern Europe).
Wade, R.H. (1990) *Governing the Market: Economic Theory and the Role of Government in East Asian Industrialization*, Princeton, NJ: Princeton University Press.

Robert Gwynne

GERALD K. HELLEINER (1936–)

Gerry Helleiner's contributions have taken in academic work, support to the creation and success of a range of organisations, and policy advice. Ultimately, he is perhaps best known for his close involvement in policy advice in African countries, above all Tanzania, and for the way that this involvement has played a central role in efforts to improve the relationship between governments of low-income developing countries and international financial institutions and aid organisations.

Helleiner's intellectual formation – like that of many of his generation – was shaped by the upheavals of European and international politics from the 1930s into the post-Second World War period. He was born in Austria in 1936, to parents whose families had fallen foul of the Nazis and who were forced to emigrate. He and his mother were on the last plane out of

Vienna before the war began. The family went first to England and then settled in Canada, where his father, Karl, lectured in economic history. Gerald Helleiner graduated in 1958 from the University of Toronto, studying political science and economics. He then went to Yale, where he did a 'straight' economics PhD. Nonetheless, his interests were already shifting to the plight and challenges of developing countries and he was appointed, in 1961, to the new Yale Economic Growth Center where he was based until 1965, including a year spent researching and teaching in Nigeria. In 1965 he moved to the University of Toronto, where he, his wife, Georgia, and three children have been based since, in a career interspersed with stays overseas.

His time at the Economic Growth Center at Yale produced a comprehensive economic history of Nigeria from 1900 to the mid-1960s (Helleiner 1966). The book was written just as it was becoming obvious how dramatically oil production would change the economy and just before the two coups of 1966 (cited in the preface) that initiated a cycle of political instability in Nigeria. The principal emphasis of the book was on the way in which extensive expansion of export-oriented peasant agriculture had been the key to Nigeria's economic dynamism, along with the stimulus to expansion given by government spending, especially on infrastructure (road and rail), education, and employment in the public sector.

Something of the wide sweep of the radar of Helleiner's interests is also shown by his pioneering work on intra-firm trade, organised, among other places, in *Intra-Firm Trade and the Developing Countries* (1981). This book argued for a greater acknowledgement of the huge significance of intra-firm trade, since a 'high proportion of the international flow in the markets for goods and services ... takes place within firms' and built on the fact that these transactions typically take place 'in consequence of central commands rather than in response to price signals and they are recorded at prices which can be arbitrarily established and do not necessarily have anything to do with market prices'. Of course an obvious source of interest in intra-firm trade as it relates to firms operating in developing countries is the scope for 'transfer pricing' by transnational firms as a ruse to evade taxation. However, the book looked at a fuller range of issues than this. Overall, the analysis was built on the argument that standard trade theory was shown by the phenomenon and significance of intra-firm trade to be an inadequate tool for understanding the reality of much international trade in goods and services.

Many critics of orthodox economic theory and policy are technically weak in economics. Gerry Helleiner has always – by contrast – sustained a firmly independent analytical policy stance underpinned by an extremely high level of technical capability as an economist broadly in the Keynesian tradition. It is perhaps this strength as an economist, combined with his enduring sympathy for the challenges faced by governments in low-income countries, that has enabled him to play such an important role in efforts to forge a realistic and progressive dialogue between 'the West' and the developing world.

This role has partly involved serving on – and sometimes helping to create – international organisations. These have included two Canadian bodies – the International Development Research Centre and the North–South Institute; acting as a member of the advisory board of the World Institute for Development Economics Research (WIDER, part of the United Nations University and based in Helsinki, Finland); providing support to the Intergovernmental Group of 24 on International Monetary Affairs; acting as Chair of the Board of Trustees of the International Food Policy Research Institute (IFPRI) from 1990, where he was credited with resuscitating an organisation at the time in crisis; and advising the African Economic Research Consortium (AERC). He was involved in the Canadian mission to support the Democratic Movement in South Africa in its formulation of post-apartheid economic policy and served on the advisory board of the Macro-Economic Research Group (MERG) that produced a policy report, *Making Democracy Work: A Framework for Macroeconomic Policy in South Africa* in 1993. He has also worked on other policy reform missions, including, twice, to Uganda and also to Guyana and Nicaragua.

But above all, Helleiner is renowned for his long-term involvement in and commitment to African economic development, especially in Tanzania. From 1966 to 1968 he took leave from Toronto to become the founding director of the Economic Research Bureau at University College in Dar es Salaam. His involvement in Tanzania lasted through the years of **Julius Nyerere**'s exercise in 'African socialism' and the fraught relationships with the IMF and World Bank and into the political and economic reform years of the 1990s. Tanzania's relationship with international lenders had broken down – not for the first time – in the mid-1990s. When some kind of *rapprochement* began to look possible, the government and the donors, with Danish assistance, agreed to appoint an independent team of experts to analyse the relationship and to advise on a way forward. This team, headed by Professor Helleiner, produced the 'Helleiner Report' in 1995. This report was not only hugely constructive in Tanzania but more broadly it became a seminal contribution in the evolution of ideas that aid would be more effective if there were greater 'national ownership' of the reform process and the disbursement of aid funds. After this report, the Tanzanian government and the donors agreed a series of initiatives to monitor the transparency and accountability of *both* parties to the aid relationship.

Helleiner's concern with the ideals of ownership was stamped by his earlier experiences in Tanzania, his sympathy with much of what Nyerere was trying to achieve, and a distaste for the high-handed behaviour of the IMF and what he saw as its unthinking and inappropriate application of neo-classical economic theory to conditions of extreme poverty. In a talk he gave on his own reflections on the Nyerere years, at a conference celebrating Nyerere's legacy after his death, much of the humane and independent approach Helleiner has taken to economic development is clear. Noting the unpopularity of Nyerere's villagisation programme, the unfeasibility of a sweeping nationalisation programme, and the over-ambition of the basic industries strategy, Helleiner nonetheless reminded the audience of Nyerere's emphasis on free education, on equity, on a leadership code for politicians, and on targeting the poorest. He argued that economic failure in Tanzania ultimately had very little to do with Nyerere's socialism; rather, it was rooted in external shocks (the oil price rises of the 1970s, the cost of war against Idi Amin in Uganda, and adverse weather) and in bad macro-economic policy management in response to these shocks.

This argument continued a theme Helleiner had long insisted on. His 'The IMF and Africa in the 1980s' (1983) was one of the earliest critiques of IMF conditionality and lending policy in Africa. The essay bore a hallmark fairness in tone and was all the more effective for the calm way it built its argument. The essay, which foresaw a period of increasing conflicts between African governments and the IMF, argued from the premise that the key to African economic crisis lay in deteriorating terms of trade combined with the rising cost of borrowed capital and, therefore, the desperate shortage of foreign exchange to cover the import requirements of stability and growth. Adjustment was inevitable.

> The question is not whether there should be a difficult and painful adjustment, on both the demand and the supply side, but how it should be undertaken. With what assistance? Over what time horizon? With the burden distributed how? With what mix of policies and what sequencing? With what terms for foreign borrowing?
> (Helleiner 2000: 11)

His critique of the IMF focused both on the inequity of the IMF's liquidity expansion and on the analytical inadequacies behind conditionality (as well as on the gormlessness of sending missions that had little or no knowledge or experience of Africa). The first argument was that IMF low-conditionality finance had not grown fast enough to match the value and significance of trade for poor countries while the terms of high-conditionality loans were growing tougher. The second argument was that the IMF's analytical models were blunt and impervious to the variety of sources of balance of payments difficulties. Given this variety, as well as the more general lack of understanding of complex economic phenomena, Helleiner argued that the 'only possible professional stance in these circumstances is of considerable humility and caution in the dispensing of advice' (*ibid.*: 17). He envisaged an alarming prospect for relations between African governments and the IMF in the coming years: 'badly prepared antagonists of modest ability employing data of dubious quality and entering upon a series of battles over very complex policy questions' (*ibid.*: 22). This was a pretty accurate vision of how the following decade unfolded in a number of cases, although over time there was typically less and less effective resistance within African governments to the model on offer.

Helleiner stressed a number of needs in this essay – more access to low-conditionality liquidity for the poorest African countries, more flexible conditional credit arrangements in the face of external shocks, a greater variety in the types of aid delivered by donors, institutional innovation to develop conflict resolution mechanisms to resolve tensions between the Fund and debtor governments, and a surge in training and research and data collection in Africa. Despite the HIPC (Highly Indebted Poor Country) initiative, some new IMF lending facilities, the possibility of a shift to a 'post-Washington Consensus' in the World Bank, a range of data collection exercises, and a shift towards prioritising 'national ownership' of aid programmes and policies, many might argue that these needs remain unmet. A gulf remains. For example, what Helleiner and many others understand by the term 'ownership' appears still to differ from how the Fund sees it. In an IMF paper on ownership, ownership was defined as a 'convergence of interests' between debtor countries and the IMF. Specifically criticising the World Bank's view of ownership, this effectively meant that the IFIs (international financial institutions) 'now want local policy makers not simply to do what it recommends but also to believe in it' (Helleiner 2000: 3).

The pragmatic and humane, moral and technically sure-footed approach that has characterised Helleiner's policy and organisational work has stamped his academic writing, which has also been prolific. He has written or edited close to fifty volumes as well as publishing many articles in refereed journals and contributing chapters to edited books. His main academic writings have been in the fields of developing country trade and finance and the role of transnational companies in developing countries. As with the rest of his work, this body of writing escapes glib 'left' or 'right' labelling, but is broadly within a Keynesian tradition drawing on elements of structuralist economics. He has also served on the boards of fifteen academic journals.

Major works

Helleiner, G.K. (1966) *Peasant Agriculture, Government, and Economic Growth in Nigeria*, Homewood, IL: Richard D. Irwin.
—— (1973) 'Manufactured Exports from Less Developed Countries and Multinational Firms', *Economic Journal* 83(329): 21–48.
—— (1981) *Intra-Firm Trade and the Developing Countries*, London: Macmillan.
—— (1983) 'The IMF and Africa in the 1980s', *Essays in International Finance No. 152*, Dept of Economics, Princeton, NJ: Princeton University.
—— (1989) 'Transnational Corporations, Foreign Direct Investment and Economic Development', in Chenery, H.B. and Srinivasan, T.N. (eds), *Handbook of Development Economics*, Vol. II, Amsterdam: North-Holland.
—— (1990) *The New Global Economy and the Developing Countries: Essays in International Economics and Development*, London: Edward Elgar.
—— (1994) *From Adjustment to Development in Africa: Conflict, Controversy, Convergence, Consensus?*, ed. by Giovanni Andrea Cornia and Gerald K. Helleiner, Basingstoke: Macmillan.
—— (2000) 'Towards Balance in Aid Relationships: External Conditionality, Local Ownership and Development', produced by Brian Tomlinson, as an amalgam of two earlier papers by Helleiner, for the Reality of Aid International Advisory Committee meeting, Costa Rica, September. Available at http://www.devinit.org/realityofaid/jcrpapers.htm.
—— (2001) *Non-Traditional Export Promotion in Africa: Experience and Issues* (edited and Introduction), UNU/WIDER, London: Palgrave.

Selected mission reports

'Financing Africa's Economic Recovery' (1988) for UN Secretary-General.
'Guyana: the Economic Recovery Programme and Beyond' (1989) report of a Commonwealth Advisory Group.
'The Helleiner Report: Report of the Group of Independent Advisors on Development Cooperation Issues Between Tanzania and its Aid Donors' (1995).

Further reading

Culpeper, R., Berry, A. and Stewart, F. (eds) (1997) *Global Development Fifty Years after Bretton Woods: Essays in Honour of Gerald K. Helleiner*, Basingstoke: Macmillan.

Christopher Cramer

ALBERT O. HIRSCHMAN (1915–)

Most of the major academic contributions of the economist, Albert Hirschman, fall within the political economy of development. In the community of development scholars, he is perhaps best known for his theory of unbalanced growth, whereby a pattern of quicker growth and more egalitarian development may eventually be stimulated by concentrating development efforts on key industries and locations in hitherto lagging regions, especially for the underdeveloped countries of the South. Hirschman excels at placing economic phenomena and processes within their broader social context. His work on economic development, collective action, the political bases of economic thought, and attitudes toward market society frequently addresses areas of thought unfamiliar to most economists. His writing has illuminated a wide range of social behaviour that lies beyond simplistic rational-choice models, and it characteristically eschews attempts to overgeneralise in favour of a more nuanced approach that pays special attention to the particular and the unique.

Hirschman is considered somewhat of a maverick within the field of economics, especially for his adoption of an interdisciplinary approach to development studies, unusual for contemporary economists. He has been an iconoclastic trailblazer in his research on economic development, frequently stepping outside his own discipline in order to better address the complexities and particularities of local development processes in various areas of the world. He is the author of numerous influential and controversial works, often adopting novel concepts and ideas in disagreement with mainstream economic theory and major figures in the field. His work commonly exhibits an impatience with disciplinary boundaries that all too frequently separate economists from other scholars in development studies, limiting their ability to address fully the multifaceted nature of development. He has been described as a playful genius, a person who loves being unconventional and does not shy away from controversy.

Over his long, varied and distinguished career, Hirschman worked as an economist with the US Federal Reserve Board; as an economic advisor in Bogotá, Colombia; and as a professor at Yale, Columbia and Harvard, where he was the Lucius N. Littauer Professor of Political Economy. He joined the School of Social Science at the Institute for Advanced Study in Princeton in 1975, becoming professor emeritus in 1985. His works have been the subject of numerous symposia in economics and related disciplines in development studies, and he has been the recipient of many awards and honours, including the 1983 Talcott Parsons Prize for Social Science from the American Academy of Sciences, the 1997–98 Toynbee Prize, and the 1998 Thomas Jefferson Medal awarded by the American Philosophical Society.

Hirschman was born on 7 April 1915 in Berlin. After attending the Sorbonne and the London School of Economics, he obtained a PhD in economic science from the University of Trieste in 1938. During and immediately following his doctoral studies, much of his life was dominated by the struggle against fascism in Europe, where he played an active role in several countries. While in Italy in the mid-1930s, he joined the underground opposition to Mussolini. In 1936 he fought with the Spanish Republican Army and later with the French Army until its defeat in June 1940. He remained in France for an additional six perilous months, taking part in clandestine operations based in Marseilles to assist political and intellectual refugees from Nazi-occupied Europe, until his escape to the USA in January 1941 to avoid capture and arrest.

Following his arrival in the USA, Hirschman's major focus in his early career concerned the structure of world trade, and especially the political implications of its asymmetries and patterns of concentration. These interests determine much of the material around which his first book (Hirschman 1945) is organised. In the immediate post-World War II era, Hirschman was particularly occupied, like many other development scholars of his generation, by economic problems (e.g. the 'dollar shortage' in countries such as France and Italy) associated with efforts to smooth European reconstruction and integration. The book became immediately notable for the interdisciplinary approach it adopted, characteristic of Hirschman's work throughout his career, situated in the 'grey zone' between economic and political theory. It asserted that economic power is often the handmaiden of political power between states. It also generated a good deal of controversy, another hallmark of Hirschman's career, in that it added to the growing scepticism of many development scholars over the way in which the Marshall Plan was being implemented in post-war Europe.

With the publication of his second book, *The Strategy of Economic Development* (1958), Hirschman became widely recognised as a leading development scholar. The analysis upon which the book is based grew out of four years of practical experience in Colombia as an economic advisor to the government and several private firms. In the book, Hirschman strongly rejected the wholesale and unquestioning importation of the formal concepts and ready-made, universal strategies of conventional neoclassical economic theory, the application of which he believed inevitably sacrifices necessary attention to local developmental particularities in order to conform to the narrow strictures of modern scientific modelling. He contended that, rather than adopting conventional generalised prescriptions for economic development, the developmental needs of particular regions and countries ought to be assessed on a case-by-case basis, paying special attention to the particularities of indigenous resources and structures to achieve the desired developmental results. Efforts to impose formal universal models and strategies, regardless of the historical and structural realities of specific development contexts, were a sure-fire recipe for disastrous development policies and programmes, with many dire economic, political and socio-cultural consequences.

In the book, Hirschman argued against both *laissez-faire* and 'rational' nationwide economic planning for developing countries. Especially in underdeveloped countries, development ought to be spurred by policies and programmes promoting 'unbalanced growth', in which governments would deploy their scarce resources strategically to foster critical disequilibria important to encouraging private investment and related entrepreneurial activities in key industries and locations in lagging regions. Previously hidden and underutilised resources would best be mobilised by targeting development efforts on key industries with strong forward/backward linkages to the local economy. Such linkages would generate 'spread' effects, whereby the initial inequalities of unbalanced growth centred on industrial complexes would give way to a more egalitarian pattern of territorial deconcentration of the benefits of modern development to surrounding peripheral areas. If focused on dynamic industrial growth poles, an initial emphasis on unbalanced growth and socio-economic/spatial disequilibria would allow new governments in the South relatively quickly to overcome the common historical legacy of polarising 'backwash' effects inherited from previous (neo)colonial spatial structures. These were dominated by outward-oriented primate cities delinked from much of the stagnant domestic economy and outlying peripheral regions.

Hirschman's concept of unbalanced growth underscores the particularities of resources and structures in peripheral regions upon which key industries could be constructed, thereby facilitating beneficial linkages to the local economy. His approach criticises conventional doctrinal models of economic development that neglect local particularities. In a later article, 'The Rise and Decline of Development Economics' (1982a), he expands this focus into a more general critique of the rise of 'mono-economics' accompanying the ascendancy of neoliberal development strategies in the South. According to Hirschman, development economists may be divided into two basic types. On the one hand, there are those who take a mono-economic perspective, whereby orthodox neoclassical theory is equally applicable in both the North and South. Although the South may have peculiar characteristics, such as higher levels of uncertainty, its economic agents make decisions and its markets function essentially according to the same logic as in the North. Most importantly for neoliberals, individuals or firms respond similarly to structures of incentives, whether they are (properly) established by the economic market or (improperly) set by the political market.

On the other hand, there are the followers of a 'duo-economic' approach who believe that standard market economics is of limited relevance to the special problems of development in the South. Without substantial theoretical modification, they see conventional economic models and strategies as incapable of addressing problems associated with underdevelopment that arise from a series of particular Southern structural conditions. These include: massive unemployment and underemployment, despite instances of high growth and industrialisation; frequent market failures based on poorly developed circulation systems, financial networks, and other economic structures; the continuing influence of culture and tradition on common forms of behaviour that detract from utility maximisation; the persistence of extreme societal polarisation and inequalities, which are aggravated by poorly articulated social, sectoral and regional structures; and a pronounced vulnerability to externally generated crises due to high levels of foreign dependency and economic concentration within the external sector.

In the article, Hirschman is especially critical of the mono-economic approach of the conventional neoclassical paradigm. He claims that neoliberals ought to remove their conceptual blinkers and methodological straitjackets in favour of a broader, more flexible vision of development capable of addressing the diverse realities of the South. The domination of development studies by 'grand theories' has generated increasing tension between the desire to formulate universally valid principles and formal models, and the need to understand the variety of actual experiences and potential alternatives of development in the South. Hirschman contends that solutions to development problems must be sought in the contextuality of development, which is a product of particular historical processes. The context of development is constantly changing in scale, over time and among societies – creating both new obstacles and new opportunities for variations. Development frameworks, especially those designed to contribute toward policy-making, need to devise ideas and methods capable of accommodating this geographical and historical diversity.

In the editor's introduction to another important book, *A Bias for Hope: Essays on Development and Latin America* (1971), Hirschman coined the term 'possibilism' to describe his position against grand theorisation and overgeneralisation. In development studies, and the social sciences in general, the search for general laws often obscures the role of the particular and unique in human pursuits, which may produce much of the unpredictability that frequently confounds the scientific modelling of society. Typically, such general formulations view development progress in Southern societies linearly, based on formulations rooted in the history of the North, and according to the application of universal laws developed in and for Northern societies. The failure of development efforts, especially those of external origin, is commonly attributed to the presence of some inescapable obstacles to development unfortunately present in Southern societies. However, the development trajectories of particular societies often take paths that are *a priori* quite unlikely and cannot be easily anticipated; obstacles may become opportunities and vice versa, confounding rules with unexpected consequences. Hirschman's work often uses this intellectual starting point to get at the many complexities of development issues and problems, always leaving open possibilities for unanticipated phenomena and events to produce novel forms of social change.

In addition to his work in development studies, Hirschman also employed his possibilist approach to address broader problems of social theory. In *Exit, Voice and Loyalty* (1970), he offers an elegant and profound rethinking of individual choice within institutions or organisations, comparing the implications of dissatisfied clients alternatively exiting from an organisation or giving voice to their complaints. In this volume and much of his more recent work, including his latest book, *Crossing Boundaries* (1998), Hirschman turns to social issues and problems which straddle the border between economics and politics. But he shows no inclination to follow the lead of many other economists who, via the reductionist framework of public choice theory, have essentially turned politics into economics. Instead, his work stresses that conventional models of economic behaviour based on rational choice cannot satisfactorily explain diverse forms of 'public-minded' behaviour such as voicing one's convictions on public affairs, participating in demonstrations or working in support of candidates for public office.

Hirschman's disdain for reductionist, ideologically driven, and invariant arguments is also clearly evident in his studies of historical views of capitalism (1977, 1982b). In recent years he has alluded to his penchant for 'self-subversion' to stimulate intellectual inquiry (1995) – constantly and systematically revisiting his principal ideas, concepts and theories in order to add needed modifications, qualifications and complications in various ways. In his studies of capitalism, a complex and highly variable dynamic of development is portrayed, one in which capitalism under certain circumstances may be a powerful civilising influence, but also may sometimes destroy a people's moral and social fabric, or may simply be too feeble to overcome the constraints of prior social forms. These competing ideological views of capitalism, in which seemingly similar processes may produce quite divergent results in different places and times, underscore Hirschman's overriding interest throughout his work to express rather than conceal the full range of human and social complexities that influence development processes, which can only be very partially addressed by using standard methods of economic science based on formal concepts and *a priori* models.

Major works

Hirschman, A.O. (1945) *National Power and the Structure of Foreign Trade*, Berkeley, CA: University of California Press.
—— (1958) *The Strategy of Economic Development*, New Haven, CT: Yale University Press.
—— (1958) *Development Projects Observed*, Washington, DC: Brookings Institution.
—— (1970) *Exit, Voice, and Loyalty: Responses to Decline in Firms, Organizations, and States*, Cambridge, MA: Harvard University Press.
—— (1971) 'Introduction: Political Economics and Possibilism', in A.O. Hirschman (ed.), *A Bias for Hope: Essays on Development and Latin America*, New Haven, CT: Yale University Press.
—— (1977) *The Passions and the Interests: Political Arguments for Capitalism Before its Triumph*, Princeton, NJ: Princeton University Press.
—— (1981) *Essays in Trespassing: Economics to Politics and Beyond*, Cambridge: Cambridge University Press.
—— (1982a) 'The Rise and Decline of Development Economics', in M. Gersovitz *et al.* (eds), *The Theory and Experience of Economic Development: Essays in Honour of Sir W. Arthur Lewis*, London: Allen and Unwin.
—— (1982b) 'Rival Interpretations of Market Society: Civilizing, Destructive, or Feeble?', *Journal of Economic Literature*, 20 (December): 1463–84.
—— (1982c) *Shifting Involvements: Private Interest and Public Action*, Princeton, NJ: Princeton University Press.
—— (1984) 'Against Parsimony: Three Easy Ways of Complicating some Categories of Economic Discourse', *American Economic Review* 74(2): 89–96.
—— (1986) *Rival Views of Market Society and Other Essays*, New York: Viking-Penguin International.
—— (1987) 'The Political Economy of Latin American Development: Seven Exercises in Retrospection', *Latin American Research Review* 22(3): 7–36.
—— (1995) *A Propensity to Self-Subversion*, Cambridge, MA: Harvard University Press.
—— (1998) *Crossing Boundaries: Selected Writings*, Cambridge, MA: MIT Press.

Further reading

Coser, A. (1984) *Refugee Scholars in America: Their Impact and Their Experiences*, New Haven, CT: Yale University Press.
Foxley, A. *et al.* (eds) (1986) *Development, Democracy, and the Art of Trespassing: Essays in Honor of Albert O. Hirschman*, Notre Dame, IN: University of Notre Dame Press.
Meier, G. (1987) *Pioneers in Development*, Oxford and Washington, DC: Oxford University Press and World Bank.

John Brohman

RICHARD JOLLY (1934–)

From his early studies of disarmament and education through to his more recent dedication to the cause of water supply, sanitation and hygiene,

Richard Jolly's unique combination of intense optimism and measured pragmatism have earned him widespread admiration and acclaim in the field of human development. As an economist with a passionate commitment to a people-centred approach, he has consistently argued that greater attention should be paid to addressing basic human needs in development. This he has always seen not only as morally desirable but also as beneficial for medium and longer-term economic growth and global security. Jolly was at the centre of debates on the relationship between economic growth and social welfare, concluding – alongside others such as **Dudley Seers** and **Hollis Chenery** – that it was not necessary to choose; instead, he argued it was necessary to pursue both growth and poverty reduction *in order to produce* sustainable economic development and stability. In a recent statement on education for all, Jolly maintained that even in periods of stagnation, huge gains in social welfare could be achieved without significant increases in expenditure, through 'determined leadership, cost-consciousness and ingenuity'.[1] This sentiment encapsulates much of his life's work.

Born on 30 June 1934, Richard Jolly graduated from Cambridge University in 1956 with a degree in Economics, and later completed a doctorate at Yale University, once again specialising in Economics. In 1969 he joined the Institute of Development Studies (IDS) at the University of Sussex as a fellow. In 1972 he followed Dudley Seers as Director, a position that he held for nine years. During this period, on secondment from the IDS, he acted as special consultant on North–South issues to the Secretary-General of the Organisation for Economic Co-operation and Development (OECD) in 1978. Also, between 1978 and 1981, he had his first experiences of working in the UN system when he acted as member and rapporteur of the United Nations Committee on Development Planning.

In 1978 the first edition of *Disarmament and World Development* was published, a volume co-edited by Jolly. This book, updated and republished in 1984, explored the potential for directing resources freed up by disarmament towards poverty-focused development through targeted restructuring policies. In this study, Jolly's willingness to tackle huge global issues with imagination and optimism is already evident, as is his rigorous exploration of examples from around the globe which disprove conventional thinking about what is possible and achievable within the given constraints of the international system. In his contribution, Jolly argued that the estimated financial costs of providing for many of the world's basic needs within the space of a decade were less that half of one year's expenditure on armaments; a fact which 'underlines the disastrous and inhuman diversion to armaments which might otherwise be turned to peace and humane progress' (Graham *et al.* 1984: 129). Contemporary policy makers would be well advised to reflect on this. Jolly was never under any illusions that the *financial* aspect was the only dimension of the transformation necessary to allow resources released by disarmament to be channelled into development. Despite the difficulties involved, he argued that there were many examples from the past where such a fundamental restructuring of production had occurred, e.g. the run down of coal, textile and electronics production in many industrialised countries. Furthermore, he pointed out that a focus on social welfare could prove a useful political tool even for authoritarian and military regimes.

The themes developed in *Disarmament and World Development* recur in later work, most notably the groundbreaking *Adjustment with a Human Face* (Cornia *et al.* 1987), of which Jolly was co-editor. This was completed during his time as Deputy Executive Director for Programmes at UNICEF, a position he held from 1981 to 1995. The context for the collection was the pattern of contradictory global forces set in motion in the early 1980s and that gave rise to a severe world recession. The Bretton Woods institutions addressed these through advancing stabilisation measures and structural adjustment policies, which sought to rectify balance of payments deficits and improve macro-economic management. At the time, UNICEF was at the forefront of renewed attempts to put basic child protection and poverty alleviation measures in place. Central to the study's argument was the idea that efforts were necessary to ameliorate the effects of adjustment policies on the most vulnerable sectors of society, particularly children. Jolly's personal contribution to this work was in the field of education, following on from themes explored in his doctoral research (Jolly 1969a). Jolly identified the goal of universal primary education as the most significant educational priority, arguing that it was possible to move towards this without any significant increase in total expenditure if existing institutions and income structures were adapted significantly.

The main thrust of *Adjustment with a Human Face* was that growth-oriented adjustment in itself was not enough to ensure protection of the vulnerable; and yet the vulnerable could be protected during adjustment if *targeted programmes* were adopted. Furthermore, the book pointed to successful examples where such programmes had been adopted and growth had not been damaged as a result: 'the issue should not be adjustment or growth, but adjustment *for* growth' (Cornia *et al.* 1987: 5). In contrast to much of the received wisdom at the time, Jolly and his colleagues maintained that a 'human face' in adjustment policy was in fact a precondition for long-term growth, insisting that a strategy 'which protects the vulnerable during adjustment not only raises human welfare but is also economically efficient' (*ibid.*: 290). These sentiments were consistent both with Jolly's long-held view that investment in human resources exhibits high economic returns and his *basic needs* approach to development.

During his time at UNICEF, Richard Jolly was also Vice-President of the Society for International Development (SID) from 1982 to 1985, and then Chairman from 1987 to 1996. In 1995, he left UNICEF to become Special Advisor to the Administrator of the United Nations Development Programme (UNDP). Here he was principal architect of the annual Human Development Report between 1996 and 2000. Jolly championed the '20/20 initiative', which called for developing and donor countries to devote 20 per cent of government budgets and 20 per cent of aid allocations, respectively, to services targeted at basic needs. His commitment to education remained strong, and in June 1996 he delivered the concluding statement at the Mid-Decade Meeting of the International Consultative Forum on Education for All in Amman, Jordan. The purpose of the meeting was to assess the advances made since the Jomtien Conference on Education in 1990. Although accelerated progress was necessary, 'For the first time in human history, the numbers of those without the ability to read or write is beginning to fall. This is unprecedented, almost certainly since the beginnings of mankind.' Education for all and greater equality, he argued, were 'the two essential factors for ensuring rapid economic growth and rapid advance in human development'.

Jolly is considered a pioneer of the concept of Human Development. In 2001 Mark Malloch Brown, Administrator of the UNDP, said of him: 'Nobody has done more to help promote the concept of human development, and provoke thoughtful discussion and debate about how best to achieve it, than Dr. Richard Jolly.'[2] He was awarded the Rome Prize for Peace and Humanitarian Action in 1999, in recognition of his work on the Human Development Report and earlier contributions to shaping programmes at UNICEF. He donated the $30,000 prize money to Jubilee 2000, the international coalition campaigning for debt reduction. Two years later he was knighted (KCMG) by the United Kingdom for 'long and distinguished service for international development'. Since retiring from UNDP in 2000, Richard Jolly has been Co-Director of the United Nations Intellectual History Project at the City University of New York, exploring – and in many cases defending in measured terms – the development legacy of the United Nations as part of a 12-volume history of its economic and social contribution. In the first co-authored volume of this series, *Ahead of the Curve,* published in 2001, Jolly is dismissive of concerns about the origin and ownership of ideas, emphasising instead their impact and spread and he is firm that the UN has played an important role in this regard. In a paper written in preparation for the Human Development Report 2003, he appealed for a more nuanced and flexible interpretation of success in terms of achieving UN goals. Too often, he argued, UN development projects are considered 'failures' because global goals are only partially or regionally met, when in fact huge progress has been achieved. If setting global goals is to be valuable and successful, he contended, 'it is important now to plan for *partial* success and *partial* failure, not for the extremes of either *total success or total failure*' (Jolly 2003: 18), especially in the cases of the least developed countries. In May 2004, during a talk at the Overseas Development Institute in London, he also defended the UN against accusations of being overpaid, resistant to reform and uncritical of developing countries.

In recent years Sir Richard has also devoted himself to the cause of clean water and sanitation for the global poor. As Chair of the Water Supply and Sanitation Collaborative Council (WSSCC) from 1997 to 2003, he argued for expenditure on water supply and sanitation (most of which goes to wealthier urban areas) to be redirected towards the rural poor. 'Improvements in water, sanitation and hygiene are at the core of human development and poverty alleviation,' he said on World Water Day in 2001. As with his earlier work in education, Jolly has argued for greater community participation in processes to ensure success. With reference to an approach termed 'VISION 21', which aimed to allow local people to plan and manage improved water and sanitation systems, he explained that 'we let communities themselves drive the process and it demonstrated that people had ideas that were practical, economic and sustainable'.[3] An enduring feature of Jolly's legacy will be his emphasis on the benefits of community involvement.

Jolly is likely to be remembered principally for his work as architect of the Human Development Reports in the late 1990s, and the way he has kept basic needs and poverty alleviation on the development agenda. Many of the measures he has advocated have undoubtedly had an impact. *Adjustment with a Human Face* not only steered thinking within the UN system but also influenced the World Bank's three-pronged approach to poverty reduction – addressing growth, human development and targeted social safety nets. It also strongly informs the current Millennium Development Goals (MDGs), which seek to reduce poverty significantly by the year 2015. The goals involving commitments to achieving universal primary education and reducing by half the proportion of people without sustainable access to safe drinking water by the year 2015 are particularly close to Jolly's heart. Nevertheless, many of his core passions have come under extensive critical review. A focus on basic human needs has been criticised by those wanting more emphasis on growth and by those wanting more focus on redistribution. Some welfare economists are sceptical about the efficiency and effectiveness of targeting, while the limits of community participation are increasingly well known and ventilated. However, while poverty elimination and redistribution of wealth might be nobler goals than poverty reduction and addressing basic needs, nearly half a century after the publication of *Planning Education for African Development,* the MDGs are unlikely to be met on target.

Even more dismal is the slow uptake of Richard Jolly's ideas on arms, conflict and well-being, a cause that he is advocating once more. Twenty years on, the relationship between development and global security has become more pressing and complex than ever. Dismal though this context is, Richard Jolly's optimism and pragmatism might not be misplaced. At the United Nations in New York on 30 April 2004, Ronnie Kasrils – South Africa's Minister of Water Affairs and Forestry – said of Jolly:

> He is a man who has brought passion, commitment and immense compassion to his work in the promotion of human development and human dignity. He is a man with vision, but with the remarkable skill of turning that vision into concrete reality.

These are respectful and touching words, not least because they come from a man who was once a senior military figure in Umkhonto we Sizwe, the armed wing of the African National Congress. If Kasrils has exchanged the pursuit of arms for the pursuit of wells, then perhaps Jolly's pragmatic optimism is well placed and he is right about starting small and thinking big.

Notes

1 'Education for All: The Vision to be Grasped' – Concluding statement by Richard Jolly at the Education for All Forum, Amman, Jordan, 19 June 1996.
2 See http://www.undp.org/dpa/frontpagearchive/2001/january/23jan01/.
3 Message of Dr Richard Jolly on the occasion of World Water Day, 22 March 2001.

Major Works

Cornia, G.A., Jolly, R. and Stewart, F. (1987) *Adjustment with a Human Face, Vols. 1 and 2*, Oxford: Clarendon Press.

Emmerij, L., Jolly, R. and Weiss, T.G. (*c.*2001) *Ahead of the Curve?: UN Ideas and Global Challenges*, Bloomington, IN: Indiana University Press.

Graham, M., Jolly, R. and Smith, C. (eds) (1984) *Disarmament and World Development*, 2nd edn, Oxford: Pergamon.

Jolly, R. (1983) *Third World Employment – Problems and Strategy: Selected Readings*, Harmondsworth: Penguin Education.

Streeten, P.P. and Jolly, R. (eds) (1981) *Recent Issues in World Development: A Collection of Survey Articles*, Oxford: Pergamon.

Further reading

Jolly, R. (1968) *Costs and Confusions in African Education: Some Future Implications of Recent Trends*, IDS Mimeo Series No. 18, Brighton: Institute of Development Studies, University of Sussex.
—— (1969a) *Planning Education for African Development: Economic and Manpower Perspectives*, Nairobi: East African Publishing House for the Makerere Institute of Social Research, East African Studies, No. 25.
—— (1969b) *The Purpose of Manpower Planning and its Implications for Planning Techniques*, IDS Communications No. 45, Brighton: Institute of Development Studies, University of Sussex.
—— (1970) *Skilled Manpower as a Constraint to Development in Zambia*, IDS Communications No. 48, Brighton: Institute of Development Studies, University of Sussex.
—— (2001) *Jim Grant: UNICEF Visionary*, Florence: UNICEF Innocenti Research Centre.
—— (2003) 'Global Goals – The UN Experience', http://www.waterweb.org/wis/wis6/papers/Global%20Goals%20v.030120.03.pdf.
UNDP Human Development Reports, 1996–2000, http://hdr.undp.org/reports/view_reports.cfm?type=1.

Jo Beall

CHARLES POOR KINDLEBERGER (1910–2003)

Charles Kindleberger was born in New York, the son of a lawyer. After studies at the University of Pennsylvania he completed a PhD in economics at Columbia University. From 1936 he worked with the Federal Reserve Bank of New York, followed by the Bank for International Settlements, the Federal Reserve Board in Washington, the Joint Economic Committee of the United States and Canada, the US Office of Strategic Services, and, from 1944 to 1948, the Department of State. In this last appointment, he played a part in organising the post-World War II Marshall Aid programme of reconstruction assistance for Europe, which subsequently became the prototype for large-scale aid programmes to developing countries. In 1948 he was appointed Associate Professor of Economics at the Massachusetts Institute of Technology (MIT), obtaining a full professorship three years later. He proceeded to confute the widespread stereotype of well-connected government officials honouring academic appointments by their discretion, by pouring out a mass of publications on economic history, international trade and development, banking and finance.

He remained at MIT, becoming emeritus professor on his retirement in 1976 (Miall 2003). By the time of his death on 7 July 2003 he had published twenty-nine books, excluding second and third editions. This essay concentrates on his most original contributions to development economics.

These, however, do not stand out of his extensive literature summaries. Moreover, his service in government and central banking gave him a practical approach to policy and the financial mechanics of economic adjustment that excludes simple theoretical solutions. Much of the time he reads like the notorious two-handed economist, concluding too often that circumstances alter cases (e.g. Kindleberger 1965: 387; 1969: 35–6). But even this conclusion is never less than elegant or erudite, and betrays a genuine commitment to finding out about a country before assessing its condition or prospects.

Perhaps Kindleberger's most distinctive contribution to the theory of economic development processes is his argument in *Economic Development* that 'economic development depends upon an open class structure and is particularly helped by the existence of a strong middle class' (Kindleberger 1965: 23). The precondition for successful industrialisation was a 'commercial revolution' because 'the growth of markets train(s) the entrepreneur' and 'eases his tasks'. Indeed,

> the commercial revolution does more. It starts the accumulation of capital. Inventories and ships are two early and easy objects of and outlets for capital accumulation. The productivity of capital becomes evident, the habit of saving spreads, and accumulations of capital capable of being converted to industrial use come gradually into existence.
>
> (*ibid.*: 163)

He concluded that

> there is a strong argument to be made for the proposition that economic development through industrialization should be preceded by commercialization, and the industrial by the commercial revolution. To short-cut the evolutionary process and attempt forthwith to turn a subsistence economy into an industrial economy may find a society badly equipped in capacity for transport and distribution.
>
> (*ibid.*)

For him, this was as much a political argument as it was an economic one. Early on Kindleberger expressed the view, widespread among the upper middle classes of his generation on both sides of the Atlantic, that the ruination of the Central European middle classes in the 1920s had paved the way for Hitler and fascism.

In Kindleberger's view, the 'prior commercial revolution' was also necessary to ensure adequate supplies of food and wage goods, as labour shifted from agriculture to manufacturing industry, in the classic **Arthur Lewis** model of development. Kindleberger considered that this 'Ricardian' model 'neglects the central concern of Ricardo – how the price of food is to be held down' (Kindleberger 1965: 182). With characteristic even-handedness he was willing to concede that the economic development of Japan after 1868 'may be put in the balance on the other side. Industrialization moved very rapidly . . .'. He concluded: 'This is a remarkable case. In my judgement, however, it provides the exception rather than the pattern' (*ibid.*: 164). But it is more difficult to see how a commercial revolution would provide a country with transport infrastructure. Like many other development economists, he argued against monumental projects, including these transport and energy projects. He felt that such projects tended to have 'a lower social marginal product of capital' and were typically 'under-priced' by governments (*ibid.*). Recent 'privatisation' of public utilities may have changed attitudes towards the supply by the private sector of such infrastructure, but has yet to prove that, unaided by the state, the private sector can maintain large capital-absorbing infrastructure. Kindleberger may have had in mind the privately constructed railways and canals of Europe and North America. He could not have been unaware of the speculative and frequently fraudulent nature of these undertakings.

In Kindleberger's view, the cultivation of commerce is, however, prone to two 'marketing pathologies'. The first is high mercantile profits with price-fixing, resulting in over-capacity and under-utilisation in the retail trade, leading to *petit bourgeois* conservatism and economic stagnation. The other pathology is 'speculative fever' (Kindleberger 1965: 164–5). His interest in the latter eventually resulted in Kindleberger's most popular work, *Manias, Panics and Crashes*.

Kindleberger considered that market competition was the best way of combating mercantile conservatism. Later, he put this forward as an argument in favour of liberal government policies in relation to foreign direct investment. Indeed, he viewed opposition to such investment as misguided, 'nationalist' and 'emotional' (Kindleberger 1969: 145). In this respect he was robustly internationalist, liberal and pro-business, while disclaiming any 'doctrinaire' attachment to private enterprise as a sufficient 'organiser' of economic development. He followed the liberal Keynesian consensus of his contemporaries in regarding government action as appropriate in the provision of public goods and industries with external economies and diseconomies. Even in competition with private companies, governments have the advantages of risk-spreading, better information, and ease of recruitment. In the process of starting economic development in the least developed countries, Kindleberger regarded government action as indispensable. **Alexander Gerschenkron** was an important influence on Kindleberger's thinking, about not only the role of governments in economic development, but also the functions of financial institutions in that process. But, at more advanced stages of economic development, Kindleberger was sceptical of étatiste planning, which he thought tended to over-centralise decision-making, resulting in excessively large projects. He felt, in particular, that governments are prone to getting diverted away from 'economic objectives' and to 'welfare redistribution' and the underpricing of governments' 'output' (Kindleberger 1965: 132).

Kindleberger's scepticism about government was explicit in his critique of the import-substitution strategies advocated by **Raúl Prebisch** and **Hans Singer** during the 1940s and 1950s. These strategies advocated the expansion of the domestic market in developing countries by government projects, in particular in transport, public utilities and government services, behind import barriers to ensure that the resulting economic growth was not dissipated in a surge of imports. On the face of it, such strategies would induce precisely the kind of domestic 'commercial' revolution that Kindleberger believed should precede successful industrialisation. However, he argued that, notwithstanding incidental benefits, such strategies inhibit technological improvements that may be obtained by importing more advanced manufactured goods. Above all, he considered that these strategies may divert resources away from exports and to the domestic market (Kindleberger 1963: 468). But he was typically sceptical too about arguments in favour of free trade, which he regarded as being supported by 'partial equilibrium' efficiency arguments, when economic development was about improving the efficiency of the economy as a whole. This consideration led him to conclude that market prices are an unreliable guide to economic decision-making where the process of economic development alters the whole structure of the economy (Kindleberger 1965: 191–3). There is a deeper meaning to his repeated conclusion that 'circumstances alter cases'.

Early on, Kindleberger was a critic of the application of Keynesian policies of aggregate demand management to developing countries, which he perceived to be especially prone to inflation that was not due to too much demand in the economy at large, but more 'structural'. By this he meant that the rapid expansion of employment in the sectors leading the development of the economy (in particular foreign trade, light industry, and retail services) would tend to raise earnings from employment more rapidly than food production. At low levels of income, spending on food takes up a high proportion of household budgets. The rise in food prices due to increased non-agricultural employment was therefore not amenable to aggregate demand management without bringing the process of economic development to a halt (Kindleberger 1965: 227–30).

Kindleberger robustly favoured foreign direct investment. During the 1960s, he urged the establishment of an international conference to provide ways of promoting it (Kindleberger 1969: 206–8). He believed that the absence of savings adequate to finance industrial projects is a major constraint on the economic development of poorer countries. Indeed, in his earliest writings on economic development he put forward a typology of stages of economic development, according to capital inflow from abroad. Poorer countries, he argued, use foreign capital inflows to augment their domestic saving, before eventually achieving industrial maturity and experiencing capital outflows (Kindleberger 1963: chap. 23). This belief in a savings constraint on economic development, in large measure, inspired his enthusiasm for the thrifty middle classes.

However, he was ambiguous in his attitude towards the financial liberalisation that came to characterise official US policy, and that of the World Bank and the International Monetary Fund, towards banking and finance in developing countries. During the 1950s and the early 1960s, Kindleberger took **Gerschenkron**'s side in arguing that government can more efficiently take the place of commercial banks early on in the development process. He criticised Gurley and Shaw, the leading advocates of such liberalisation in the 1950s, for underestimating the degree to which investment can be financed in advance of saving by monetary expansion (Kindleberger 1965: 234). Nevertheless, by 1970, Kindleberger was arguing in favour of more lending by foreign and domestic commercial banks in developing countries, because it 'depoliticises' commercial lending decisions, and because private finance 'adjusts to reality' more quickly than official, or government, lenders (Kindleberger 1970).

By 1976, Kindleberger was convinced that the expansion of international bank lending in developing countries had gone too far. He argued that financial liberalisation exposes developing countries to integration with international capital markets that are prone to illiquidity, which can inflict crisis on countries without robust financial systems. He urged the establishment first of domestic banking systems before developing countries opened up their financial markets to foreign financial capital (Kindleberger 1976). It was another six years before the 1982 international debt crisis convinced international bankers and economists that financial liberalisation can have painful consequences. In 1978, Kindleberger published what is arguably his most famous book, *Manias, Panics and Crashes.* In subsequent editions, lending to developing countries, in particular from the 1970s onwards, featured as a speculative mania leading to crisis. He advocated prompt central bank lending to overcome such financial crises, and urged the creation of an international 'lender of last resort' to avoid them in the future.

Kindleberger was arguably the most influential American development economist of his time. His writings on economic development are authoritative and judicious. Yet, his cosmopolitan studies of banking and economic history, his ideas on economic development ventured little beyond the mainstream US academic literature of the mid-twentieth century. In particular, the European tradition of viewing economic development as a process of changing social and political, as well as economic, structures was foreign to an American who professed himself to be an economist in a somewhat narrower sense (Kindleberger 1969: vi). His admiration for the thrift of the middle classes echoes that of John Stuart Mill a century earlier. His advocacy of a commercial revolution preceding industrialisation was rooted in his studies in the economic history of Western Europe and his native USA. Alternative routes to development seemed problematic to him. Japan, he wrote, must be an 'exception'. The saving constraint he identified in developing countries has a distinctly pre-Keynesian air, while his notion that 'mature' economies export capital is at odds with the reliance of the USA and the UK on capital imports at the start of the twenty-first century. These limitations were redeemed by the scale of his output, his work on Marshall Aid, his early recognition that financial liberalisation leads to crisis for developing countries, and his exposure of the hubris of banking and finance at the end of the twentieth century.

Major works

Kindleberger, C.P. (1953) *International Economics*, Homewood, IL: Richard D. Irwin, 2nd edn 1958, 1963.

—— (1958) *Economic Development*, New York: McGraw Hill, 2nd edn 1965.

—— (1969) *American Business Abroad: Six Lectures on Direct Investment*, New Haven, CT: Yale University Press.

—— (1970) 'Less Developed Countries and the International Capital Market', in Markham, J.W. and Papanek, G.F. (eds), *Industrial Organisation and Economic Development: Essays in Honor of E.S. Mason*, Boston, MA: Houghton Mifflin; reprinted in Kindleberger, C.P. (1981) *International Money: A Collection of Essays*, London: George Allen and Unwin.

—— (1976) 'International Financial Intermediation for Developing Countries', in R.I. McKinnon (ed.), *Money and Finance in Economic Growth and Development: Essays in Honor of Edward S. Shaw*, New York: Marcel Dekker; reprinted in Kindleberger, C.P. (1981) *International Money: A Collection of Essays*, London: George Allen and Unwin.

—— (1978) *Manias, Panics and Crashes: A History of Financial Crises*, Basingstoke: Macmillan, 3rd edn 1996 (2nd edn 1989).

Further reading

Kindleberger, C.P. (1950) *The Dollar Shortage*, New York: The Technology Press of Massachusetts Institute of Technology and John Wiley.

—— (1964) *Economic Growth and France and Britain 1851–1950*, Cambridge, MA: Harvard University Press.
—— (1984) *A Financial History of Western Europe*, New York and Oxford: Oxford University Press, 2nd edn 1993.
—— (1987) *International Capital Movements*, Cambridge: Cambridge University Press.
Miall, L. (2003) Obituary: 'Charles P. Kindleberger; economic historian and progenitor of the Marshall Plan', *The Independent*, 10 July: 16.

Jan Toporowski

SIR WILLIAM ARTHUR LEWIS (1915–91)

William Arthur Lewis was a pre-eminent figure in the emergence of development economics from the late 1940s until his death in 1991. He made a singular contribution to its elaboration as a distinct branch of economics and as a body of ideas informing the formation of public policy. His work on problems of balanced sectoral growth, 'unlimited' surplus labour, agricultural productivity and planning created an important framework and agenda which shaped much subsequent research. His stature was perhaps most succinctly summed up when Harberger called him 'the patriarch' of development economics.

Arthur Lewis was born in St Lucia in 1915, the fourth of five sons of parents who were both school teachers. His father died when he was seven and his mother thus became the most powerful early influence in his life, imparting her strong religious convictions and her love of music to her son. Lewis left school at fourteen, having completed his school certificates, to work as a civil service clerk in the St Lucia Department of Agriculture. At seventeen, he won an island scholarship which took him to the London School of Economics (LSE). He had hoped to be an engineer but the realities of the colonial West Indies meant that there was no employment for a black engineer. Instead he opted for a business administration degree, confessing later that 'I had no idea in 1933 what economics was' (Lewis 1979). Nevertheless, he graduated in 1937 with 'the highest marks ever obtained in the history of the School' (Downes 2003: 2). This opened the door to a PhD in industrial economics (which he received in 1940) and to a lectureship at the LSE, making him its first black member of staff. Between 1938 and 1947 (by which time he had become Reader in Colonial Economics), Lewis had short attachments at the Board of Trade and the Colonial Office, allowing him research access to 'reports from the colonial territories on agricultural problems, mining, currency questions, and the like'. During this period he also became a member of the executive of the League of Coloured People, editing its journal, *The Keys*. He also published a pamphlet on economic and social problems in the West Indies for the Fabian Society, an association which would yield a number of publications. Although he professed little interest in politics, describing himself as 'a social democrat who was interested in poverty eradication and economic and social development' (Downes 2003: 5), this combination of scholarship and civic activism characterised his life.

Lewis's LSE research produced two significant books which introduced themes that were to characterise his work. The first, *Economic Survey, 1919–39* (1949) situated economic development in the context of world economic history, viewing the world economy as a single interdependent system. The second, *The Principles of Economic Planning* (1949), written for the Fabian Society, set out his case for 'planning through the market', identifying the 'conditions under which the market would need to be supplemented by intelligent state action' (Bhagwati 1979: 15), and in the process developing a powerful argument against both *laissez-faire* and, particularly, 'planning by direction'. The latter he considered inefficient, inflexible, undemocratic and 'procrustean', criticisms that would later be taken up by many other commentators, many of them having little in common with Lewis.

Lewis's most important intellectual achievements came after he moved to the University of Manchester in 1947 to become Stanley Jevons Professor of Political Economy. There, for the next ten years, he made a defining contribution to the elaboration of the discipline of development economics. That contribution is encapsulated in two publications. The first was a paper entitled 'Economic Development with Unlimited Supplies of Labour', published in *Manchester School* in May 1954. The second was an elaboration of it (and much else) in a broad treatise entitled *The Theory of Economic Growth* (1955). Wrestling with the problems of poverty and poor rates of growth in developing countries, Lewis felt increasingly that neoclassical explanations for wages, profits and savings did not work in such countries and that relative prices could not be explained within that framework. He thus took his point of departure from classical economic theory. In his autobiographical statement for the Nobel committee, he recalled:

> From my undergraduate days, I had sought a solution to the question of what determines the prices of steel and coffee. The approach through marginal utility made no sense to me. And the Heckscher-Ohlin framework could not be used, since that assumes that trading partners have the same production functions ... Another problem that troubled me was historical. Apparently during the first fifty years of the industrial revolution, real wages in Britain remained more or less constant while profits and savings soared. This could not be

squared with the neo-classical framework, in which a rise in investment should raise wages and depress the rate of return on capital.

One day in August 1952, walking down the road in Bangkok, it came to me suddenly that both problems have the same solution. Throw away the neo-classical assumption that the quantity of labour is fixed. An 'unlimited supply of labour' will keep wages down, providing cheap coffee in the first case and high profits in the second case. The result is a dual (national or world) economy, where one part is a reservoir of cheap labour for the other.

(Lewis 1979)

The 'unlimited' supplies of labour were made possible, first, by the subsistence economy of the 'traditional' agrarian sector in most developing countries and, second, by the considerable underemployment and low productivity of that sector. In a system of production where the marginal productivity of cultivators approached zero, it was possible for large ('unlimited') supplies of labour to move to the capitalist sector of the economy without significantly raising industrial wages (since real wages were set by the consumption levels of the traditional sector rather than by supply and demand) or reducing profits, and without significantly reducing agrarian production. Thus 'dualism' meant that high profits and rapid industrial growth would not be limited by the same factors as in developed economies.

What would limit the prospects for growth, Lewis concluded, was the inability of the agrarian sector in such dual economies to trade with the industrial sector so that industrialisation would need either a rapid increase in agricultural productivity or export markets. Using a Ricardo–Graham model, Lewis argued that the terms of trade (say between North and South) were determined by relative labour productivities in food. Given the low productivity of the traditional agrarian sector in developing countries, there would be what Emmanuel was later to call 'unequal exchange'. **Albert Hirschman** (1982: 382) describes it thus: 'as long as "unlimited supplies of labour" in the subsistence sector depress the real wage throughout the economy, any gains from productivity increases in the export sector are likely to accrue to the importing countries . . .'. This led Lewis to the conclusion that there was a need for balanced growth: for rapid industrialisation and for a rapid increase in agrarian productivity so as to improve both the terms of trade and to reduce rural poverty. There was also a need for protection to permit this process to take place.

These two models, one explaining the problems and possibilities produced by 'dualism' and the other by the terms of trade between economies with different production functions, set the agenda for debate, discussion and further research in development economics for a generation. As Findlay (1980: 3) observes:

> His own subsequent work, and in fact a large part of the literature of development economics, can to a large extent be seen as an extended commentary on the meaning and ramifications of this central idea. Few other instances come readily to mind of an entire field being so dominated by a single paper.

For twenty-five years from 1957, Lewis elaborated and reprised these ideas in various ways. A volume on development planning appeared in 1966, extending the ideas first explored in 1949 and an edited collection, *Tropical Development, 1880–1913*, published in 1970, re-examined various issues of economic history and development in a global economy. In the years after Manchester, Lewis became as much involved in administration and policy as in academic research. From 1959 to 1963 he served as the first West Indian principal of the University College of the West Indies, helping to extend its reach beyond Jamaica to Barbados and Trinidad, and negotiating its autonomy from London University to become the independent University of the West Indies. In 1967 he was appointed Chancellor of the University of Guyana and in the mid-1970s he also lectured part-time at UWI in Barbados. Among a number of roles as adviser to the leaders of developing countries, Lewis served as UN economic adviser to President **Kwame Nkrumah** in Ghana until 1963 (helping to produce Ghana's second national development plan), as deputy managing director of the UN Special Fund, and as president of the Caribbean Development Bank from 1970 to 1974. In 1963, he became James Madison Professor of Political Economy at Princeton, a tenure which lasted until his retirement in 1983. His approach to economics and economic history gave him a broad perspective and the capacity to look beyond the narrow models that dominated neo-classical economics. They also prompted him to address related issues that lay outside the disciplinary boundaries of economics. Thus, his *Politics in West Africa* (1965) explored the relationship between democracy and development in West Africa. And, in 1982 he delivered the W.E.B. Du Bois Lectures, which were published as *Racial Conflict and Economic Development* (1985). Lewis was knighted in 1978 and received the Nobel Prize for Economics (jointly with Theodore Schultz) in 1979. In his paper for the Nobel Prize committee, **Jagdish Bhagwati** argued that Lewis had made a singular contribution on three levels – to political economy, historical analysis and the modelling of development problems.

In more sardonic vein, **Albert Hirschman** (1982: 377) observed that the Nobel Prize for economics 'is often split between one person who has developed a certain thesis [here Lewis] and another who has labored mightily to prove it wrong' [Schultz]. This comment highlights the fact that Lewis – and development economics in general – came under persistent attack from the start from both left and right ('the strange alliance of Neo-Marxism and Monoeconomics against Development Economics', Hirschman calls it). For both sides, the relative failure of indicative planning in countries like Ghana provided confirmation of their rejection of development economics. For many in the neo-classical orthodoxy, the Lewis models became 'a privileged target'. The rise of neo-liberalism and the new economic history (which in part grew out of Lewis's ideas, according to Bhagwati) demonstrated little sympathy for his views on dualistic markets or state intervention. For dependency and 'development-of-underdevelopment' theorists, too, Lewis's contention that economic exchanges between rich and poor countries could, in some circumstances, become antagonistic, missed the point. Rather, they insisted, this was the essential character of the relationship, an iron law of centre–periphery inequality in the global capitalist economy which made the hopes of balanced growth an illusion. Lewis (1984) directly rejected this dependency argument: it was appropriate, he suggested, as a description of what happened in the colonial context, in the late nineteenth and early twentieth centuries, but it made little sense in the second half of the twentieth century when independent governments sought to restructure these relationships. Nor is it likely, for all his courteous manner, that he would have had much time for the Washington Consensus and neo-liberal agenda.

Even so, there is little doubt that the tide has run against Lewis and the other important figures of development economics over the last twenty-five years. The spiralling debt crisis and the increasing inequality and instability characterising the global system have undermined ideas of promoting balanced growth. What remains is the debate about whose fault it is. For all that, Lewis's personal and intellectual standing would seem to have endured, perhaps one final testament to his stature as one of the leading economists of his time.

Major works

Lewis, W.A. (1949) *Economic Survey, 1919–39*, London: Allen & Unwin.
—— (1949) *The Principles of Economic Planning*, a study prepared for the Fabian Society, London: Allen & Unwin.
—— (1954) 'Economic Development with Unlimited Supplies of Labour', *Manchester School* 22: 139– 91.
—— (1955) *The Theory of Economic Growth*, London: Allen & Unwin.
—— (1965) *Politics in West Africa*, London: Allen & Unwin.
—— (1966) *Development Planning: The Essentials of Economic Policy*, London: Allen & Unwin.

—— (1970) *Tropical Development, 1880– 1913*, London: Allen & Unwin.
—— (1978) *Growth and Fluctuations, 1870– 1913*, London: Allen & Unwin.
—— (1978) *The Evolution of the International Economic Order*, Princeton, NJ: Princeton University Press.
—— (1985) *Racial Conflict and Economic Development*, the W.E.B. Du Bois Lectures 1982, Cambridge, MA: Harvard University Press.

Further reading

Bhagwati, J.N. (1979) 'W. Arthur Lewis: An Appreciation', reprinted in M. Gersovitz *et al.* (eds) (1982), *The Theory and Experience of Economic Development: Essays in Honor of Sir W. Arthur Lewis*, London: Allen & Unwin, pp. 15–28.
Downes, A.S. (2003) 'William Arthur Lewis: 1915–1991 – A Biography', paper prepared for D. Rutherford (ed.) (2004), *The Biographical Dictionary of British Economists*, Bristol: Thoemmes Press.
Findlay, R. (1980) 'On W. Arthur Lewis's Contributions to Economics', reprinted in M Gersovitz *et al.* (eds) (1982), *The Theory and Experience of Economic Development: Essays in Honor of Sir W Arthur Lewis*, London: Allen & Unwin, pp. 1–14.
Gersovitz, M., Diaz-Alejandro, C.F., Ranis, G. and Rosenzweig, M.R. (eds) (1982), *The Theory and Experience of Economic Development: Essays in Honor of Sir Arthur Lewis*, London: Allen & Unwin.
Hirschman, A.O. (1982) 'The Rise and Decline of Development Economics', in Gersovitz *et al.* (eds), pp. 372–90.
Ingham, B. (1991) 'The Manchester Years, 1947–1958: A Tribute to the Work of Arthur Lewis', Salford Papers in Economics, University of Salford.
Lewis, W.A. (1979) Autobiographical statement to the Nobel committee, Nobel Lectures, Economics 1969–80, http://nobelprize.org/economics/laureates/1979/lewis-autobio.html.
—— (1984) 'Development Economics in the 1950s', in Meier, G.M. and Seers, D. (eds), *Pioneers in Development*, New York: Oxford University Press for The World Bank, pp. 119–37.

Morris Szeftel

MICHAEL LIPTON (1937–)

Michael Lipton has made distinguished contributions to development research across a wide front. Most importantly, however, he is the principal contemporary theorist and advocate of the role of agriculture in development, and one of the most significant leaders of research into the causes and character of poverty. He has for long championed the role of small-scale agricultural producers, emphasising how the agricultural sector generally, and small farmers most specifically, are harmed by what he describes as 'urban bias'. For this he has sometimes been castigated (most memorably in Byres 1979) as a populist. But he is no dewy-eyed romantic harking back to the agrarian past. He is unusual among scholars working on development issues for his deep engagement with the hard sciences, especially plant breeding and nutrition. And he certainly does not envisage that all agricultural production is necessarily, and always, carried out most efficiently on small farms.

Lipton was born in London in 1937, into a German Jewish family that had left Hamburg in 1933. He went to Haberdashers' Aske's School in Borehamwood, north London, and from there to Balliol College, Oxford, to read Politics, Philosophy and Economics, graduating with First Class honours, winning several university prizes for economics, and subsequently taking a prestigious Fellowship at All Souls' College in Oxford. He was taught as an undergraduate by **Paul Streeten**, and it was as a result of some notes that he wrote for Streeten on **Gunnar Myrdal**'s *Beyond the Welfare State* (1960) that Myrdal subsequently invited him to work on parts of what was to become *Asian Drama* (Myrdal 1968). Ultimately published in three large volumes, *Asian Drama* was in its time a pioneering study in which an institutional approach was applied to the analysis of the development problems of South Asia. 'Mr Lipton', Myrdal writes,

> wrote the first draft of Appendix 10 on climate and prepared an intensive analysis of national accounting in the South Asian countries ... [and his] ... most difficult and time-consuming assignment was to carry out a lengthy comparative analysis of the development plans in the countries of the region.
>
> (Myrdal 1968: xv)

It was this experience that drew Michael Lipton into development economics and gave him his life-long interest in the development of South Asia, in particular.

According to his own account, he taught himself development economics as a young lecturer at Sussex University, and he studied mathematics and econometrics for a year at MIT in 1963–4, alongside the future Nobel Prize winners, **Joseph Stiglitz** and George Akerloff. Shortly after this, in 1965, he took the very unusual step for an economist of going to live for seven months in a village in the western Indian state of Maharashtra. In the work that he had done for *Asian Drama*, he had come to realise the inadequacy of existing micro-level research for understanding problems of agricultural development, and he set about making up for this lack. His experiences in the village (called Kavathe) gave rise to a seminal article on 'The Theory of the Optimising Peasant' (1968a) and to an important critique of India's agricultural development policy that appeared in a major book, *The Crisis of Indian Planning*, published in the same year (Streeten and Lipton 1968, see Lipton 1968b). That year, 1968, is a remarkable one in

Lipton's career, for he published (at the age of only thirty-one) not just these two outstanding contributions to understanding of peasant agriculture and of its role in development, but also a book on British economic performance (Lipton 1968c), and a book of chess problems. Chess, as well as poetry-reading (he enjoys, for example, the work of Emily Dickinson) and music, have remained passions for Michael Lipton but – unfortunately – he has not written on British economic policy since the 1960s. The Kavathe experience taught him not only about the particular rationality of small peasant farming but also stimulated his thinking about the significance of the urban–rural gap, and the idea of 'urban bias' appeared for the first time in his paper for *The Crisis of Indian Planning* (1968b). It appeared again in another seminal paper, on land reform, published in 1974, and it was finally elaborated, controversially, in the book *Why Poor People Stay Poor* in 1977. Meanwhile he had sought to put his ideas about redistributive land reform to practical effect in his contributions to an influential ILO report on the development of Sri Lanka, in 1971 (ILO 1971). All of this work was done from the Institute of Development Studies at the University of Sussex, where he was appointed as a professorial fellow in 1967, and remained a leading figure for almost thirty years. His role in the establishment of academic development studies was also reflected in his contributions as a managing editor of the still very young *Journal of Development Studies* – though he staked out a more cautious view of the role of 'Interdisciplinary Studies in LDCs' than some others have advocated, in an article published in the *JDS* in 1970. At the time, Lipton was directing another project that resulted from his Kavathe experience, the Village Studies Project (VSP), which had the ambitions of generating general theory from systematic comparison of village studies, and of situating economic activity in developing countries in its wider social context. Though the case for analysing institutions that are situated between the conventional ones for economics of the household on the one hand, and the state on the other, is a strong one, the VSP was not entirely successful, partly because of the very uneven quality of the existing village surveys on which it was based. It gave rise to several interesting outputs (for example Lipton *et al.* 1976; Connell *et al.* 1976), but it was left to others to make more sense of 'villages' as institutions (see, most recently, Krishna 2002).

Underlying much of Michael Lipton's work are ideas that derive from the Kavathe experience and from his appraisal of Indian agriculture in the 1960s. At that time thinking about peasant agriculture was influenced especially by the (later, Nobel prize-winning) Chicago economist, Theodore Schultz, whose book, *Transforming Traditional Agriculture* (1964), elaborated the view that peasant farmers are 'efficient but poor'. Schultz's argument was that peasant farmers, by maximising their profits, allocate their resources efficiently. But they are poor, and their resources are severely limited – so achieving growth of output must involve changing the conditions of production through the application of new technology and new inputs (such as chemical fertilisers). This is what the Indian government was trying to do by the later 1960s, concentrating those resources that were devoted to agriculture on areas that were thought to have a high potential for such modern cultivation. Lipton's 'theory of the optimising peasant', however, held that it would be irrational for small peasant farmers to attempt to maximise profits, for such a strategy would take no account of the considerable risks and uncertainty associated with agricultural production. In their actual agricultural practices, Lipton showed, poor farmers seek to insure themselves against failure, and they may well sacrifice potential output or profit maximisation in the process. They have, he said, a 'survival algorithm', pursuing – in terms of formal game theory – an *optimising* 'minimax strategy', rather than trying to maximise. The aim of public policy, therefore, should be to reduce the riskiness of agriculture. The dedication of resources to irrigation, for instance, rather than provision of chemical fertilisers to farmers operating in more favoured areas, would enable much broader-based, more egalitarian agricultural development. Given appropriate policies the relative efficiency of small-scale, household producers in the agricultural sector – reflected in the common observation of the existence of an inverse relationship between farm size and the productivity of land, and deriving from the fact that this mode of production reduces many of the 'transaction costs' involved, and so encourages intensive application of labour – can become the basis for dynamic and also well-distributed growth. At the time, thinking about development policy was still dominated by the idea that economic growth, necessarily involving industrialisation in the longer run, was best pursued in poor countries through planned, import-substituting industrialisation. This meant that there was a strong tendency, as Lipton noted in the Indian case, for agriculture to be neglected in the allocation of public investment. Development policy was characterised by an urban, industrial bias – a bias against what could be shown to be more efficient and more equitable ways of using public resources, through investment, in the first instance, in small-scale agriculture. Urban bias was underpinned, however, by a powerful political alliance in which the urban classes, bourgeoisie and proletarians with interests alike in cheap agricultural goods and the transfer of resources out of agriculture (another aspect of urban bias) drew in also the more powerful people in the rural economy through the transfer of some subsidies and the offer of places in the political sun. This alliance worked against the possibility of redistributive land reform that, in many rural economies, had the potential to create both a more just and a more efficient rural economy.

This is the case that Lipton set out in detail in the 1974 paper – revisited in 1993a – and that he tried to see implemented in Sri Lanka. Urban bias, therefore, militated against a potentially more effective strategy for development, based on the transformation of the rural economy as a whole – an 'agriculture first' strategy for development.

Lipton has maintained his advocacy of small-scale producers and the role of agriculture in development, while devoting most of his research in the last twenty years to understanding the causes and conditions of poverty (see Lipton 1983a–c, 1985, 1993a, 1993b). The two themes were brought together in an authoritative book, *New Seeds and Poor People* (1989), on the impact of the 'modern seeds' (or Higher-Yielding Varieties) that had brought the 'Green Revolution'. There had been a spate of influential work in the 1970s that was critical of the effects of the introduction of the new seeds, as contributing to the further impoverishment of large numbers of rural people (see especially Griffin 1974). Lipton himself had been a little sceptical in the early days (1968b: 118) – but he later showed that much of the criticism of the Green Revolution had been premature and that, most significantly because of its effects on both labour and product markets, the new technology, in South Asia certainly, had brought real benefits for poor people. The cultivation of the new seeds had absorbed more labour and helped to tighten rural labour markets, imparting a push upwards to rural wages, while increased output of cereals had tended to bring prices down. The two factors together had benefited the poor, a majority of whom depend on wages and on purchasing food rather than producing it themselves. Lipton has continued to emphasise the importance for poverty outcomes of the ways in which labour markets function, and his continued championing of small-scale agriculture is because of the ways in which this mode of agricultural production can generate productive employment while the process of the structural transformation of the economy takes place. Ultimately, such small-scale production may well be replaced by larger-scale commercial production, but it is mistaken to attempt to short-circuit this historical process.

Interested in the relevance of Asian experience for Africa, and vice versa, Lipton took up work on rural labour in Africa, initially in Botswana in the later 1970s, and subsequently with his wife, Merle, in South Africa; he began to engage more thoroughly with nutrition science (reflected first in Lipton 1983c); he has continued to work on economic demography (Lipton 1983a; and Lipton and Eastwood 1999); he undertook studies of aid effectiveness; and most recently he has entered into the controversy about the potentials and problems associated with the applications of biotechnology. He is firmly of the view that this technology has enormous potentials in regard to the livelihoods of poor people, and that much of the criticism – like that of the Green Revolution three decades ago – is muddle-headed. He remains an active researcher and, with his champion chess-player's style of argument, he is as engaging and formidable in debate as ever.

Major works

International Labour Office (1971) *Matching Employment Opportunities and Expectations: A Programme of Action for Ceylon*, Vol. 2, *Working Papers*, Paper 10.

Lipton, M. (1968a) 'The Theory of the Optimising Peasant', *Journal of Development Studies*, 4: 327–51.

—— (1968b) 'Strategy for Agriculture: Urban Bias and Rural Planning', in Streeten, P. and Lipton, M. (eds), *The Crisis of Indian Planning: Economic Planning in the 1960s*, London: Oxford University Press.

—— (1968c) *Assessing Economic Performance: Some Features of British Economic Development 1950–65 In Light of Economic Theory and Principles of Economic Planning*, London: Staples Press.

—— (1970) 'Interdisciplinary Studies in LDCs', *Journal of Development Studies* 7: 5–18.

—— (1974) 'Towards a Theory of Land Reform', in Lehmann, D. (ed.), *Agrarian Reform and Agrarian Reformism*, London: Faber, pp. 269–315.

—— (1977) *Why Poor People Stay Poor: A Study of Urban Bias in World Development*, London: Temple Smith.

—— (1983a) *Demography and Poverty*, Washington, DC: The World Bank.

—— (1983b) *Labour and Poverty*, Washington, DC: The World Bank.

—— (1983c) *Poverty, Undernutrition and Hunger*, Washington, DC: World Bank.

—— (1985) *Land Assets and Rural Poverty*, Washington, DC: World Bank.

—— (1989) *New Seeds and Poor People*, London: Unwin Hyman (with R. Longhurst).

—— (1993a) 'Land Reform as Commenced Business: The Evidence Against Stopping', *World Development* 21(4): 641–57.

—— (1993b) *Poverty and Policy*, Washington, DC: World Bank (M. Lipton and M. Ravallion).

—— (1993c) *Including the Poor*, Washington, DC: World Bank (with J. van der Gaag, eds).

Lipton, M. and Eastwood, R. (1999) 'The Impact of Change in Human Fertility on Poverty', *Journal of Development Studies* 36(1): 1–30.

Lipton, M. *et al.* (1976) *Migration from Rural Areas: The Evidence from Village Studies*, Oxford: Oxford University Press.

Further reading

Byres, T.J. (1979) 'Of Neo-populist Pipe-dreams', *Journal of Peasant Studies* 6: 210–44.

Connell, J. *et al.* (1976) *Assessing Village Labour Situations in Developing Countries*, Delhi: Oxford University Press.

Griffin, K. (1974) *The Political Economy of Agrarian Change*, London: Macmillan.

Krishna, A. (2002) *Activating Social Capital*, New York: Columbia University Press.

Myrdal, G. (1960) *Beyond the Welfare State: Economic Planning in the Welfare States and its International Implications*, London: Duckworth.
—— (1968) *Asian Drama*, 3 vols, London: Penguin.
Schultz, T. (1964) *Transforming Traditional Agriculture*, New Haven, CT: Yale University Press.

John Harriss

REVEREND THOMAS ROBERT MALTHUS (1766–1834)

Not many people widely quoted in the contemporary development literature have given their name to a familiar concept and a word, as has Malthus. 'Malthusian' connotes a pessimistic and negative long-term prospect: a 'Malthusian scenario' is a frequent phrase anticipating an overuse of resources driven by excess population and rapid population growth, with the likelihood of serious demographic and economic collapse and 'crisis' (Graham and Boyle 2002). Similarly, the 'Malthusian trap' is often invoked to argue that population growth is too rapid to permit savings and investment in national economic growth and thereby to raise standards of living; 'neo-Malthusian' is a term widely applied to views that identify the contemporary relevance of his ideas, first developed over 200 years ago, on the relationship between population and resources, essentially that population growth and sustainable resource use are fundamentally incompatible in the long term.

Today we would call Malthus a political economist, though he was also an Anglican clergyman. He studied mathematics at Cambridge during a period much affected by the intellectual ferment of the Enlightenment and over the period of the French Revolution, when the nature of society and the relationship between poverty and inequality were dominant in public debate. Malthus was prominent in that debate, principally in two famous essays in 1798 and 1803. The second essay, with several subsequent editions, including as a contribution to the *Encylopaedia Britannica* in 1824 (Malthus 1830), was the more reasoned and is more widely quoted. It was first published just before the beginning of Malthus's long period as Professor of Political Economy (from 1805 till his death) at the East India Company's Staff College, Haileybury, the major training institution for rulers of British India throughout the nineteenth century. However, there is not much about India in the essays and he himself never visited India (or went out of Europe) (Flew 1970).

Malthus's outstanding and long-lasting contribution to debates on the nature of development has been to argue that, whereas population had the potential to grow geometrically, the resource base to support that population with food and other essentials could only grow much more slowly. Since he assumed a relatively fixed relationship between population and resources, taking resources as givens so that they could not really be created, ultimately population must outstrip resources unless its growth could be controlled. Population control, Malthus argued, could be, and had largely been, achieved by preventive measures – i.e. positive checks controlling the birth rate, essentially through sexual restraint outside and within marriage (but not the use of contraception). However, where these failed (and he saw them failing in the England of his lifetime) then the balance would inevitably be set by direct constraints – i.e. negative checks on the death rate, by 'natural' factors, such as disease, famine and (not quite 'natural') warfare.

Population growth in most human societies is usually controlled by preventive measures – humans do not breed to the limits of their fecundity, and social institutions such as marriage and custom will affect family size and childbearing patterns. However, Malthus believed, given the inequality and poverty of his own times (just after the French Revolution and the Napoleonic wars, a time of rapid industrialisation in western Europe), that the poor were showing insufficient restraint and that they were poor as a result of having large families that they could not support. As a result their mortality was raised, due to malnutrition and disease. The poor were 'victims of their own passion'! These ideas represented a response to apologists for the French Revolution, commentators we would now see as developing socialist or egalitarian arguments, who claimed that poverty and degradation were due primarily to inequalities in wealth rather than the absolute lack of resources to support any population, i.e. Malthus's position (Wrigley 1986). The debate on population between Malthusian ideas and socialist, **Marx**ian ideas persists today (Woods 1989).

Today we can appreciate some of the essential failings but also the enduring strengths of Malthus's arguments. The most obvious weakness is in the argument that resources were largely given and could not really be created beyond an arithmetic ratio. That this is not the case had become evident in the early/mid-nineteenth century, with the great increase in the availability of land, largely in the New World. This revolutionised the global food supply, with densely populated Europe being increasingly supplied with cheaper food from overseas sources: the American plains, Australia and Argentina. New resources, of technology (e.g. refrigerated ships) as well as land, were being created and were able to support the burgeoning European and global populations at rising levels of living.

Technology continued to revolutionise food production throughout the twentieth century, most evidently in the very substantial increases in food production associated with Green Revolution technologies from the

1970s, to the extent that the erstwhile bread-basket 'Malthusian' cases of India and Bangladesh are no longer major food-deficit countries, despite their continuing – though now falling – high rates of population growth (Dyson 1996). Malthusian attention in the food debate has now moved to Africa, where there are serious food deficits. However, are these more appropriately attributed to the rapid population growth of the last half-century (the Malthusian view) or rather to inappropriate national and global food policies and inadequate resources being allocated to agricultural development? Is the contemporary African food crisis Malthusian or a political/economic crisis?

The prominence of Malthusian ideas in Development Studies peaked during the 1960s and 1970s, a period of rapid population growth, especially in the global South, and substantial economic growth in richer countries. Population growth was seen to be a constraint on development, with many poor countries caught in a Malthusian trap. In 1990, when fertility rates were not thought to be generally declining in African countries and before HIV/AIDS was taken to be the serious demographic problem that it now is, the World Bank's Annual Report could argue that 'Population growth is the biggest single development problem for Africa'. The neo-Malthusian view is very much associated with the American commentators Lester Brown (1998) and Paul Ehrlich (Ehrlich and Holdren 1971; Ehrlich and Ehrlich 1990) and Norman Myers (1998) in the UK, who, quite separately, in a series of influential books and articles from the 1970s, argued the case for the necessity for preventive checks. These views were important for the growth of the family planning movement, later 'softened' by the late 1980s and 1990s into a concern for reproductive health and mother-and-child programmes, and all integrated into global thinking through United Nations declarations and objectives. These developed a widely accepted neo-Malthusian agenda in that the main problems were seen to lie on the population side of the equation, and solutions were to be sought in restricting population growth rather than in expanding production. However, it is not a matter of either/or: there is clearly a need to consider both simultaneously, as in the case of the World Bank which has been a strong force in the development of Green Revolution technologies as well as extending major support for population programmes.

One further influential strand in the neo-Malthusian argument at that time was associated with the Club of Rome (Meadows *et al.* 1972). In a series of alarmist simulations of current resource depletion and increasing pollution, this largely European-based group predicted a serious overburdening of the global system by the year 2000, to the extent that economic development would be thrown sharply into reverse. Globally, population growth and increasing consumption of physical resources with polluting technologies seemed no longer sustainable.

These neo-Malthusian scenarios have proved to be largely unfounded. Today, the world is much richer than thirty years ago, global agreements have been promulgated to control pollution, new technologies have brought new resources into the economic system, population growth rates are everywhere falling, and there is a global awareness of the need to balance population and resources at all scales. As a result of greatly increased geological knowledge and higher oil prices, driven upwards by OPEC principally for political reasons, there is now a world surplus of oil, and not the imminent fuel crisis forecast by the Club of Rome. New resources are being used (e.g. renewable energy from wind and solar power), while others (coal, copper) are much less in demand. As Julian Simon (1981) argued persuasively in his polemic on population, the long-term prices for energy and mineral resources have been falling, as a sign of long-term surplus and sustainability rather than depletion and degradation. Simon's position diametrically opposes that of Malthus: for him people are themselves resources, creators and managers of natural producers of resources, rather than just mouths to feed. Simon (1981) agrees that low population growth is preferable to high population growth, but argues that it may also be preferable to zero population growth. Investment in people through education and training and better health, as well as better access to technology and resources, will lead to enhanced productivity for the poor as for the rich.

The perspective about population being both cause and effect – common to both Malthus and **Marx** – is also shared with the Danish agricultural economist, **Ester Boserup** (Boserup 1990; Lee 1986). She is the most prominent contemporary advocate of the view that population growth can act as a stimulus rather than an impediment to economic change. For her, population growth can be a critical stimulus for agricultural intensification, with rising outputs per unit of input of land, labour and technology. Where there is no population growth, there is no pressure to raise outputs, to introduce new technologies or to raise labour inputs. We are now aware, particularly through the work of Michael Mortimore and collaborators, first in the Kano Close-Settled Zone in Nigeria, but more generally in West Africa (Mortimore 1998), and subsequently in East Africa, most famously in Machakos, Kenya, that there can be 'more people, less erosion' (Tiffen, Mortimore and Gichuki 1994), that long-term environmental improvement and productivity gains are possible with the stimulus of population growth, combined with commercialisation of the economy. In the light of their evidence, any neo-Malthusian interpretation of the current African environmental, economic and political plight therefore needs serious reassessment.

Given this range of qualifications and counter-arguments, why does the neo-Malthusian discourse remain prominent in Development Studies? We could cite four sets of reasons:

Empirical evidence: There is ample empirical evidence that high rates of population growth, consistent over several decades, have been a major constraint on development in some circumstances, and Malthus himself adduced wide-ranging empirical evidence in his essays (Wrigley 1986). However, alternative development scenarios may always be suggested that seek to restore any population/resource imbalances by concentrating on the resource side of the equation rather than population. How do we know that population is the problem even in difficult environmental situations? More specifically, is Africa over- or under-populated? This is a multivariate problem that cannot be answered by considering the population factor independently.

The long-term perspective: One school of thought sees the Malthusian scenario as merely having been postponed as a result of new lands and new technologies, and that in the long term there is an essential logic about the need to achieve a balance between population and resources. There are limits to technology, and so there must be limits to population growth. The threat of 'standing room only', of a very large global population living at low and falling levels of consumption is a worst-case scenario. Current UN projections are that the global population will grow from 6 billion in 2000 to stabilise at 9 billion in 2100. Can this population be sustained at current (and hopefully rising) levels of consumption globally?

Environmentalist discourses: The widely accepted need to strive towards 'sustainable development' does require a balance between population and resources. Much of the attention is towards having more efficient and less polluting methods of resource use rather than lower consumption. However, 'deep Greens', usually a vocal minority in the environmental movement, argue for a fundamental incompatibility between consumption growth and long-term sustainability, essentially the Malthusian position. A solution must involve preventive checks on population growth as well as consumption changes.

Global poverty and inequality: Discussion of Malthusian scenarios is most common for Southern countries. It is typically a view from the outside – and has to be seen in the broader context of global inequality. The Malthusian threat has receded in the rich North; indeed there may be more demographic threat from population ageing and below-replacement fertility than from population growth, or even from excess consumption. But in the poor South, in a globalising world of increasing economic inequalities, population growth appears from the outside to continue to be a problem. Do the poor continue to be victims of their own passion? As in Malthus's time, a critical discourse for all of humankind revolves around production and distribution, the ownership and use of resources, human rights and environmental justice, as much as consumption, the number of consumers and their levels of living.

Major works

Malthus, T.R. (1798) *An Essay on the Principle of Population, as it Affects the Future Improvement of Society, with remarks on the speculations of Mr. Godwin, M. Condorcet, and other writers*, London: J. Johnson; reprinted in London: Macmillan and the Royal Economic Society.

—— (1803) *An Essay on the Principle of Population, as it Affects the Future Improvement of Society; or a view of its Past and Present Effects on Human Happiness; with an Inquiry into our Prospects respecting the Future Removal or Mitigation of the Evils which it Occasions*, 2 vols, London: J. Johnson.

—— (1830) *A Summary View of the Principle of Population*, London: John Murray, from *Encyclopaedia Britannica* (1824), reprinted in D.V. Glass, *Introduction to Malthus*, London: Watts.

Further reading

Boserup, E. (1990) *Economic and Demographic Relationships in Development/Ester Boserup*, Baltimore, MD: Johns Hopkins University Press.

Brown, L.R. (ed.) (1998) *State of the World, 1998*, Washington, DC: Worldwatch Institute.

Dyson, T. (1996) *Population and Food: Global Trends and Future Prospects*, London: Routledge.

Ehrlich, P.R. and Ehrlich, A.H. (1990) *The Population Explosion*, New York: Simon and Schuster.

Ehrlich, P. and Holdren, J. (1971) 'The Impact of Population Growth', *Science* 171: 1212–17.

Flew, A. (1970) 'Introduction', in T.R. Malthus, '*An Essay on the Principle of Population' and 'A Summary View of the Principle of Population'*, London: Penguin, pp. 7–55.

Graham, E. and Boyle, P. (2002), 'Population Crises: From the Global to the Local', chap. 13 in Johnson, R.J., Taylor, P.J. and Watts, M.J. (eds), *Geographies of Global Change: Remapping the World*, Oxford: Blackwell, pp. 198–215.

Lee, R.D. (1986) 'Malthus and Boserup: A Dynamic Synthesis', in D. Coleman and R. Schofield (eds), *The State of Population Theory*, Oxford: Blackwell, pp. 96–130.

Meadows, D.H., Meadows, D.L., Randers, J. and Behrens, W.W. (1972) *The Limits to Growth*, London: Pan Books.

Mortimore, M. (1998) *Roots in the African Dust: Sustaining the Dry Lands*, Cambridge: Cambridge University Press.

Myers, N. (1998) 'Global Population and Emergent Issues', in N. Polunin (ed.), *Population and Global Security*, Cambridge: Cambridge University Press, pp. 17–46.

Simon, J.L. (1981) *The Ultimate Resource*, Princeton, NJ: Princeton University Press, and Oxford: Martin Robertson.

Tiffen, M., Mortimore M. and Gichuki, F. (1994) *More People, Less Erosion: Environmental Recovery in Kenya*, Chichester: Wiley.

Woods, R. (1989) 'Malthus, Marx and Population Crises', chap. 6 in Johnson, R.J. and Taylor, P.J. (eds), *A World in Crisis? Geographical Perspectives*, Oxford: Blackwell, pp. 151–74.

Wrigley, E.A. (1986) 'Elegance and Experience: Malthus at the Bar of History', in Coleman, D. and Schofield, R. (eds), *The State of Population Theory*, Oxford: Blackwell, pp. 46–64.

W.T.S. (Bill) Gould

MAO ZEDONG (1893–1976)

Mao was co-founder of the Chinese Communist Party, first president of the People's Republic of China, prominent strategist of guerrilla warfare and people's struggle and a major communist theoretician. His ideas were important for many revolutionary struggles, particularly in Third World countries during the decades of the 1950s–1970s. His ideas about social change and economic development turned out to be largely disastrous for China but some of them, nevertheless, had an impact on development thinking.

Mao Zedong was born on 26 December 1893 in Shaoshan village, near Changsha, the capital of Hunan province. He was the son of a middle-income farmer and his father wanted him to become a farmer as well. He was forced to marry at the age of fourteen but his wife died some years later. The young Mao showed great interest in reading and learning, something that he maintained into old age. At sixteen, he began studying law and classical Chinese texts, first through private lessons and later in schools. He developed an interest in Chinese history and literature, especially of the Tang dynasty (eighth/ninth centuries), geography, social sciences and philosophy. He also read Western thinkers and philosophers.

Around the turn of the century, profound changes were taking place in China. Western powers had forced the empire to grant them territorial concessions in the south, most notably in Shanghai. Nationalists were revolting against the imperial Qing dynasty that prevented any modernisation of China's economy and rigid social structure. In 1911 the empire collapsed and China became a republic with Sun Yat Sen as its first president. In these turbulent years of revolution, Mao served for six months in the nationalist army. However, the Republic did not succeed in bringing unity and peace to the country. The country was divided among military warlords who wielded absolute power in their respective areas.

In 1918 Mao graduated from the Changsha teachers' training college. He became a history teacher and later established a cultural book society.

For a short while, he published a weekly magazine in which he explained his idea that China was on the eve of a social revolution. He also organised a strike by secondary school pupils. He went twice to Beijing, where he learned about Marxist socialist theories. His activities attracted the attention of revolutionaries and he was invited to Shanghai where the Comintern, the international socialist union dominated by the USSR, initiated the Chinese Communist Party (CCP), which joined the Guomindang nationalist party of General Chiang Kai-Shek.

Mao became an organiser of peasant and industrial unions in Hunan province. There he married his second wife. In 1926 and 1927 Mao made several visits to the countryside where he investigated the peasant rebellions on which he wrote in his famous *Report of an Investigation into the Peasant Movement in Hunan*. He explained that the peasantry constituted a potential revolutionary force and that 'a revolution is not a dinner party, or writing an essay, or painting a picture' (1971, vol. 1: 28). In the meantime, the CCP had increased in force. Chiang and the warlords considered the party a threat and in 1927 Chiang forcefully eliminated the CCP in Shanghai. Naturally this led to a split between the Guomindang and the CCP.

Strong disagreement existed within the CCP regarding the strategies to be pursued. Mao represented the view that the establishment of soviets in rural areas and the building up of the Red Army were indispensable in order to capture the cities and, ultimately, the country. From 1928 until 1931 he and his comrades established and consolidated the Jiangxi soviet in the border area of the southern provinces of Jiangxi and Fujian. Mao's second wife stayed behind in Changsha, where she was beheaded by the nationalist troops in 1930. Meanwhile, in Jiangxi, Mao met his third wife. Like his second wife, she seems to have had an independent character. This corrresponded with Mao's ideas about the role of women in society, which, seen against the background of Chinese traditions and Confucianist thoughts, can be considered as progressive. The first law he enacted in the first Soviet Republic of China gave men and women equal rights in marriage and divorce. It would also be the first law enacted in the People's Republic twenty years later.

In Jiangxi, Mao's star was rising. In 1931, he was elected as the first Chairman of the Soviet. When Chiang's troops forced the Red Army to leave Jiangxi, Mao led the army to Yan'an in Shaanxi, North-West China, a distance of some 9,600 km. This journey became famous as the Long March of 1934–6. It included about 90,000 people, of whom only 20,000 survived. For eleven years Yan'an became the headquarters of the CCP. Mao studied Marxist philosophy and wrote some important essays in which he applied Marxist thinking to Chinese reality. In *On Practice* (Mao 1937) he stressed the necessity to investigate empirically the socio-economic conditions in areas where the CCP wanted to be active. In *On Contradiction* (Mao 1937), contrary to classic Marxist thinking that the economic base determines the political and cultural upper structures of the society, he argued that the power of the human will could also determine the material base.

For health reasons, Mao's third wife went to the Soviet Union in 1936. She would return after the proclamation of the People's Republic but the couple had already divorced in 1938. In Shaanxi Mao met his fourth wife, Jiang Qing, who later became one of the leading figures of the Cultural Revolution. Like his earlier wives, she was quite independent but the marriage was a failure and the couple ended by living separately. Mao emerged as the major Communist leader and in 1943 became chairman of the Politbureau and CCP's Central Committee. With his growing power, he became more dominant and less flexible towards people holding other ideas. A personal cult developed around his life and he was called 'the Great Helmsman of the Chinese Revolution'. In the introduction to the manifesto of the CCP at the 7th Party Congress in 1945, it was stated that the CCP must take the thoughts of Mao Zedong as guidelines for all its activities, and the history of the CCP was rewritten to give him a more prominent place.

During the second Sino-Japanese War (1937–45), both the nationalists and the CCP had to fight against the Japanese while continuing the civil war. The civil war ended with the flight of Chiang Kai-Shek to Taiwan. On 1 October 1949, Mao proclaimed the People's Republic of China. He became the first president of the republic, as well as chairman of the CCP and president of the military commission. He was re-elected to these positions in 1954.

Mao and his comrades inherited a country of 9.6 million km^2 devastated by decades of revolts and war. China's population was 600 million, the majority of whom were peasants living in extreme poverty and accustomed to suppression by local landlords, regionally based warlords and Japanese occupation forces. The first tasks of the CCP were to restore order and security, and to rebuild the country. Land reform proved to be crucial. Tenants and the rural landless received land while the landlords were executed. Nowhere in history had such a large-scale land reform ever taken place. It was only logical that, after this period, China's development policy became based on the only socialist model known, namely the Stalinist approach to development. Thus, the major features of the First Five Year Plan (1953–7) were central planning, industrialisation – stressing large-scale heavy industry – and mechanisation of the countryside.

The model was not uncontested. A real break with the Stalinist model came in 1958, when Mao launched the Great Leap Forward policy in an attempt to decentralise the economy by establishing self-reliant communities. These communes each numbered 50,000 or more people and were meant to co-ordinate the local economy. This concerned the organisation of the smaller units of work brigades and teams for farm work, mass mobilisation for infrastructural works, and rural industrialisation. The aim of the policy was social as well, namely to eliminate differences in class, education, sex and age. It lasted three years and coincided with natural disasters in various parts of the country, such as droughts, typhoons and floods. The policy and natural disasters created chaos and starvation in the countryside. The real size of the disaster will never be known but estimates of its human victims range between 20 and 40 million.

This disastrous adventure revealed the existence of a serious struggle within the CCP. Mao believed in the priority of politics above economics; self-reliance; radical equality; and mass mobilisations for permanent revolutions. His opponents considered economic growth as the first priority and adhered to the classical Marxist-Leninist idea of the party as a vanguard of the proletariat. Among others, Mao had the support of Lin Biao, army general and Minister of Defence. The latter ordered the composition of a 'Little Red Book' containing quotations from the work of Mao, and subsequently he ordered every soldier to use it as a guide book. In the Red Army distinctions were abolished. Students followed with campaigns against 'elite thinking'. In 1966, this resulted in the launching of the Great People's Cultural Revolution and the organisation of students in Red Brigades to spread the thinking of Chairman Mao. They swarmed over the country, attacking intellectuals and party officials, destroying cultural heritages and forbidding 'capitalist habits' such as the reading of non-revolutionary literature. The Cultural Revolution disrupted the country totally and at the end of 1967 the army had to restore order.

However, the party struggle continued. The so called 'Gang of Four', with Mao's wife, Jiang Qing, as one of its members, set the tone and continued their campaigns against bourgeois habits. Mao died in September 1976. One month later the Gang of Four was arrested and with that the Cultural Revolution came to its real end. Although officially Mao's ideas have never been denounced, since his death China has followed a different development policy. In the Socialist Market Economy, households have been restored as the basic production unit, private ownership in industry is encouraged, and foreign investment in Special Economic Zones is welcomed. The new policies have led to economic growth but also enlarged the difference between the new rich and the poorer parts of the population, especially in the countryside. Western China also now lags far behind the east and southeast of the country.

Three tendencies can be discerned in Mao's thinking. The first is that he succeeded in adapting **Marx**ism–Leninism to Chinese realities. This applies in particular to his idea that peasants and not the industrial proletariat constituted the revolutionary class. Second, as a leader, Mao was merciless with his opponents within and outside the party. Third, he was a visionary, utopianist thinker. Mao believed in permanent criticism and in the power of the will of people, who did not need modern technology to change the world.

Mao's ideas were important for many revolutionary struggles, particularly in Third World countries. In Vietnam, rural guerrilla warfare led to victory in 1975, but proved catastrophic in Cambodia. Today, rebel movements in the Philippines, Nepal and Peru still refer to Maoism as their source of inspiration.

Indirectly, Mao's thoughts and the Maoist development model in China, as perceived by many Western intellectuals, stimulated 'alternative' models and strategies in development thinking. Concepts such as bottom-up approaches, urban bias, decentralisation, self-reliance and primary health care that are commonplace today, find their origin in debates of the 1960s and 1970s. Maoism certainly contributed significantly to these debates.

Major works

Mao Zedong (1966) *Quotations from Chairman Mao Tse-Tung*, Peking: Foreign Languages Press (also known as the Little Red Book).
—— (1971) *Selected Readings of Mao Tsetung*, Beijing: Foreign Languages Press.
Among others, these selected readings contain major essays such as:
Mao Zedong (1926) *Analysis of the Classes of the Chinese Society*.
—— (1927) *Report of an Investigation into the Peasant Movement in Hunan*.
—— (1936) *Problems of Strategy in China's Revolutionary War*.
—— (1937) *On Practice*.
—— (1937) *On Contradiction*.
—— (1939) *The Chinese Revolution and the Chinese Communist Party*.
—— (1942) *Talks at the Yenan Forum in Literature and Art*.

Further reading

Li Zhisui (1994) *The Private Life of Chairman Mao – The Memoirs of Mao's Personal Physician*, New York: Random House.
Short, P. (1999) *Mao – A Life*, New York: Henry Holt and Company.
Spence, J. (1999) *Mao*, New York: Viking/Penguin.

Ton van Naerssen

KARL MARX (1818–83)

Karl Marx was a revolutionary political and economic philosopher who, together with Friedrich Engels, produced a materialist theory of the evolution of societies. The theory was powerful enough that it influenced both social theory and the geopolitical reality that theory tries to understand. Marx was born in Trier and educated in philosophy at several German universities. He served briefly as editor of the Cologne newspaper *Rheinische Zeitung* in 1842 and 1843 but was forced to resign because of his radicalism and left for Paris, and then Brussels, where he wrote *The German Ideology* in 1847, his first sustained statement of an overall philosophy. Marx and Engels also organised a network of revolutionary groups into the Communist League, for which they wrote a statement of principles called *Manifesto of the Communist Party* (1847). After being arrested and banished, Marx finally settled in London in 1849, where he was active in the First International Workingmen's Association formed in 1864. A series of works, including *A Contribution to the Critique of Political Economy* (1859), *Wages, Price and Profit* (1865), and the first volume of *Das Kapital* (1867, translated into English as *Capital* volume 1, 1887), formed the core of Marx's economic writings. Volumes 2 and 3 of *Capital*, edited by Friedrich Engels, were published posthumously in 1885 and 1894, and translated in 1907–9. *Capital: A Critique of Political Economy* is regarded as Marx's mature statement on capitalist development.

Marx and Engels were Enlightenment modernists, who believed in social progress and the perfectibility of humankind. They were optimists in the sense of trust in the transformative potential of science and the material plenitude made possible by technological advance. Yet they were also highly critical of the particular social and political form taken by modernism – that is, capitalism. They saw capitalist development as a process of human emancipation but also as alienation from nature, as a process of human self-creation, but one directed by a few powerful people, as progress in material life, but progress motivated by socially and environmentally irrational drives, like competition. Marxian analysis thus became not the scholastic pursuit of truth for its own sake, and certainly not legitimisation theory for the rich and famous, but a theoretical guide to radical political practice aimed at completely changing society. Development, they thought, should meet the needs of the working class and should be socially and democratically directed. Marx and Engels came to liberate modernism, not to praise it.

In their early years, Marx and Engels were followers of the German philosopher, G.W.F. Hegel. Hegel thought that a kind of transcendental reason, the Absolute Idea, or World Spirit, was the hidden source of human consciousness, and the guiding force behind historical development. Yet, whereas in religion, Spirit is all-knowing from the start, in Hegelian idealism the Spirit's uncertainties were sources of the oppositions and crises inherent in every real thing. In the Hegelian dialectic, the negative element (the anti-thesis) is the living source of dynamic process. For Hegel, negation asserts itself as opposition and criticism against everything that endeavours to persist. The antagonisms between contradictory elements (thesis and antithesis) are exacerbated to the point that they can no longer co-exist, precipitating a crisis in which the contrary elements are reabsorbed into a higher and qualitatively different unity (the synthesis). In this way, the World Spirit progresses from notion to notion, each uniting in itself a new stage of reality, spiritual and material. The dialectic is Hegel's way of showing how development follows a complex yet rationally determined path (Cornu 1957; Hegel 1967).

Marx followed Hegel's dialectical method but disagreed with Hegel's teleological and spiritual idealism. In opposition to idealism, Marx was a dialectical *materialist.* As Marx and Engels put it at the time:

> In direct contrast to German philosophy [Hegel] which descends from heaven to earth, here we ascend from earth to heaven. That is to say, we do not set out from what men say, imagine, conceive, nor from men as narrated, thought of, imagined, conceived, in order to arrive at men in the flesh. We set out from real, active men, and on the basis of their real life-process we demonstrate the development of the ideological reflexes and echoes of this life-process … Life is not determined by consciousness but consciousness by life.
>
> (Marx and Engels 1981: 47)

Marx and Engels concurred with Hegel that the driving force in development was conflict and opposition. But Marx began his analysis of conflict by examining not transcendental consciousness, but real economic conditions. From the time of *The German Ideology*, Marx and Engels (1970) saw struggle within social relations as the driving force of history. However, Marx at first understood social relations as a 'form of intercourse' between people, with 'intercourse' including all kinds of relations, such as trade and commerce, as well as the class relations that he eventually came to focus on. By comparison, Marx's conception of the development of the material productive forces (human labour power and means of production) was more advanced. These different levels of conceptual understanding lent the tenor of technological determinism to Marx's first overall statement about social development. As Marx (1969b: 503–4) put it:

> In the social production of their life, men enter into definite relations that are indispensable and independent of their will, relations of production which correspond to a definite stage of development of their material productive forces. The sum total of these relations of production constitutes the economic structure of society, the real foundation, on which arises a legal and political superstructure and to which correspond definite forms of social consciousness. The mode of production in material life conditions the social, political and intellectual life process in general. It is not the consciousness of men that determines their being, but, on the contrary, their social being that determines their consciousness. At a certain stage of their development, the material forces of society come in conflict with the existing relations of production, or – what is but a legal expression for the same thing – with the property relations within which they have been at work hitherto. From forms of development of the productive forces, these relations turn into their fetters. Then begins an epoch of social revolution. With the change of the economic foundation the entire immense superstructure is more or less rapidly transformed .... In broad outlines Asiatic, ancient, feudal, and modern bourgeois modes of production can be designated as progressive epochs in the economic formation of society.

In this statement, the development of the forces of production drives change in the entire 'mode of production of material life' following a conception influenced still by Hegelian idealism (the self-development of the productive forces standing in for the unfolding rationality of the World Spirit). By the time of Marx's *Capital* (1976), however, the forces of production were subordinated to the relations of production, and these were thought of more exactly as class relations within the labour process.

For Marx, the most essential aspect of social relations was control over the productive forces available in any society. In class societies, the production of existence was controlled by a ruling elite. This created a fundamental social cleavage, a class relation, between owners of the productive forces and the labourers who performed the work necessary for life to continue. The aspect of this most crucial for issues of economic development was the extension of the working day beyond the labour time necessary for simply reproducing the worker and his or her family. So, for Marx, societies were 'exploitative' when uncompensated surplus labour, or its products, were taken from the direct producers, the exploitation process being an arena of the conflict and opposition that Marx, as a dialectical thinker, focused on. In class struggle, the dominant elite used a combination of economic, political and ideological forces against the dominated, and the dominated resisted through overt means like organisation (unions, political parties), and hidden means, like reluctant compliance (putting a spanner in the works). For Marxists, while workers produce the means of economic growth, they do not control the development process. This makes capitalism both highly productive and very unstable, a dangerous combination.

The other main social relation is within the capitalist class, among capitalists competing on markets. Capital itself came from surplus made in earlier periods of history, from merchant's capital (in which profits were made by buying cheap and selling dear), from savings and hoarding, but mostly from that fearsome source of 'primitive accumulation', the raiding of non-capitalist societies for their accumulated wealth, their resources and their people:

> The discovery of gold and silver in America, the extirpation, enslavement and entombment in mines of the indigenous population of that continent, the beginnings of the conquest and plunder of India, and the conversion of Africa into a preserve for the commercial hunting of blackskins, are all things which characterize the dawn of the era of capitalist production. These idyllic proceedings are the chief moments of primitive accumulation.
>
> (Marx 1976: 915)

Capitalists were originally commercial farmers or small manufacturers who put accumulated money into productive use as capital to make profit. Under market conditions, they had to produce commodities at low prices. Competition forced capitalists to extract the maximum surplus value from workers. Competition forced capitalists to re-invest most surplus in improving the forces of production, especially tools, machines and infrastructures in an attempt at making production more efficient. Competition also forced the adoption of innovative types of organisation (corporations as compared with family firms, multinationals as compared with national corporations). Thus under capitalism, investment, productivity and development are not a matter of will, but result from compulsion forced by competition – it is a case of develop or lose in the race to survive. Each technological or organisational change forced by competition then has multiple effects running through the entire economy, society and culture – hence the theme of 'progress' in modern history. For Marx, development occurred unevenly in social terms of class (the owning class becoming richer) and geographical terms of space (some countries becoming richer than others), hence the theme of uneven development in Marxism. Development was a contradictory process altogether (Marx 1976; Harvey 1982; Becker 1977; Weeks 1981; Peet with Hartwick 1999).

For Marx, social transformation essentially involved a shift from one mode of production (for example, feudalism) to another (for example, capitalism). Marx envisioned such transformations as violent episodes undertaken when all the productive possibilities of old social orders had been exhausted. Crises in material development sharpened and intensified ongoing social struggles, lending struggle the potential for broad social change. The new social system, produced through class struggle, did not materialise out of thin air, nor from utopian desire alone, but was formed out of the embryonic relations already present in the dying body of the old society. Marx saw capitalism as a necessary stage in history, developing technology and destroying old beliefs in gods, ghosts and goblins. Following from this, Marx saw capitalism and imperialism as progressive aspects of the development of non-capitalists societies, like India and China, making possible a better society in the future, a viewpoint that made Marx vulnerable to criticisms of Eurocentrism. The main point is that Marx favoured modern technology and social development because these made a better life possible for the mass of ordinary people. He just wanted the development process to be owned, controlled and directed democratically, by working people. Only then could the full potential of modern development be fully realised – as the satisfaction of needs, as education for all and as sustainablity. For Marx, this would require a further transformation in mode of production, from capitalism to socialism.

Marx is important in two main ways. First, his doctrines were supposedly the political-economic basis on which the Communist countries (the Soviet Union, the East European bloc, China, etc.) were established. These countries used a development model founded on growth of the productive forces, understood narrowly as heavy industries like steel, chemicals and engineering, the idea being to increase productivity as quickly as possible. Consumer goods industries were less developed within a political culture stressing hard work for the state and sacrifice for the future. Second, Marx's writings form the intellectual basis for a range of subsequent theories such as dependency and world systems theories, as well as constituting a philosophy of development in and of themselves. This body of ideas continues to influence radical social theory up to the present time.

Major works

Marx, K. (1969a) *Wages, Price and Profit* (1865 original), in Karl Marx and Frederick Engels, *Selected Works*, Vol. 2, Moscow: Progress Publishers, pp. 31–76.

—— (1969b) Preface to *A Contribution to the Critique of Political Economy* (1859 original), in Marx, K. and Engels, F., *Selected Works*, Vol. 1, Moscow: Progress Publishers, pp. 502–6.

—— (1976) *Capital: A Critique of Political Economy* (1887, 1907–9 original), Harmondsworth: Penguin.
Marx, K. and Engels, F. (1847–8 original) *Manifesto of the Communist Party*, in Marx, K. and Engels, F., *Selected Works*, Vol. 1, Moscow: Progress Publishers, pp. 108–37.
—— (1970 and 1981) *The German Ideology* (1847 original), New York: International Publishers.

Further reading

Becker, J. (1977) *Marxian Political Economy*, Cambridge: Cambridge University Press.
Cornu, A. (1957) *The Origins of Marxian Thought*, Springfield, IL: C.C. Thomas.
Harvey, D. (1982) *The Limits to Capital*, Oxford: Blackwell.
Hegel, G.W.F (ed.) (1967) *The Phenomenology of Mind*, New York: Harper Publishers.
Peet, R. with Hartwick, E. (1999) *Theories of Development*, New York: Guilford.
Weeks, J. (1981) *Capital and Exploitation*, London: Edward Arnold.

Richard Peet

MANFRED MAX-NEEF (1932–)

Manfred Max-Neef, the son of German immigrants, is a Chilean economist who has had a significant influence on grassroots movements and local development projects worldwide. Since teaching at the University of California, Berkeley, in the early 1960s, he has held several academic posts in the USA and Latin America, including Rector of the Universidad Bolivariana and subsequently of the Universidad Austral de Chile in Valdivia (1994–2002). He has also worked for international organisations like the Food and Agricultural Organisation (FAO) and the International Labour Organisation (ILO). He is the founder and Executive Director of the Centre for Development Alternatives (CEPAUR) in Santiago, Chile, which is dedicated to transdisciplinary research and action projects in order to reorient development by stimulating local self-reliance, satisfying fundamental human needs, and promoting human-scale development. He has also been an executive council member of the Club of Rome. In 1983 Max-Neef was awarded the Right Livelihood Award in recognition of his work on development alternatives. The prize is often described as the Alternative Nobel Prize, and is awarded for outstanding vision and work on behalf of our planet and its people. In 1993 he stood as an independent candidate in the Chilean Presidential election and achieved a respectable minority vote. His work and writing continue to influence and inspire grassroots organisations and local movements in search for alternative, sustainable development.

Max-Neef's thinking originates to a large extent in his frustration with classical, conventional economics. The discipline of economics, he argues, has become obsessed with being a science, with providing technical solutions to human problems and with abstract measurements such as GNP, growth rates, and balance of payments. As a result, economists have become 'dangerous people' (1992: 19), whose theories and models cast a powerful spell on policy-making, but have proved dismally incapable of responding adequately to poverty and human misery. Instead, conventional economics makes the majority of poor people *invisible*, as it assigns no value to work carried out at the subsistence and domestic level. This reflects the discipline's obsession with gigantism, efficiency, measurements and abstractions. In the process, economics has, in the words of Max-Neef, lost 'a good deal of its human dimension' as well as its connection to moral philosophy (1992: 20). For these reasons, Max-Neef writes, 'I severed my ties with the trends imposed by the economic establishment, disengaged myself from "objective abstractions", and decided to "step into the mud" ' (1992: 21). He became a self-described 'barefoot economist', someone who has a profound distrust of grandiose solutions imposed from the top down and whose main concern is instead with local action and small-scale, bottom-up development.

Max-Neef's ideas and experiences as a 'barefoot economist' are set out in his most well-known book, *From the Outside Looking In* (1992). Since its first publication in 1982 by the Dag Hammarskjöld Foundation in Sweden, the book has been translated into five languages and has achieved the status of a classic text on grassroots development projects. In the book, Max-Neef details his personal experiences with participatory grassroots projects in two Latin American settings, interspersing his empirical account with more theoretical reflections on the nature of development. The result is a moving and passionate book, where the voices of the poor speak with clarity and force and where conventional theories and models are shown to be lacking. His recollection of his involvement with an integrated development project among peasants in Ecuador is particularly interesting and powerful, and demonstrates both the willingness and capacities of poor people to organise and participate, given the right circumstances and opportunities. The participants' own accounts of their development problems (reprinted verbatim in the book) provide a forceful antidote to the widespread perception that poor people are ignorant of their situations, and speak, as Max-Neef puts it, with a voice so vivid and so true 'that no expert with all his formal knowledge could have improved upon them' (1992: 71). Sadly, the project also demonstrates the intrinsically political character of participation and empowerment, in that all the project's documents and reports from the peasants' meetings were confiscated by

Ecuador's military authorities. Shortly afterwards, the project was closed, presumably because its success in organising and empowering the peasants was seen as too dangerous by the political elite.

This experience leads Max-Neef to a profoundly *political*, as opposed to technical, vision of development. To assume 'goodwill on the part of governments to really improve the lot of the invisible sectors [the poor] is naïve', he argues, and in this sense development represents the interests of the dominant class (1992: 114). In the national accounts, where growth rates and import/export ratios reign supreme, the poor are expendable, or at best, they can wait patiently until the effects of economic growth 'trickle down'. This scepticism of top-down, grandiose solutions leads Max-Neef to put his trust in local action: 'if national systems have learned to circumvent the poor, it is the turn of the poor to learn how to circumvent the national systems' (1992: 117). His vision for a better future accordingly arises from small-scale action, where the people concerned actively participate and where self-reliance and ecological harmony are guiding principles. It is only in such environments, he argues, that 'human creativity and meaningful identities can truly surface and flourish' (1992: 117). Drawing on traditions of thought stretching back to Aristotle, Max-Neef argues that large cities lead to alienation, passivity and indifference, whereas small environments allow for a dynamic equilibrium between nature, human beings and technology, as in such settings people have the opportunity to feel directly responsible for the consequences of their actions (1992: 132). As a firm believer in the relationship between theory and praxis, Max-Neef put his ideas to the test in Tiradentes, a small city in Brazil. The project was an effort to revitalise the city, counter the trend of hyper-urbanisation and improve the livelihoods of local people. Described in *From the Outside Looking In*, the project succeeded with relatively few resources not only to increase people's incomes, but also their sense of well-being, community and control, hence confirming Max-Neef's belief that local action holds the key to development.

Max-Neef's conviction that 'big problems require a great number of small solutions' (1992: 204) is given further elaboration in his second book, *Human Scale Development* (1991). The book's main postulate is that 'development is about people, not about objects', and proposes an 'ethics of well-being' against the logic of economics which dominates modern life (1991: 16, 64). Human-scale development is based on:

> the satisfaction of fundamental human needs, on the generation of growing levels of self-reliance, and on the construction of organic articulations of people with nature and technology, of global

> processes with local activity, of the personal with the social, of planning with autonomy and of civil society with the state.
>
> (1991: 8)

As an aid to achieving this form of development, the book proposes a radically different conception of human needs and wants. Whereas conventional economics commonly assumes that human needs are infinite, and that human beings are driven by an insatiable desire for material possessions, Max-Neef asserts that fundamental human needs are 'finite, few and classifiable'. What is more, these needs are 'the same in all cultures and in all historical periods' (1992: 18). What changes over time and from culture to culture is the manner in which people satisfy their needs. Apart from the fundamental need of subsistence or survival, there is no hierarchy of needs, as suggested for example by the psychologist Abraham Maslow, but instead needs are constant and complementary. This in turn allows for a reclassification of poverty, which can no longer be defined simply in terms of income below a certain level, as all unfulfilled needs reveal a human poverty, such as, for example, poverty of affection or of understanding.

Max-Neef identifies nine fundamental human needs; subsistence, protection, affection, understanding, participation, leisure, creation, identity, and freedom, with the possibility that transcendence (a process that he doesn't define or elaborate on) may in time become a tenth universal need (1992: 27). Needs are also defined according to the existential categories of being, having, doing and interacting, and on the basis of these Max-Neef develops a 36-cell matrix designed to help small communities enhance their awareness of their deprivations and potentials. The matrix is filled with examples of so-called satisfiers, that is, the means by which various needs are fulfilled in a specific community. Satisfiers have different characteristics, and while some are relatively straightforward (food satisfying the need for subsistence, formal education the need for understanding), others are more complicated. Some satisfiers can be destroyers of the very need they are said to fulfil, an example being excessive military expenditure which is supposed to provide protection, but instead promotes insecurity and inhibits subsistence, participation, affection and freedom (1991: 33). Others are 'synergic' satisfiers, meeting more than one particular need simultaneously. Through this redefinition of human needs and the development of the needs/satisfiers matrix, Max-Neef aims to assist communities in realising their potential for local self-reliance and develop a strategy for the actualisation of human needs.

Human Scale Development represents a departure from the prevailing economic rationale and the equation of development with attaining the material living standards of the most industrialised countries, and suggests instead that an abundance of goods and economic resources can exist side by side with poverty of, for example, affection and creativity. In fact, the valorisation of material possessions has, according to Max-Neef, led to alienated societies, where life 'is placed at the service of artifacts, rather than artifacts at the service of life' (1991: 25). While human-scale development does not exclude conventional development goals such as economic growth, it differs from such approaches in that the aims of development are considered not only as points of arrival, but as part of the process itself. Fundamental human needs must be realised throughout the entire process of development, and in this way the realisation of needs becomes, instead of a goal, the motor of development (1991: 53).

Max-Neef has made a noticeable contribution to development thinking and practice through his pioneering critique of development economics, his early advocacy of grassroots participation and empowerment, and his redefinition of human needs. His arguments for development alternatives and community participation remain relevant, perhaps even more so as civil society, participation and poverty reduction have become the contemporary buzzwords of mainstream, top-down development models. One prominent vehicle for disseminating his ideas has been *Development Dialogue*, the journal of the Dag Hammarskjöld Foundation, with which he had a close association over many years. Overall, it is thus possible that his emphasis on participation and grassroots involvement has had a belated influence on international institutions like the World Bank, as seen, for example, in its recent large-scale 'Voices of the Poor' project. For Max-Neef, this would not necessarily signify the success of his ideas, but rather the ability of the holders of power to conceal their vested interests and pour old wine into new bottles (1992: 115; see also Wisner 1988). For him, human-scale development can arise only from local settings, not from distant development organisations and predefined development models. Accordingly, his own writings do not so much propose an alternative development model as provide flexible tools whereby communities can arrive at their own solutions to their individual problems. Unlike so-called post-development writers, however, who also focus on grassroots movements and local knowledge, Max-Neef remains within the language of development, utilising the same terms and phrases as orthodox models. Many post-development thinkers, by contrast, argue for alternatives *to* development, as opposed to development alternatives, and this entails a change of the discourse that produced the problem of development in the first place (see Esteva 1987; Escobar 1992).

Like the economist **E.F. Schumacher**'s embrace of the dictum 'small is beautiful', Max-Neef's vision of human-scale development is open to the charge of romanticism. Yet, for someone who describes his political philosophy as 'human eco-anarchism' (1992: 55) and who wishes to revitalise our capacity to dream (1991: 3), this is not necessarily a criticism. That said, Max-Neef denies being a smallness fanatic, recognising that small can also be ugly and evil, but maintains that smallness of scale allows for a degree of responsibility and human interaction that is impossible in larger environments (1992: 143, 160). The relationship of these local communities to their national and global surroundings remains largely outside the parameters of Max-Neef's writings; the challenge is to 'think small and act small, but in as many places as possible'. The path, he argues, 'must go from the village to the global order', an insight that remains a powerful inspiration for local action and grassroots movements (1992: 117).

Major works

(available in English; he has written more extensively in Spanish)

Ekins, P. and Max-Neef, M.A. (eds) (1992) *Real-Life Economics: Understanding Wealth Creation*, London: Routledge.

Max-Neef, M.A. (1992) *From the Outside Looking In: Experiences in 'Barefoot Economics'*, London: Zed Books.

Max-Neef, M.A. with contributions from Antonio Elizalde and Martin Hopenhayn (1991) *Human Scale Development: Conception, Application and Further Reflections*, New York and London: The Apex Press.

Further reading

Escobar, A. (1992) 'Reflections on "Development": Grassroots Approaches and Alternative Politics in the Third World', *Futures* 24(5): 411–36.

Esteva, G. (1987) 'Regenerating People's Space', *Alternatives* 12(1): 125–52.

Wisner, B. (1988) *Power and Need in Africa*, London: Earthscan.

Rita Abrahamsen

TERENCE GARY McGEE (1936–)

Terence ('Terry') McGee is a development geographer who has made key contributions to our understanding of urbanisation and development and the structure of cities, particularly as regards the rapidly growing conurbations of Southeast Asia. He is best known for his explanation of the role in development of Southeast Asian cities characterised by poor infrastructure, rapid growth due to rural–urban migration, political instability and extensive poverty. He was a pioneer in the 1970s in emphasising the significant position in the urban economy of the 'informal sector' (the extensive range of unregulated enterprises and activities in the black market or underground economy that operate without any formal or official recognition by government). Up to this point there had been hardly any acknowledgement of its overwhelming contribution to economy and society in some of the world's largest cities, the way people are employed, how and where they live and how so many of the economic and social relationships in the Asian city can be explained.

McGee was born on 5 January 1936 in Cambridge, New Zealand, a small agricultural town on the North Island situated in the heart of the Waikato farming district where he gained early experience in milking cows and mending fences, as well as the normal New Zealand aptitude for ball skills at rugby and cricket. At the age of ten he moved with his family to Auckland where he was at school on the city's North Shore and where, in 1954, he enrolled at Auckland Teachers' College while also attending BA classes part-time in history and geography at the University of Auckland. Graduating with a teaching diploma in 1954, he transferred to Victoria University of Wellington, New Zealand, where he majored in geography and gained his BA in 1957. In Auckland, McGee attributes his early appreciation of history and its later importance in his writing to Willis Airey (1897–1968), a leading New Zealand historian of Russia, and Keith Sinclair (1922–93), poet and expert on New Zealand history. In Wellington his mentors were Keith Buchanan (1920–98), brilliant lecturer and utopian socialist who wrote widely on Asia; Harvey Franklin, an economic geographer who introduced him to the work of the institutional economist, Clarence Ayres (1891–1972), an American who wrote the book, *The Theory of Economic Progress*, in 1944 and who propounded a theory of 'institutional lag', whereby technological changes were always several steps in front of change in socio-cultural institutions; and Ray Watters, a human geographer who had adopted ethnographic and anthropological approaches in his research on the Pacific.

He continued with his MA studies in 1958–61, where the courses taught by Buchanan, Franklin and Watters were influential and where Buchanan introduced him to the French regional geographers in Asia, notably the work in Vietnam by Pierre Gourou (1900–99) and Charles Robequain (1897–1936). Another important influence was members of the Chicago ecological school of sociologists, including Louis Wirth (1897–1952), whose approaches towards urbanism as a way of life he used in his Master's thesis on Indian migration to the City of Wellington, New Zealand, which he attained with First Class Honours in 1961.

His academic career began alarmingly when he took up a position as Temporary Assistant Lecturer in the Department of Geography at Victoria University to teach a class on the geography of Europe, a region which he had not at that time visited and which he only accomplished with the help of lecture notes that Franklin had left behind when he went on leave. But

1959 was also the year he gained his first continuing appointment as Lecturer in geography at the University of Malaya in Kuala Lumpur. He was to remain there for six years and recognises this as a very formative period in his life. McGee went to Asia expecting to teach but considers he learned far more than he ever imparted (Presidential Address to Canadian Institute of Geographers 1991). While in Malaysia he enrolled in a PhD at Victoria University on Malay migration to Kuala Lumpur City, a subject that followed on from his Master's thesis. The research involved an extensive household questionnaire survey, staying as a guest of Malay families and travelling widely throughout the country to track down migration source areas. This period also presented him with an opportunity to explore Southeast Asia extensively and it was on one such expedition that the idea for a book on the Southeast Asian city was conceived.

He returned to New Zealand to become a Lecturer in geography at Victoria University in 1965 and completed both the book and his PhD thesis. *The Southeast Asian City: A Social Geography of the Primate Cities of Southeast Asia* was published in 1967 and his graduation followed in 1968. McGee propounded the novel view that Asian cities were characterised by a form of 'pseudo urbanisation', typified by conspicuous consumption, extraction of surplus from the countryside and a drag on development. This was a perspective that progressively diminished in importance as the developing world became more closely integrated with the global economy in the last decades of the century. The book was the first in a distinguished list of titles on Asian urbanisation. A forthcoming book will study the extended metropolitan regions of Asia and be jointly authored with some of his former PhD students. This was also a busy period in his life with marriage, the birth of his son and building a house in Otaki (near Wellington, New Zealand) taking centre stage.

One of McGee's most influential papers on the economic structure of cities was jointly written in 1968 (with Warwick Armstrong, a colleague in Wellington) before returning to Asia, to a Senior Lectureship at the University of Hong Kong. This paper, heavily influenced by members of the Latin American dependency school of development thinkers (notably the Argentine economist, **Raúl Prebisch**, and American, **Andre Gunder Frank**, was titled 'Revolutionary Change and the Third World City: A Theory of Urban Involution' (*Civilisations* 1968). It was the beginning of a collaboration with Armstrong that was to be renewed in Canada in the early 1980s and a precursor to their influential 1985 book, *Theatres of Accumulation: Studies of Urbanization in Asia and Latin America.* In Hong Kong McGee was soon promoted to a personal Chair in 1970 and continued his interest in the informal sector with the first of a series of major publications on street vendors (*Hawkers in Hong Kong: A Study of Policy and Planning in* *the Third World City*, 1974). He also began a productive relationship with the Canadian International Development Research Centre which was to result in further titles on the same topic during the 1970s.

In 1973 McGee took up a Senior Fellowship in Human Geography at the Australian National University (ANU) in Canberra, Australia. His five years at the ANU were marked by innovative new work on the political economy of development and collaborative research in the Pacific with academic colleagues (notably David Drakakis-Smith (1942–99), with whom he had worked in Hong Kong, and Gerard Ward). Publications of particular note at this time included 'The Persistence of the Proto-proletariat' (McGee 1974), and 'Conservation and Dissolution in the Third World City (McGee 1979). It was also a time saddened by the death of his first wife and the added responsibilities of being a single parent. In 1978 he moved to Canada to take up the joint positions of Professor of Geography and Director of the newly created Institute of Asian Research at the University of British Columbia in Vancouver. In spite of a heavy administrative workload, McGee continued to contribute effectively to Asian studies through fieldwork in Malaysia investigating urban female workers in the new semi-conductor factories and their rural links expressed through sending part of their incomes back to home villages. This interest in urban–rural dynamics was to lead, late in the 1980s, to some of McGee's most important work – the identification of new extended metropolitan regions in Asia characterised by the outward expansion of urban activity into the intensely crowded agricultural hinterlands of cities like Jakarta on the Indonesian island of Java. McGee termed this novel metropolitan form *desakota,* taken from the *bahasa* Indonesian names for village and town. He argued that it was an Asian form of urbanisation distinctive from that of the developed countries (apart perhaps from Holland). This was an ecological viewpoint in the tradition of cultural geographers like Carl Sauer (1889–1975) and led in 1991 to the publication of *The Extended Metropolis: Settlement Transition in Asia* (ed. with N. Ginsberg and B. Koppell).

A number of objections have been raised to the idea of extended metropolitan regions (EMRs) being distinctive or peculiarly Asian. Briefly, these critiques have argued that the book was over-generalised in its claims and that it neglected the present evidence and inevitability of growth in market systems in Asia that will replicate patterns found in the developed countries (Tang and Chung 2000; Dick and Rimmer 1998). McGee has responded by saying that the idea was never based on Asian as opposed to Western traits but was one which stresses the influences on the urbanisation process of the context in which it operates. Government policy, foreign investment and various local factors, like the roles of municipalities and community organisations, explain much of the character and change found in

EMRs such as Jabotabek (Indonesia), Bangkok (Thailand) and the Pearl River Delta (China).

Besides contributing extensively to the academic debate on urbanisation processes in Asia, McGee's work has had a practical and applied focus. During the 1990s he forged strong links with the National Centre for Humanities and Social Sciences in Hanoi (Vietnam) as part of a long-term project looking at Socialist Transition Economies and developing research capacity in Vietnamese research institutions. Further investigations examined the fiscal crisis in Southeast Asia at the end of the 1990s for the Canadian International Development Agency and identified a list of urban indicators for the Asian Development Bank. He reached compulsory retirement age in 2001, leaving his positions at the University of British Columbia to become a Professor Emeritus and spend part of each year teaching in Australia and researching a new book and further papers on Asian urbanisation.

McGee has made fundamental contributions to the study and understanding of modern urbanisation processes. His work, with its emphasis on human and institutional features of urban life and change in Southeast Asia, has followed a consistent path, enduring through the 1970s when geographers all around him were caught up in what was then termed a 'quantitative revolution' based on the primacy of numerical techniques, to emerge in the new century with his ideas intact and equally relevant to the contemporary debate. A measure of the resilience of his ideas is also seen in the way his formulation of theory contributed to new concepts in allied fields other than his own. For example, views about the character of retailing in the informal sector (hawking), an activity so typical of Asian cities, were adapted by him to explain the persistence of shantytown features in these same cities (*Development and Change*, 1979). Though this was the only paper on housing he has ever written, it remains a key contribution in understanding the nature of the squatting process and the determination of urban management strategies to cope with its effects.

Major works

Armstrong, W. and McGee, T.G. (1985) *Theatres of Accumulation: Studies of Urbanization in Asia and Latin America*, London and New York: Methuen.

Ginsberg, N., Koppell, B. and McGee, T.G. (eds) (1991) *The Extended Metropolis: Settlement Transition in Asia*, Honolulu: University of Hawaii Press.

McGee, T.G. (1967) *The Southeast Asian City: A Social Geography of the Primate Cities of Southeast Asia*, London: Bell.

—— (1971) *The Urbanization Process in the Third World: Explorations in Search of a Theory*, London: Bell.

—— (1974) *Hawkers in Hong Kong: A Study of Policy and Planning in the Third World City*, Hong Kong: Centre of Asian Studies, University of Hong Kong.

McGee, T.G. and Robinson, I. (eds) (1995) *The Mega-Urban Regions of Southeast Asia: Policy Challenges and Response*, Vancouver: University of British Columbia Press.

Further reading

Dick, H.W. and Rimmer, R. (1998) 'Beyond the Third World City: The New Urban Geography of Southeast Asia', *Urban Studies* 35(12): 2303–21.

McGee, T.G. (1974) *The Persistence of the Proto-proletariat: Occupational Structures and Planning for the Future of Third World Cities*, Los Angeles, CA: School of Architecture and Urban Planning, University of California (reprinted in Abu-Lughod, J. and Hay, R. (eds) (1977) *Third World Urbanization*, Chicago, IL: Maaroufa, pp. 257–70.

—— (1979) 'Conservation and Dissolution in the Third World City: The Shantytown as an Element of Conservation', *Development and Change* 10(1): 1–22.

—— (1991) 'Presidential Address: Eurocentrism in Geography: The Case of Asian Urbanisation', *The Canadian Geographer* 35(4): 332–42.

—— (2002) 'Reconstructing the Southeast Asian City in an era of Volatile Globalisation', *Asian Journal of Social Science* 30(1): 8–27.

McGee, T.G. and Armstrong, W. (1968) 'Revolutionary Change and the Third World City: A Theory of Urban Involution', *Civilisations* XVIII(3): 353–78.

Tang, W.-S. and Chung, H. (2000) 'Urban–Rural Transition in China: Beyond the *Desakota* Model', in Li, S.M. and Tang, W.-S. (eds), *China's Regions, Polity and Economy: A Study of Spatial Transformation in the Post-Reform Era*, Hong Kong: The Chinese University of Hong Kong Press, pp. 276–308.

John P. Lea

GUNNAR MYRDAL (1898–1987)

A prolific writer and active policy advisor, Gunnar K. Myrdal had a profound effect on development thinking through the early development decades of the 1950s and 1960s. Trained as an economist, Myrdal's writings engaged with sociology and gained a wide and influential audience of planners, governments and academics through his long career. Moreover, the issues he raised about the institutional and social milieu for favourable development are still debated by contemporary researchers. Born in Stockholm in December 1898, Gunnar Myrdal was associated with that city throughout his life, despite his extensive work and travel around the globe, particularly in the USA, Switzerland and India. Educated at Stockholm University, he gained a law degree in 1923, followed by a PhD in economics in 1927. His PhD focused on the role of expectations in price formation, although he is best known for his later work on macro-economics and social factors. Following a Rockefeller travelling fellowship in the USA in 1929–30, and an equally short period in Geneva as associate professor at the

Institute of International Studies (1930–1), he returned to Stockholm where he held posts at the University of Stockholm. As professor of Political Economy (1933–50) and then as professor of International Economics (1960–7), his international work was associated particularly with the directorship of the Swedish Institute for International Economic Studies. His public role was furthered by his appointment as executive secretary to the UN Economic Commission for Europe (1947–57). In addition to acting as advisor to the Swedish government on economic, social and fiscal policy, Myrdal was a member of the Swedish Senate, becoming the Minister of Commerce, and a member of the population, housing and agricultural commission. He died in Stockholm in May 1987.

Myrdal was part of an iconoclastic generation of development economists whose work extended Keynes' economics and state interventionism. Learning from wealthy Western and Eastern European countries, this group adapted Keynesian demand management and drew on Schumpeter's work in viewing development as a structural transformation process. Rejecting neo-classical economics, most of these so-called structuralists saw the lack of industrialisation as the reason for countries' poverty (see Martinussen 1997: 50ff. for an overview). Alongside Myrdal, writers including Richard Nurske, Thomas Balogh, Paul Rosenstein-Rodan, **Raúl Prebisch** and **Hans Singer** highlighted issues such as dual economies, labour surplus, unbalanced growth, the vicious circle of poverty, unequal exchange and dependency. In their policy prescriptions, they argued for rapid industrialisation and redistribution with growth. Compared with this generation of structuralists, however, Myrdal's work considered non-economic phenomena more systematically (Martinussen 1997: 73). Arguably, Myrdal's direction of a major study of the situation of the African-American population funded by the Carnegie Corporation from 1938 to 1943 contributed to his understanding of the legal, social and cultural factors involved in the creation and perpetuation of mass poverty. His aim in this study was, he stated, to determine 'the social, political, educational, and economic status of the Negro in the US, as well as defining opinions held by different groups of Negros and whites as to his "right" status' (*An American Dilemma: The Negro Problem and Modern Democracy*, 1944: x). This represents his first major work, the themes of which would recur in later studies. Also, in comparison with many structuralists at the time, Myrdal believed that the accumulation of capital could assist in economic development, which drew him closer to the 'modernisation' thinkers in this regard. Myrdal's influence on development thinking can be seen in the four key fields of regional development, dependency approaches, the state, and non-economic factors.

With regard to regional planning, Myrdal contributed several key concepts that had a wide impact on development studies thinking and on planning. He was one of the first writers to offer an explanation for the contrast between a dynamic part of a developing country – usually the capital city – and stagnation in the rest of the country. He attributed regional inequalities to the 'spiral of cumulative causation' which results in positive spread effects in the core area, and negative backwash effects in the periphery (*Economic Theory and Under-developed Regions*, 1957). Myrdal's 1957 book identified a similar process to **Albert Hirschman** but his approach was more humane, and had more influence (Brookfield 1975: 99). By means of the 'process of cumulative causation' found in free market economies, economic activities became concentrated in certain centres, to the detriment of other sectors of the economy and country and are reinforced by movements of capital, goods and people. The concepts of spread and backwash effects had an immediate impact on debates in regional development and urbanisation as planners struggled to account for the location and impacts of urban centres (Reitsma and Kleinpenning 1985: 202). Although neither Hirschman nor Myrdal theorised about the best initiators of positive spread effects, a commonly agreed impetus was urban industrial development, which had been vital in Europe. By the 1960s, however, Myrdal had concluded that industrialisation could not absorb all underemployed people and argued for the need to develop agriculture in order to create employment and increase productivity and efficiency. Compared with Hirschman's optimism that growth, although initially unbalanced geographically, would equalise over time, Myrdal believed that this would be unlikely unless strategic state intervention took place. It was only after the work of Myrdal and Hirschman that uneven development in the development process became firmly established, demanding further examination and theorisation. For example, **John Friedmann**'s work on core–periphery built on Myrdal and Hirschman, although he was able to be more precise in the treatment of regional economics (Brookfield 1975: 101).

Myrdal also engaged with a growing body of theory examining the international context for inequalities in wealth between nations, particularly between core European and North American countries and former colonies in South America, Africa and Asia. He noted that economic, social and political dependency relations existed between regions at the international level (Reitsma and Kleinpenning 1985: 246). Influenced by the earlier writings of Latin American economists and planners such as **Raúl Prebisch** and the influential United Nations ECLA (Economic Commission on Latin America) school, Myrdal regarded underdevelopment as intrinsically related to development elsewhere through the pervasive effects of negative backwash effects. Differing from **Walt Rostow**'s stage model of development independent of international relations, Myrdal took the structuralist argument that post-Second World War international trade could not act as the growth engine for developing countries as it would only cause the 'trickle up' of benefits to industrialised countries (Martinussen 1997: 78–9). However, he crucially extended the structuralist account by his argument of circular and cumulative causation, in which an outflow of surplus, the brain drain and the destruction of domestic crafts were all indicative of backwash effects. Underdevelopment was also tied to the vagaries of the global economy, argued Myrdal, with backwash effects stronger in a depression than a boom (Brookfield 1975: 100). Developing a theme to be further explored by later neo-Marxists, Myrdal drew attention to colonial legacies which, through their law and order policies and tax collection, were oriented to the interests of colonial powers, not poor countries (Martinussen 1997).

Economics in the 1950s and 1960s assumed that market failure was the norm in developing countries, meaning that the state would have to mobilise domestic and foreign savings for investment in directed industrial development, with state plans (see Corbridge 1995). In line with mainstream thinking, Myrdal viewed the state as 'knowledgeable and compassionate' (Lal 1985), although he remained critical of certain state economic policies such as those in India (Toye 1993: 100). The state envisaged by Myrdal (1960) was a highly organised and centralised one with features such as collective bargaining, democratic participation and a welfare state. States were important institutions, to be created by policies and political determination (Martinussen 1997: 227). However, according to Myrdal's research, a developmentally favourable state was not necessarily found in poor countries. Myrdal assumed that most developing country states were 'soft', and would be exploited by powerful individuals and groups (Myrdal 1968: chap. 18; 1970). Myrdal first used the term 'soft state' to describe a country the government of which is incapable of instituting radical reform policies or which fails to follow through with policies (Reitsma and Kleinpenning 1985: 332). His work on India and South Asia in *Asian Drama* (a majestic combination of theory and policy) develops these aspects for the first time, analysing the detrimental impact of inefficient political development agencies. Another feature of soft states was the weak formulation of law, permitting both corruption and the continuation of exploitative systems such as sharecropping (Martinusssen 1997: chap. 16). As the interests of the powerful control the state and its economic performance, the goal is to envisage a strong state structure which, according to Myrdal, can enforce social discipline or generate a tolerance of authoritarianism. Rather than focus on political structure, Myrdal's approach thus interpreted soft states as 'a problematic absence of the right cultural context for development planning' (Toye 1993: 124). This interpretation was long seen as highly ethnocentric, although it is in line with some recent conservative interpretations of culture and development (compare Berger and Huntington 2002). In the 1958 Storr lectures at Yale, Myrdal outlined the kind of state he envisaged for developing countries, where, although responding to international crises, national planning acts as a positive internal force, enhanced by democratisation, economic nationalism and equalisation.

As much a sociologist as an economist of development, Myrdal worked for many years to understand the impact of non-economic factors on development and poverty. Through his critique of capital-centric theories, he moved beyond economic theory to a more general theory of societal stagnation and transformation (Martinussen 1997: 80). Criticising previous planning models for their production and income orientation, he highlighted the need to consider attitudes to work and life, and political and cultural institutions (Myrdal 1968, Appendix 3, 4; also Martinussen 1997: 230). His multi-factor model included six categories that ranged from output and incomes, and conditions of production, to attitudes towards work and life, institutions and policies. He viewed these aspects of the development process as interrelated, tending to shift in similar directions with the result of either development or underdevelopment (Martinussen 1997: 80). His 1968 book, *Asian Drama,* described attitudes to life and work including those obstructing economic growth (such as low levels of work discipline, punctuality, orderliness) as well as superstitious beliefs and irrationality, and readiness for change. Myrdal's belief in the circular causation of poverty, which rested partially on the poor's attitudes, was not unique at this time; the work of Oscar Lewis and others analysed the – now discredited – idea of a 'culture of poverty' (Lewis 1959).

Although widely influential, Myrdal's work was not without its critics. While easily summarised, his work failed to acknowledge others' work in similar fields and at times stated the obvious (Brookfield 1975: 100). During the 1980s 'counter-revolution' against Keynesian and state-centred development, Myrdal's work was strongly criticised for its *dirigisme* dogma (Lal 1995 [1985]: 56). Neo-liberals particularly disagreed with Myrdal's assumption of an omniscient central state authority, and his implicit goal of egalitarianism (Lal 1995 [1985]). Similarly, while recognising Myrdal's concern for human capital formation, Toye criticises his erroneous claim that 'productive consumption' – that is, investment in education and health – 'proved that the standard macro-economic distinction between consumption and investment was worthless in policy making' (Toye 1993: 97–8). By emphasising the cultural predilection to comply with government planning rather than the political structures of states, Myrdal's soft state theories were later perceived as 'embarrassingly close to Peter Bauer's linking of development prospects to cultural characteristics of ethnic groups' (Toye 1993: 124). African-American critics focus on his evasion of a class critique and his faith in the American way of life, although they acknowledge the depth of statistical information presented in *American Dilemma* (Reed 2001). Nevertheless, Myrdal's analysis of structures and power – albeit superficial – laid the 'groundwork for a fruitful analysis of interactions between the state and economic development' (Martinussen 1997: 227).

Myrdal's work continued to have a resonance in development thinking and policy throughout his life. In 1974, he shared the Nobel Prize for Economics with Friedrich A. von Hayek 'for their pioneering work in the theory of money and economic fluctuations and for their penetrating analysis of the interdependence of economic, social and institutional phenomena' (also Martinussen 1997: 227). In his collaborations with development thinkers such as **Ester Boserup** and **Michael Lipton**, and his extensive friendships with scholars and politicians, Myrdal's influence was felt through succeeding generations. His wife, Alva Myrdal (1902–86), was a highly significant figure in her own right, working variously for UNESCO, as a Swedish MP and Cabinet Minister, an Ambassador, and a disarmament negotiator (for which she won the Nobel Peace Prize). In their different fields, Gunnar and Alva Myrdal were prime examples of a social democrat commitment to equality and public service allied with concrete information. In his attention to detail and his sensitivity to the multifaceted character of mass poverty, Gunnar Myrdal's work contributes to an understanding of the human cost of underdevelopment through his descriptions of widespread un(der)employment, poor use of resources, and income inequalities. In his view that inequality was morally unacceptable and that the agenda of development was to guarantee social equity, Myrdal came close to what later became termed a basic needs strategy (Martinussen 1997: 82).

Major works

Myrdal, G. (1944) *An American Dilemma: The Negro Problem and Modern Democracy*, New York and London: Harper and Brothers.

—— (1956) *An International Economy*, London: Routledge.

—— (1957) *Economic Theory and Under-developed Regions*, London: Duckworth.

—— (1960) *Beyond the Welfare State: Economic Planning in the Welfare States and its International Implications*, London: Duckworth.

—— (1963) *Challenge to Affluence*, London: V. Gollancz.

—— (1968) *Asian Drama*, 3 vols, London: Penguin.

—— (1970) *The Challenge of World Poverty: A World Anti-Poverty Program in Outline*, London: Penguin.

Further reading

Berger, N. and Huntington, S.P. (eds) (2002) *Many Globalizations: Cultural Diversity in the Contemporary World*, Oxford: Oxford University Press.
Brookfield, H. (1975) *Interdependent Development*, London: Methuen.
Corbridge, S.E. (1995) *Development Studies: A Reader*, London: Arnold.
Dickenson, J. *et al.* (1996) *Geography of the Third World*, London: Routledge.
Lal, D. (1985) 'The Misconceptions of Development Economics', reprinted in Corbridge (ed.), *Development Studies: A Reader*, London: Arnold.
Lewis, O. (1959) *Five Families: Mexican Case Studies in the Culture of Poverty*, New York: Basic Books.
Martinussen, J. (1997) *Society, State and Market*, London: Zed Books.
Reed, Jnr, A. (2001) 'Race and Class in the Work of Oliver Cromwell Cox – American Writer', *Monthly Review* 52(9) (February): 23–32.
Reitsma, H.A. and Kleinpenning, J.M.P. (1985) *The Third World in Perspective*, Maastricht: Van Gorcum.
Toye, J. (1993) *Dilemmas of Development*, Oxford: Blackwell, 2nd edn.

Sarah Radcliffe

KWAME FRANCIS NKRUMAH (1909–72)

The *Osagyefo* (meaning the redeemer, as he was popularly known to supporters), Dr Kwame Nkrumah, the man who led Black Africa's first independent nation, Ghana, was born in Nkroful, in Ghana's Western Region close to the Ivorian border, in mid-September 1909, the son of the senior wife of an Nzima goldsmith. Some of his detractors would later cast aspersions on his Ghanaianness because of this geographical accident of birth. According to a contemporary, he was a shy child who was highly influenced by his mother, emerging as a very determined, highly motivated and disciplined individual (Davidson 1973). Nkrumah believed that Africans must not only shed the yoke of colonial inferiority but must be prepared to take up responsibilities, as a prerequisite for retrieving power from the coloniser. One of the earliest challenges the young Nkrumah had to face was that of education, in a colony where the vast majority of children dreamed of attending school but never had the opportunity to do so. After finishing his elementary education in 1926, he became a teacher in Half Assini, a small coastal town near the Ivorian border, from where at the age of seventeen he was recruited by the Reverend A.G. Fraser, the founder of Achimota College, to become a student there. This he did for the next four years. After Achimota, he taught in a number of schools before leaving for the USA in 1935.

Nkrumah's choice of the USA for further studies was influenced by the rejection of the snobbery he saw 'in the British way of life' as well as the influence of people like Nigerian nationalist leader, Nnamdi Azikiwe, and the realisation that many Black pioneers were themselves luminaries of US educational institutions (Davidson 1973). During his sojourn in the USA, he attended Lincoln University, Pennsylvania and got to know many of the Black emancipators in the US, men such as W.E.B. Du Bois, and Marcus Garvey, who helped in the formulation of his Pan-Africanist ideals. Despite this growing period of intellectual enlightenment, life in America was not easy for the young Nkrumah, who had to toil to pay his way through university, endured racism and imperialism, factors which impelled him to return to his native land to lead the anti-colonial struggle.

Armed with Western liberal philosophy, Nkrumah became convinced about the need for Africa to both overturn the colonial system and struggle for continental unity. These two projects would come to dominate the thinking and life of Nkrumah, and he was convinced that the latter could not be achieved without attaining the first objective. Thus his famous maxim: 'Seek ye first the political kingdom and all things shall be added', which points to his ability to draw connections between the colonial subjugation and the wretched plight of the mass of the African people (Mohan 1967). Once political independence had been won, Africa and its leaders would then embark on the task of continental unity, and a clear anti-Balkanisation strategy, which he saw as the way out of colonial bondage and neo-colonialism. In his political treatise, *Africa Must Unite* (1968), Nkrumah argued that Ghana's independence would be meaningless without the complete liberation of the African continent.

In May 1945, Nkrumah left New York homeward bound via London, where he met George Padmore, the West Indian journalist, and the publisher T.R. Makonnen from British Guiana as well as attending the Fifth Pan-Africanist Congress in Manchester in the same year. During the two years he spent in London, Nkrumah made an immediate impact on London's small Black community as vice-president of the West African Students' Union, as well as endearing himself to the city's intellectual community through lecturing and studying for the Bar. He was also instrumental in setting up a secret group, which emerged from the Manchester Congress among activists of the West African National Secretariat, called The Circle, with policies designed to foster unity and national independence in West Africa.

During his stay in London he received an invitation from his old friend from Lincoln University, Ako Adjei, and Dr J.B. Danquah to return home to assume the role of secretary general of the newly formed party, the United Gold Coast Convention (UGCC), which was about to be launched. After much soul-searching, he left for the Gold Coast in November 1947, arriving to find a country heaving with social and constitutional changes, including a shoppers' strike and restive young men, many of whom had remained unemployed since being demobilised in 1945. Nkrumah soon fell out with the leaders of the UGCC after being accused of being a communist, which led him to form the Convention People's Party (CPP) in 1949. The guiding principle of the CPP was Positive Action, which involved mass action in order to achieve the specific goal of political liberation, and Tactical Action, a strategy involving neo-colonial accommodation with the colonial power in alliance with the petty bourgeoisie in order to defeat the chiefs and the middle class. Through the vehicle of the CPP, Nkrumah won independence for Ghana in 1957 and became the leader of Black Africa's first independent sovereign state. This provided him with the opportunity to put some of his ideas into practice for his country's development. This has been referred to as Nkrumahism (Mohan 1967).

Nkrumahism was not a systematic or coherent ideology; indeed, Mohan described it as 'a personalised choice of slogans and formulas' with a 'hopeless attempt to superimpose an apparent order over a mass of contradictions' (Mohan 1967: 211). Nonetheless, one can unravel several strands to it: the philosophy of 'Consciencism'; 'socialism in one country'; Pan-Africanism; Positive and Tactical Actions; political and economic strategies for national liberation and continental unity. However, as Mohan has argued, the doctrine was riddled with contradictions.

Consciencism was a 'third way' philosophical outlook designed by Nkrumah to aid Africa's social and political revolution. It is an eclectic ensemble of the triple heritage of the African continent which will enable African society to digest the Western, the Islamic and the Euro-Christian elements in Africa, and develop them in such a way that they fit into the African personality (Nkrumah 1964: 79). The African personality is defined as a cluster of humanist principles underlying traditional African society. In Nkrumah's world the need for a new philosophical or ideological outlook for Africa was premised not just because of the syncretic nature of post-colonial Africa, but because the essence and purposes of capitalism are contrary to African society and the African *Weltanschauung*. For him capitalism was the denial of the African personality and conscience of Africa. Not only was capitalism unjust, but it was also unworkable and alien to Africans (*ibid.*).

Thus in order to resituate Africa's humanist and egalitarian principles, which had been transgressed by colonialism, socialism was the answer. With regard to domestic policies, in the early days the CPP was riddled with anti-communist psychosis (Mohan 1967), designed to neutralise the radical wing of the Party and depoliticise the trade unions in order to gain colonial accommodation. Part of the CPP's project of modernity was to foster national unity (as a prelude to continental unity) and in order to achieve this Nkrumahism directed its attack on the chiefs, perceived as the symbols of the *ancien régime* and sources of 'tribalism' and regionalism. However, after 1951 Nkrumah defended chieftaincy in the country (Mohan 1967), provided that they delivered the vote in a new form of post-colonial decentralised despotism. Paradoxically, the emergence of a bourgeoisie, that genre of modernity *sui generis,* was to be discouraged at all costs, as they were seen as a comprador class and as harbingers of neo-colonialism. In reality, Nkrumah was concerned about the emergence of a class that might challenge the CPP for state hegemony. Mohan has argued that the rejection of the middle class put paid to the rise of an incipient Ghanaian capitalism and transformed the large farmer-chiefs into the political opponents of the CPP's dominance (Mohan 1967). National liberation, it was argued, should be constructed with the petty bourgeoisie (the famous Veranda Boys) as the vanguard and a depoliticised worker-peasant movement (Mohan 1967). The surplus that was so necessary for social and economic reconstruction was to be squeezed from the peasantry operating within a modernised agricultural sector managed by the *apparatchiks*, via the state-controlled marketing boards.

If an indigenous bourgeoisie was to be denied, then the task of social reconstruction and industrialisation was left to the state. Nkrumah invested huge amounts in both human and physical capital: schools and universities, state farms, harbours and roads, including the Accra Tema highway and factories. The Nkrumahist state was a Bonapartist one: highly centralised, autonomous of any class or group, but a developmental one with the government playing a major role in industrial transformation. To ensure national unity and stability, the *sine qua non* for economic development, not only must *dirigisme* remain supreme, but also the ruling party must have total monopoly of power as all counter-hegemonic groups were proscribed. This alienated large sections of Ghanaian society, leading to several assassination attempts.

The economic core of Nkrumahism was proto-**Marx**ist, tinctured by Leninism on a framework of African socialism, which denied the class nature of African societies by celebrating its egalitarian past as if these were monolithic (Ottaway and Ottaway 1981). Nkrumahism had a continental dimension, which was meant to 'bring home Pan-Africanism' from its diasporan origin to Mother Africa and transform it from a rather nebulous movement, concerned vaguely with black nationalism, into an expression of African nationalism. In 1963, Nkrumah called for Africa to unite (Nkrumah 1963) as the main continental body, the Organisation of African Unity, was inaugurated in Addis Abba, the Ethiopian capital. His message appeared to have been ignored by African leaders. In 1965, he published his *Neo-colonialism: The Last Stage of Imperialism*, pointing to the role that transnational corporations play in draining surplus from Africa, hence fostering the continent's underdevelopment. Following the United Nations' débâcle in its peace-keeping activities in the Congo, which led to the death of that country's leader, Patrice Lumumba, Nkrumah called for an African High Command in order to prevent internal and external rogue elements from destabilising budding African governments (Zack-Williams 1997a). In order to eliminate all vestiges of colonialism in Africa, Nkrumah's Ghana became a haven for liberation fighters from several African countries.

Nkrumahism demanded that Africa should remain neutral or non-aligned: supporting neither the Western alliance nor the Warsaw Pact. Thus to counterbalance Western influences, Ghana developed close economic and political relationships with the Eastern bloc nations in addition to those already existing with the West. The country also played a major role in the Non-Aligned Movement (NAM), an assembly of radical Third World nations established at Bandung, Indonesia, in 1955. In this way, not only was Ghana able to play off one superpower against another, but also Ghana was able to attract resources from the East (Zack-Williams 1997b) on relatively more favourable terms than could be obtained from the advanced capitalist nations. Nkrumah was removed from power on 22 February 1966 while en route to Vietnam in search of the peace that he felt might bring an end to the brutal war taking place in that country. It was then that senior military officers in cahoots with foreign cold warriors removed him from power after depressing the price of cocoa, Ghana's main export.

He returned to Africa, having gained asylum in nearby Guinea Conakry where he was declared joint President of Guinea with Ahmed Sekou Touré. He remained in Guinea for the remainder of his life, apart from a visit to Bulgaria for medical treatment and where he died in 1972. By the time of his death, Kwame Nkrumah was not just the *Osagyefo* of Ghana but, as *The Legon Observer* (Ghana, 18 May 1972) noted: 'To the black man in all parts of the world, Nkrumah gave a new pride.' Hence his remains, initially buried in his home town of Nkroful, were later reinterred with ceremonial pomp in a grand mausoleum constructed adjacent to Accra's independence arch and stadium.

Major works

Nkrumah, K. (1957) *Ghana: The Autobiography of Kwame Nkrumah*, London, Edinburgh: Thomas Nelson.

—— (1961) *I Speak of Freedom: A Statement of African Ideology*, London, Melbourne, Toronto: Heinemann.

—— (1963) *Africa Must Unite*, London: Panaf.

—— (1964) *Consciencism: Philosophy and Ideology for Decolonisation and Development with Particular Reference to the African Revolution*, London: Heinemann.
—— (1965) *Neo-colonialism: The Last Stage of Imperialsim*, London: Thomas Nelson & Sons.
—— (1970) *Class Struggle in Africa*, London: Panaf.
—— (1990) *The Conakry Years: His Life and Letters*, London: Panaf.

Further reading

Birmingham, D. (1998) *Kwame Nkrumah: The Father of African Nationalism*, Athens, OH: Ohio University Press.
Boateng, C.A. (1995) *Nkrumah's Consciencism: An Ideology for Decolonisation and Development*, Dubuque, IA: Kendall/Hunt.
Bretton, H.L. (1966) *The Rise and Fall of Kwame Nkrumah: A Study of Personal Rule in Africa*, London: Pall Mall Press.
Davidson, B. (1973) *Black Star: A View of the Life and Times of Kwame Nkrumah*, London: Allen Lane.
Milne, J. (1990) *Kwame Nkrumah: The Conakry Years, His Life and Letters*, London: Panaf, Zed Press.
Mohan, J. (1967) 'Nkrumah and Nkrumahism', in Miliband, R. and Saville, J. (eds), *Socialist Register 1967*, London: Merlin Press.
Ottaway, D. and Ottaway, M. (1981) *AfroCommunism*, New York and London: Africana Publishing.
Rathbone, R.J.A.R. (2000) *Nkrumah and the Chiefs: The Politics of Chieftaincy in Ghana, 1951–60*, Oxford: James Currey.
Sherwood, M. (1996) *Kwame Nkrumah: The Years Abroad, 1935–47*, Legon: Freedom Publication.
Zack-Williams, A.B. (1997a) 'Peacekeeping and an 'African High Command', *Review of African Political Economy* 71(3): 131–7.
—— (1997b) 'Labour, Structural Adjustment and Democracy in Sierra Leone and Ghana', in Siddiqui, R.A. (ed.), *Sub-Saharan Africa in the 1990s: Challenges to Democracy and Development*, Westport, CT and London: Praeger, pp. 57–69.

Alfred Babatunde Zack-Williams

JULIUS KAMBARAGWE NYERERE (1922–99)

The attempt to apply the philosophy of *Ujamaa* characterises the thinking and political work of Julius Kambaragwe Nyerere during his presidency of Tanzania. Nevertheless, Nyerere's thinking was not guided solely by the ethos of African tradition in which the philosophy of *Ujamaa* was encapsulated, for he was a modernist and a practical politician who tried to synthesise the best of his people's ideas with those of the Western world in pursuance of transformation and the development of his country.

In so doing, he exhibited the highest level of moral commitment. Many critics have called his *Ujamaa* strategy an economic failure. However, it is more useful to interrogate his philosophy and practice holistically in order to distil that which was valid, at the time, from that which proved unworkable but which could be revisited and modified in order to inspire future generations. In the words of Ali Mazrui,

> When he was president of the United Republic of Tanzania, Julius K. Nyerere's vision was bigger than his victories; his perception was deeper than his performance. In global terms, he was one of the giants of the twentieth century. And like all giants he had both great insights and great blind spots. While his vision did indeed outpace his victories, and his profundity outweigh his performance, he did bestride this narrow world like an African colossus.
>
> (Mazrui 2002: 1)

Mazrui, previously a great critic of Nyerere, adds that with *Ujamaa,* which means 'familyhood', Nyerere attempted to 'build a bridge between indigenous African thought and modern political ideas'. By making it the organising principle of Tanzania's entire economic experiment from the Arusha Declaration of 1967 to the mid-1980s, Nyerere turned it into a foundation for African socialism (*ibid.*: 5). Thus an assessment of his political and development thinking must be blended with the consideration of wider issues with which he was involved throughout his political life.

Julius Nyerere was born on 13 April 1922 in Butiama village in the Musoma District of Tanganyika's Mara Region. In 1947 he obtained his diploma in education from Makerere University in Kampala. Thereafter he taught at St Mary's Secondary School in Tabora for two years before studying for his Master's degree in history, geography and economics at the University of Edinburgh. On returning to Tanganyika in 1952, he taught at Pugu Secondary School, Kisarawe in Dar es Salaam between 1953 and 1954.

His political life began when he resigned from teaching to help found the Tanganyika African National Union (TANU) on 7 July 1954, immediately becoming its president. In 1960, he became prime minister of Tanganyika under 'self-government' and continued to hold the position when Tanzania became independent on 12 December 1961. Nyerere soon resigned his position and handed over the government to his deputy, Rashid Mugume Kawawa, so that he could return to consolidate the party among the ordinary membership. When he felt that work to be done, he took charge of the government once more in December 1962, after the declaration of a republic, and became Tanganyika's president until his eventual resignation in November 1985.

The inspirations of an outstanding thinker and politician like Julius Nyerere cannot be traced to a single influence or a particular phase (Green

1995: 81–2). However, several trends and directions are indicative of how his thinking about rural development, with which his *Ujamaa* ideology was most associated, evolved.

His approach was based on a strong attempt to find inspiration in traditional African society. Beyond that, he tried to mould these ideas into a new compact of economic and political theory. Some have seen the origin of his ideas in Fabian socialist ideology and Catholic social teachings (Ibhawoh and Dibua 2003: 62), while others regard his political ideas as parallel to those of Rousseau (Stöger-Eising 2000: 34–5). However, Irene and Roland Brown have argued that Rene Dumont's book, *False Start in Africa*, had a 'particular significance' for Tanzania, especially Dumont's explanations of rural poverty as forming the root cause of lack of industrialisation (Brown and Brown 1995: 9–11).

Nyerere also read **Marx**ism and scientific socialism and later came to develop close relations with the People's Republic of China, with which he entered into several bilateral economic co-operation programmes. However, he always tried to marry new ideas gleaned from his prodigious reading to the conditions on the ground in which African traditional ideas and norms played a significant role. Joseph Warioba, one-time prime minister and confidant of Nyerere, believes that Nyerere's ideas about self-reliance were inspired very much by his conversations with ordinary people and their lived experiences. He adds that Nyerere believed very much in the idea of 'man-centred' development (interview, Arusha, 28 September 2004).

We have to delve deeper to understand how he integrated traditional and intellectual ideas into a personal philosophy and political practice. For this purpose, the main texts are the *Arusha Declaration* and its two sequels, 'Socialism and Rural Development' (1975a) and *Education for Self-Reliance* (1967b). The declaration, adopted as official doctrine by the Tanganyika African National Union (TANU) in February 1967, officially placed Tanzania on a socialist path and an associated strategy of self-reliance (Nyerere 1975b). It argued that agriculture formed the basis of Tanzania's economic development, and that only through the hard work and intelligence of the people could production levels be raised (Nyerere 1977).

Nyerere's *Ujamaa* philosophy and its ideological base are set out in *Socialism and Rural Development* (1975a). The 'traditional African family' is identified as the unit through which the principle of *Ujamaa* was to be examined. In his view, *Ujamaa* constituted the 'Third Way' to modern development. According to Nyerere, the African traditional family 'lived and worked together because that is how they understood life'. The results of their joint efforts were divided unequally among them, but this was done according to well-defined customs and traditions (Nyerere 1975a: 1).

Although Nyerere, the traditionalist, extolled the virtues of what he regarded as traditional African values, Nyerere, the modernist, simultaneously recognised their limitations. He pointed out that the result of these practices 'was not the kind of life we really wish to see existing throughout Tanzania'. Quite apart from the failure of individuals to live up to the ideals of the social system, he noted that two other factors prevented traditional society from 'flowering'. The first was gender inequality: 'It is impossible to deny that the women did, and still do, more than their fair share of the work in the fields and in the homes.' They were made to suffer from inequalities, which had nothing to do with their contribution to the family welfare. This, he concluded, 'was inconsistent with the socialist conception of the equality of all human beings and the right of all to live in security and freedom'. He added: 'If we want our country to make a full and quick progress now, it is essential that our women live in terms of full equality with their fellow citizens who are men' (*ibid.*: 3).

The other limitation of traditional society, according to Nyerere, was the prevalence of poverty. Here, Nyerere began to 'graduate' from the traditional world of the African family to modern neo-colonialism. He accepted that the traditional system offered an attractive degree of equality, albeit at a low level that could be raised. He blamed the prevalence of poverty on two factors, namely 'ignorance' and the scale of operations of the family unit. In his view, the principles of mutual respect, sharing and joint production, and work by all were, and are, the foundation of human security, of practical human equality and of peace between members of a society. They can also form the basis of economic development 'if modern knowledge and modern techniques of production are used' (*ibid.*: 3). Here, Nyerere stood alongside all modernists.

This recognition of the need to move rural production forward in modern conditions constituted the basis of *Ujamaa Vijijini* (Ujamaa Villages). The implementation programme set out in the Arusha Declaration began to confront the realities in rural areas and the way that peasants looked at the same problem. To be sure, the strategy of rural development was not entirely Nyerere's idea. It was suggested by the World Bank in their first Mission report, which formed the basis of the *First Five Year Plan* adopted at Tanganyika's independence in 1961 and which came into effect in 1964 (Nabudere 1981: 98).

This plan had adopted the 'transformation approach' to promote 'Village Settlement Schemes', which would depend on the injection of foreign capital into rural development. The plan also had a 'socialistic' ring to it, envisaging the creation of 60 'settlement villages' with each 'scheme' involving 250 individual family farms. It was hoped that the schemes would bring about 'a relatively abrupt transition' to modern techniques with regard to land use, land tenure, patterns of agricultural production and a change in economic attitudes. Such schemes could 'also be relied upon ... to relieve incipient land hunger and population pressure in certain areas' (Government of Tanganyika 1964 I: 21). In the event, very little foreign capital was attracted; instead, there was capital flight, leading to the imposition of exchange controls that showed signs of reversing this trend (Loxley 1972: 104). The programme proved too costly and unproductive, while the individual traditional family unit survived, recording 'impressively' increased production, diversification and productivity (Ngotyana 1972: 125).

By 1965 panic arose in official circles regarding the lack of foreign capital inflow to support the plan. From this perspective it appeared that African socialism of the *Ujamaa* type would be the answer. The Arusha Declaration's main innovations were an expansion of the Settlement Schemes as new villages and a programme to nationalise enterprises aimed at saving foreign exchange. However, when the *Ujamaa Vijijini* policy was implemented, the peasants and pastoralists immediately resisted what they saw as a derogation of their rights to ancestral lands. From a policy of persuasion, the government resorted to using force to implement its objectives and this soon precipitated the collapse of the entire programme. Faced with this failure and the pressure to adopt structural adjustment programmes by the World Bank and the International Monetary Fund, Nyerere stood down as president of Tanzania in 1985, creating room for reforms that the new president, Ali Hassan Mwinyi (1985–95), was forced to adopt in line with the demands of the Bretton Woods institutions in order to obtain new loans to 'revive' the economy.

Nyerere's decision to resign was very unusual in Africa, especially within one-party states. However, his leadership had a long record of modesty. Although the son of a chief, he strove to live a simple life. He placed little significance on owning money; his salary and those of his ministers were very modest. He hated corruption and tried to banish it from his government. The party and army insisted on building him a larger house in Butiama ahead of his retirement. However, he slept in it for only two weeks before falling sick. He went to London for treatment for leukaemia, and died soon afterwards at St Thomas's Hospital, aged 77. He was buried at his new house in Butiama during a state funeral.

Nyerere dedicated the last years of his life to South–South co-operation. Serving as chair of the South Commission proved a creative experience, marking the beginning of his 'international career', which lasted more than thirteen years and put Nyerere's vision, unique capacities and political acumen in the service of the Third World and its development. The Commission's landmark report, for which Nyerere chose the title, *The*

Challenge to the South, comprised the first comprehensive 'Southern' analysis of global development challenges. It argued forcefully that intellectual and technical preparedness was vitally important if developing countries were to confront the complexities of the modern age and the challenges posed by the North's continuing global dominance. Thus, one of the report's principal priorities was the need to establish permanent and adequate institutional support for collective global action by the South. This was not taken up directly but in September 1995, the South Centre became an intergovernmental organisation – the first South–South institution with a comprehensive global development mandate.

Nyerere served as chancellor of the University of East Africa (1963–70), University of Dar es Salaam (1970–85) and Sokoine University of Agriculture (1984); was awarded numerous honorary degrees; and received prestigious awards, including the Nehru Award for International Understanding, the Third World Prize, Nansen Medal for Outstanding Service to Refugees and Lenin Peace Prize.

With typical modesty, Nyerere declined to write an autobiography and did not encourage others to produce a biography. Two schools of thought have emerged in the debate about his contribution (Ibhawoh and Didua 2003: 70). The first argues that Nyerere's strategies were a complete failure, wasting economic resources in the name of slavish adherence to ideology (Nursey-Bray 1980; Von Freyhold 1979; Hyden 1980; Coulson 1985; Green 1995). The second school, while conceding that the economic performance of *Ujamaa* was modest, nevertheless points to achievements in social welfare such as the provision of health and educational facilities; a movement towards greater social equality in income distribution, the maintenance of political stability and a strong sense of Tanzanian national identity (Legum and Mmari 1995; Pratt 1999; Ishemo 2000; Mwakikagile 2002).

Nyerere's legacy therefore goes beyond developmentalism. It includes his contribution to the struggle for Pan-Africanism and the creation of an East African federation, his role as Chairperson of the Liberation Committee of the Organisation of African Unity and the Front Line States (the forerunner to the Southern African Development Community). His Pan-African contribution led to Nyerere being acknowledged as the fourth 'greatest' African of all time in *New African* magazine (2004), which described him as 'A great leader who refused to allow the trappings of power to corrupt him. He was respected by his country, Africa and the rest of the world.'

Major works

Nyerere, J.K. (1966) *Freedom and Unity/Uhuru na Umoja: A Selection from Writings and Speeches 1952–1965*, London and Dar es Salaam: Oxford University Press.

—— (1967a) *The Arusha Declaration*, Dar es Salaam: Government Printer.

—— (1967b) *Education for Self-Reliance*, Dar es Salaam: Government Printer.

—— (1968) *Freedom and Socialism/Uhuru na Ujamaa: A Selection from Writings and Speeches 1965–1967*, London and Dar es Salaam: Oxford University Press.

—— (1973) *Freedom and Development/Uhuru na Maendeleo: A Selection from Writings and Speeches 1968–1973*, London and Dar es Salaam: Oxford University Press.

—— (1975) *Speeches in Parliament*, Dar es Salaam: Government Printer, 18 July.

—— (1975a) 'Socialism and Rural Development', in Cliffe, L., Lawrence, P., Luttrel, W., Migot-Adholla, S. and Saul, J.S. (eds), *Rural Cooperation in Tanzania*, Dar es Salaam: Tanzania Publishing House.

—— (1975b) *The Development of Ujamaa Villages*, Presidential Circular No. 1 of 1969, Dar es Salaam: Government of Tanzania.

—— (1977) *The Arusha Declaration Ten Years After*, Dar es Salaam: Government Printer.

Further reading

Brown, I. and Brown, R. (1995) 'Approach to Rural Mass Poverty', in Legum, C. and Mmari, G. (eds), *Mwalimu: The Influence of Nyerere*, Trenton, NJ: Africa World Press; London: James Currey; Dar es Salaam: Mkuki Na Nyota.

Coulson, A. (1985) *Tanzania: A Political Economy*, Oxford: Oxford University Press.

Dumont, R. (1966) *False Start in Africa*, London: Deutsch.

Government of Tanganyika (1964) *First Five Year Plan*, Dar es Salaam: Government Printer.

Green, R.H. (1995) 'Vision of Human-Centred Development: A Study in Moral Economy', in Legum, C. and Mmari, G. (eds), *Mwalimu: The Influence of Nyerere*, Trenton, NJ: Africa World Press; London: James Currey; Dar es Salaam: Mkuki Na Nyota.

Hyden, G. (1980) *Beyond Ujamaa in Tanzania: Underdevelopment and an Uncaptured Peasantry*, Berkeley, CA: University of California Press.

Ibhawoh, B. and Dibua, J.I. (2003) 'Deconstructing Ujamaa: The Legacy of Julius Nyerere in the Quest for Social and Economic Development in Africa', *African Journal of Political Science* 8(1).

Ishemo, S.I. (2000) 'A symbol that cannot be substituted': The Role of Mwalimu J.K. Nyerere in the Liberation of Southern Africa, 1955–1990', *Review of African Political Economy* 27(83): 81–94. (Other tributes appeared in *ROAPE* 26(81) and 26(82), 1999.)

Legum, C. and Mmari, G. (eds) (1995) *Mwalimu: The Influence of Nyerere*, Trenton, NJ: Africa World Press; London: James Currey; Dar es Salaam: Mkuki Na Nyota.

Loxley, J. (1972) 'Structural Change in the Monetary System of Tanzania', in Saul, J. and Cliffe, L. (eds), *Socialism in Tanzania*, Vol. II, Nairobi: East Africa Publishing House.

Mazrui, A.A. (2002) *The Titan of Tanzania: Julius K. Nyerere's Legacy*, Binghamton, NY: Institute of Global Cultural Studies, Global Publications, Binghamton University, State University of New York.

Mmari, G. (1995): 'The Legacy of Nyerere', in Legum, C. and Mmari, G. (eds), *Mwalimu: The Influence of Nyerere*, Trenton, NJ: Africa World Press; London: James Currey; Dar es Salaam: Mkuki Na Nyota.
Mwakikagile, G. (2002) *Nyerere and Africa: End of an Era; The Biography of Julius Kambarage Nyerere (1922–1999), President of Tanzania*, USA: Protea Publishing.
Nabudere, D.W. (1981) *Imperialism in East Africa, Volume 1: Imperialism and Exploitation*, London: Zed Press.
New African (2004) '100 Greatest Africans of All Time', 432 (August/September): 12–32.
Ngotyana, R. (1972) 'The Strategy of Rural Development', in Saul, J. and Cliffe, L. (eds), *Socialism in Tanzania*, Vol. II, Nairobi: East African Publishing House.
Nursey-Bray, P.F. (1980) 'Tanzania: The Development Debate', *African Affairs* 79(314): 55–78.
Pratt, C. (1976) *The Critical Phase in Tanzania, 1945–67: Nyerere and the Emergence of a Socialist Strategy*, Cambridge and New York: Cambridge University Press.
—— (1999) 'Julius Nyerere: Reflections on the Legacy of Socialism', *Canadian Journal of African Studies* 33(1): 137–52.
—— (2000) 'Julius Nyerere: The Ethical Foundation of his Legacy', *The Round Table* 355: 355–64.
South Commission (1990) *The Challenge to the South: The Report of the South Commission*, Oxford: Oxford University Press.
Stöger-Eising, V. (2000) 'Ujamaa Revisited: Indigenous and European Influences in Nyerere's Social and Political Thought', *Africa* 70(1): 118–43.
Von Freyhold, M. (1979) *Ujamaa Villages in Tanzania: An Analysis of a Social Experiment*, New York: Monthly Review Press; London, Ibadan and Nairobi: Heinemann Educational Books.

Dani W. Nabudere

RAÚL PREBISCH (1901–86)

Economist, teacher, central banker, civil servant and institution builder, Raúl Prebisch was a pragmatic but persistent advocate of a more egalitarian system of international trade between the more developed countries of the North (which he referred to as the 'centre') and the less developed countries of the South (the 'periphery'). In economic circles and beyond, he is best known for what has been labelled as the 'Prebisch–Singer' thesis concerning the terms of trade between the developed and developing countries. He is certainly the most influential Latin American development economist and probably its most eminent.

Prebisch was born on 17 April 1901 in Tucumán, Argentina and died on 29 April 1986, shortly after his eighty-fifth birthday, in Santiago, Chile. His father was of German origin and his mother came from a traditional family of Tucumán, a provincial city in the interior of Argentina. He studied economics at the University of Buenos Aires, the main national university, graduating in 1922. From 1925 to 1948 he lectured in political economy at the School of Economics in the same University. From 1930 to1935 he was initially Under-Secretary of Finance and Agriculture and later the advisor to the Ministers of Finance and Agriculture. More importantly, he was one of the principal founders of Argentina's Central Bank, becoming its first Director in 1935, a position he held until 1943 when political changes forced him to resign. He was then invited to assist the Mexican Central Bank and went on to advise other central banks in Latin America. In 1949 he was hired as a consultant by the recently established UN Economic Commission for Latin America (ECLA),[1] located in Santiago de Chile, only to be catapulted the following year into the headship of the organisation with the title of Executive Secretary, a position he kept until 1962. From 1962 to 1964 he served as Director-General of the UN's Latin American Institute for Economic and Social Planning (ILPES), also based in Santiago, and which he was instrumental in establishing.

Before joining ECLA, Prebisch was already one of the most highly regarded economists in Latin America. But his rapid rise to the top has much to do with a document he wrote a few months after his arrival and which he presented to the institution's conference in 1949. This took place in Havana in front of a select audience, including the various country representatives of the organisation. This magisterial work, 'The Economic Development of Latin America and its Principal Problems',[2] created quite a stir among the delegates, who were most impressed not only by its ideas but also by Prebisch's forceful personality, persuasiveness and vision. Years later, the prominent development economist, **Albert Hirschman** (1961: 13), was to refer to this document as 'ECLA's manifesto'. Indeed, given its polemical character, it was a sort of manifesto and this explains why it was circulated under his own name, something which was highly unusual since UN rules dictate that documents should be anonymous and bear only the imprint of the institution. It urged Latin American countries to industrialise and launched an attack on the international division of labour in which the central countries specialised in the production of manufactures while the peripheral countries specialised in the production of raw materials such as agricultural products and minerals. Thus the central countries largely exported industrial products to the peripheral countries while the latter exported primary products to the former. His proposal was highly unusual at the time; even more so was his critique of the neo-classical international trade theory which had hardly been questioned before.

Conventional international trade theory argued that the current economic specialisation of the developed countries in the production of industrial commodities and of the developing countries in primary commodities worked to their mutual benefit as each enjoyed comparative advantages in these respective areas. Furthermore, this theory argued that international trade would help to reduce the income gap between rich and poor countries. Prebisch (1949), however, observed that incomes grew faster in the centre than in the periphery. The main reason for this widening gap was, in his view, the secular deterioration of the prices of primary goods in comparison with industrial products in the world market. To support this argument he relied on historical price data from a 1948 UN study, which did not indicate authorship but had been written by **Hans Singer**, who was then working at the UN in New York. This unexpected long-run tendency for prices of primary products to deteriorate relative to the prices of manufactured goods meant that the periphery had to export an increasing quantity of raw materials in order to continue importing the same amount of industrial commodities. Thus, Prebisch argued, the prevailing international trade system, by confining the periphery to the production of primary commodities, favours mainly the centre. He set out to explain the reasons for this trend as it contradicted existing international trade theory.

His thesis regarding the deterioration of the periphery's terms of trade reveals Prebisch's distinctive approach to the problems of development and underdevelopment as compared with neo-classical theories. Prebisch's (1959, 1964) analysis of this process deals both with demand and supply conditions of commodity markets. On the demand side, Prebisch argues that the terms of trade deteriorated against the periphery because of the different income-elasticity of demand for imports by the centre and periphery or, as he put it, due to the 'dynamic disparity of demand' between centre and periphery. This means that the centre's imports of primary products from the periphery rise at a lower rate than its national income, while the periphery's imports of industrial commodities from the centre grow at a faster rate than its income. As for the supply arguments, or the 'cycle version' of the periphery's deterioration of the terms of trade, these are related to the differential effect of the world economic cycle on centre and periphery. The capitalist economic system evolves in a cyclical fashion with the centre being the initiator of these cycles, which provoke an adaptive cyclical response by the periphery. During an economic upswing, the terms of trade generally turn in favour of primary producers but during a downswing they turn against them to an even greater degree. This results in the long-run deterioration of the periphery's terms of trade, especially as the downswings tend to last longer than the upswings.[3]

This differential impact of the world economic cycle on centre and periphery is explained by the differential behaviour of prices, profits and wages in both the centre and periphery during the phases of the cycle. During an economic upswing wages and prices grow substantially in the centre while they hardly rise in the periphery because of the availability of surplus labour. Meanwhile during a downswing the fall in wages and prices at the centre is limited due to the existence of trade union power and the oligopolistic structure of industry. In the periphery, by contrast, the downswing greatly reduces prices and wages as producers are able to compress wages substantially on account of surplus labour and the non-existence and/or weakness of trade unions.

Although Prebisch did not discover independently that the terms of trade of primary products were in secular decline, as he relied on the previous study of **Singer**, he was the first to provide an economic cycle analysis of this phenomenon. In his seminal essay, Singer (1950) partly advances arguments similar to those of Prebisch, although his analysis was done independently. Thus, the hypothesis on the deterioration in the terms of trade is known in the economic literature as the 'Prebisch–Singer hypothesis'. In condemning the deterioration of the commodity terms of trade, Prebisch is not arguing against international trade, nor has he ever suggested delinking as did some *dependencia* advocates like **Andre Gunder Frank** and **Samir Amin**. On the contrary, he sees international trade and foreign capital as essential elements for raising productivity and growth in the periphery. The Prebisch–Singer hypothesis created much controversy and was much criticised, particularly by neo-classical economists. But the emerging consensus is that it has stood the test of both time and new statistical techniques remarkably well, although the reasons for it may have shifted. Thus even today, barring a few exceptional cases, the gains from trade continue to be distributed unequally between those countries exporting mainly primary products and those exporting mainly manufactures.

Prebisch was the guiding light of ECLA which, under his inspirational leadership, became one of the most notable UN regional organisations. It had a major influence on development thinking and policy, especially in Latin America, mainly during its heyday of the 1950s and 1960s. The 'ECLA manifesto' became the cornerstone of what was to become known as 'structuralism' or the Latin American structuralist school of development. Prebisch assembled at ECLA a first rate team of mostly young social scientists, but mainly economists, from various Latin American countries who all made, to a greater or lesser extent, a contribution to the development of structuralism and the core–periphery model. The structuralist view of development stresses the crucial significance of the differences in the economic, social and political structures between the core and peripheral countries. This core–periphery model argues that the world economy is divided into an industrial centre and an agrarian periphery. While the core or developed countries have an internally homogeneous structure with similar levels of productivity in different sectors, the periphery or underdeveloped countries are characterised by heterogeneity as the various sectors of the economy have major differences in productivity arising from different levels of technology. This duality is also replicated within each of the economic sectors. Thus the periphery is characterised by 'structural heterogeneity', or internal inequality, which is reinforced by international trade. Furthermore, unequal exchange between the centre and the periphery also reinforces the external inequality between the core and the periphery. Structuralism is thus one of the first attempts to construct a theory of development and underdevelopment.

This structuralist analysis led Prebisch to propose a new development strategy for the periphery, in particular Latin America, so as to counteract, and possibly overcome, the negative impact of unequal exchange on the periphery. It entailed a radical change in the export-led development model based on primary products, or what he called the 'outward-oriented' development path, which most periphery countries had been following. In its place he argued for a new 'inward-oriented' development strategy based on a process of import-substitution-industrialisation (ISI). This required protectionist measures such as a set of duties on manufacturing imports as well as a tax on primary exports so as to induce a switch of resources from primary production for exports to industrial activities which at first would be directed at the domestic market but later would also be geared to exports. He also proposed allowing union activity in the primary export sector to push up wages, to defend primary commodity prices through concerted international action, and to press for the reduction or elimination of protection for primary commodities in the centre (Prebisch 1959: 263). The main thrust of his argument was aimed at changing the periphery's structure of production and developing an industrial sector which would lead to higher rates of productivity growth and to a greater ability to retain the 'fruits of technological progress', leading thereby to higher rates of growth and incomes which could reduce the gap between the core and periphery. While the ISI process did not fulfil all of its promises, which Prebisch was one of the first to point out, he insisted that industrialisation was a key route to development of the periphery. He was also a firm believer in the importance of the State for achieving development.

It was only natural for Prebisch to move from ECLA to the wider stage. His mission for a fairer international economic system led him to propose the creation of a new UN agency. On the strength of his report to the UN Secretary-General (Prebisch 1964), as well as his lobbying skills, the UN Conference on Trade and Development (UNCTAD) was established in 1964. He was appointed its first Secretary-General, a post he held until 1969. In his report he advocated a fairer international trading system which would tackle the problem of primary commodity prices and assist developing countries to shift to industrial exports by tariff preferences in the developed countries, among other measures. Prebisch's ideas had a major influence in shaping the demands of many developing countries who were proposing a New International Economic Order (NIEO) in the 1970s.

Upon his 'retirement' he remained closely associated with ECLA in Santiago. It was only in the last years of his prolific life that Prebisch set out to develop systematically his ideas on what he termed 'peripheral capitalism' in a series of articles originally published in *CEPAL Review*, the journal of ECLA, which he helped to create in 1976. He became its first Director, a position he held for several years (e.g. Toye and Toye 2004). These articles form the basis of what was to become his major last work (Prebisch 1981). Although schooled in the neo-classical tradition, his acute mind and pragmatic sense soon led him to question it. His ideas shaped structuralism and had a major influence on dependency and world systems theory. The last years of his life coincided with the rise of neoliberalism of which he became a major critic. As an institution builder and activist, he was able to transform some of his ideas into action but his vision for a fairer international economic system remains to be fulfilled.

Notes

1 Now the Economic Commission for Latin America and the Caribbean (ECLAC).
2 The document was published in English a year later, see Prebisch (1950).
3 *Economic Survey of Latin America 1949*, New York: United Nations, 1951. Although the authorship is assigned to the Secretariat of the Economic Commission for Latin America (ECLA), it is known that Prebisch wrote the whole of Part One, entitled 'Growth, Disequilibrium and Disparities: Interpretation of the Process of Economic Development' comprising five chapters, pp. 3–87. This text is Prebisch's most elaborate attempt to present his principal theses.

Major works

Prebisch, R. (1950) *The Economic Development of Latin America and its Principal Problems*, Lake Success, NY: United Nations. (Trans. of *El Desarrollo Económico de la América Latina y Algunos de sus Principales Problemas*, Santiago: Comisión Económica para América Latina (CEPAL), 1949.)
—— (1959) 'Commercial Policy in the Underdeveloped Countries', *American Economic Review* 49(2): 251–73.
—— (1963) *Towards a Dynamic Development Policy for Latin America*, New York: United Nations.
—— (1964) *Towards a New Trade Policy for Development: Report by the Secretary-General of the United Nations Conference on Trade and Development*, New York: United Nations.
—— (1976) 'A Critique of Peripheral Capitalism', *CEPAL Review* 1: 9–76.
—— (1981) *Capitalismo Periférico: Crisis y Transformación*, Mexico City: Fondo de Cultura Económica.

—— (1984) 'Five Stages in My Thinking on Development', in Meier, G.M and Seers, D. (eds), *Pioneers in Development*, New York and Oxford: Oxford University Press, pp. 175–90.

Further reading

Furtado, C. (1988) *La Fantasía Organizada*, Buenos Aires: Editorial Universitaria de Buenos Aires. This biographical essay written by an ECLA insider, and one of Latin America's most original development thinker after Prebisch, provides a fascinating insight into the origins and developments of Prebisch's ideas at ECLA.

Gurrieri, A. (ed.) (1982) *La Obra de Prebisch en la CEPAL*, 2 vols, Mexico City: Fondo de Cultura Económica.

Hirschman, A. (1961) 'Ideologies of Economic Development in Latin America', in Hirschman, A. (ed.), *Latin American Issues: Essays and Comments*, New York: Twentieth Century Fund.

Love, J.L. (1980) 'Raúl Prebisch and the Origins of the Doctrine of Unequal Exchange', *Latin American Research Review* 15(3): 45–72.

Mallorquín, C. (1998) *Ideas e Historia en Torno al Pensamiento Económico Latinoamericano*, Mexico City: Plaza y Valdés Editores.

Rodrí guez, O. (1980) *La Teoría del Subdesarrollo de la CEPAL*, Mexico City: Siglo Veintiuno Editores.

Singer, H.W. (1950) 'The Distribution of Gains Between Investing and Borrowing Countries', *American Economic Review* 40(2): 473–85.

Toye, J. and Toye, R. (2004) *The UN and Global Political Economy: Trade, Finance and Development*, Bloomington, IN: Indiana University Press.

Various authors (2001) 'Tribute to Raúl Prebisch', *CEPAL Review* 75: 1–108.

Cristóbal Kay

WALTER RODNEY (1942–80)

Walter Rodney was one of the most influential writers on history, development and underdevelopment to have emerged from the Caribbean. His intellectual and political work has had worldwide impact. A quarter of a century after his untimely death (Rodney was assassinated in his native Guyana at the age of thirty-eight), Rodney is still widely read. In particular, his book *How Europe Underdeveloped Africa* (first published in 1972) remains influential.

Rodney was born in Georgetown, Guyana in March 1942. At this time, one key aspect of the colonial economy was booming, with the allies consuming all that could be produced of British Guyana's main export, bauxite (the ore for aluminium). Two-thirds of allied aircraft would be made from Guyanese ores. Yet wartime shortages of consumer goods and disruption to Atlantic shipping meant that many in Guyana had to struggle to make ends meet or found themselves out of work. Just months before Rodney was born, the Americans commenced the construction of an airbase close to Georgetown. This was one of eight sites in the western hemisphere where the British permitted the USA to establish military facilities following a September 1940 agreement between British Premier Churchill and US President Roosevelt. In exchange Britain received fifty surplus American destroyers.

In other words, Rodney was born at the cusp of a shift from Britain and Europe to the USA in terms of where the centre of Western economic and political power was seen to reside. The relative eclipse of British Empire and the rise of American superpower soon became evident. Coming too were the struggles for national liberation in Africa and Asia, the birth of the Third World amidst enhanced global polarisation and contest between capitalism and communism and the rise of programmes and strategies for 'development'. These momentous geopolitical shifts and contests constituted the backdrop to his intellectual and political formation.

Whilst Rodney grew up in Guyana and entered high school in 1953, he travelled to Jamaica in 1960 to study at the University of the West Indies (then still the University College of the West Indies). Rodney majored in history then moved to London in 1963 for graduate studies at the history department of the School of Oriental and African Studies (SOAS). In 1966 he earned a doctorate for a thesis on 'The History of the Upper Guinea Coast, 1545–1800'. While at SOAS, Rodney participated in reading and study groups on **Marx**ism. Rodney was far from an orthodox Marxist, however, and he subscribed to no formal party-line. At the same time as Rodney made his way through the school and university system, other Caribbean intellectuals were formulating radical conceptions of the region's dependence, fragmentation and marginality. This work certainly influenced Rodney, although his key contribution was concerned with the related problem of African underdevelopment.

Rodney was an Afro-Guyanese from a working-class family, and African and Black liberation, along with class struggle, were the other inputs into his political, intellectual and personal development. According to Lewis's (1998: xvii) survey of Rodney's intellectual and political thought, we should:

> situate his intellectual and political evolution in the transatlantic diasporic locations of Guyana, Jamaica, London and Tanzania. These geographically dispersed locations were, of course, linked by the history of the British Empire and capitalism. What emerges from Rodney's work is not only a critique of empire and capitalism in general but a dissection of the domestic political elite that assumed

political authority from the colonisers in Africa and the Caribbean as well as an analysis of the process of recolonisation.

On completion of his doctorate, Rodney took up a position in the history department of the University College of Dar es Salaam, Tanzania, then chaired by Terence Ranger (who later became a distinguished historian of Southern Africa). Tanzania was then embarking on a path of 'African socialism' under **Julius Nyerere**. It proved a fertile environment for Rodney and other radical scholars engaged in debates on the making of Africa's underdevelopment. However, Rodney sought to return to the Caribbean with the ambition to develop African studies back in Jamaica. He was prevented from doing so by the Jamaican government who banned him in 1968, after he returned there briefly at a time of considerable political turmoil. As part of a combined crackdown on radical intellectuals (and the banning of texts, such as the writings of Malcolm X) and on the grassroots Rastafari movement, Rodney was declared *persona non grata*. A collection of his reflections and exchanges in Jamaica was published in London in 1969 under the title *The Groundings with my Brothers*. In these Rodney (1969: 59) asserted that:

> The Government of Jamaica, which is Garvey's [Marcus Garvey, born 1887 in Jamaica, died 1940 in the UK, was a founding figure of Black Liberation/Nationalism] homeland, has seen it fit to ban me, a Guyanese, a black man, and an African. But this is not very surprising because though the composition of that Government – of its Prime Minister, the Head of State and several leading personalities – though that composition happens to be predominantly black, as the Brothers at home say, they are all white-hearted.

Fortunately, Rodney was able to continue working in Dar es Salaam, where he stayed until 1974. The pull of the Caribbean remained, however, and in 1974 Rodney returned to Guyana. Although denied a university position there (on account of his growing differences with and critiques of Burnham's authoritarian government), he continued to produce academic writings and relied on guest lecture tours in North America and Europe and external research grants to survive financially. By this time his international reputation had grown, in large part owing to the broad readership that his *How Europe Underdeveloped Africa* was gaining. This was, however, little protection against the agents of Burnham's regime who assassinated Rodney on 13 June 1980, largely in response to his political activity directed against the incumbent regime.

Rodney's most influential book is a statement of what became known as dependency theory. It is thus marked by many of the controversies (and limits) of that perspective. To Latin American conceptions of dependency with their emphasis on class relations and imperialism, Rodney injected and insisted on the roles of racial categories and identities. On Rodney's return to Jamaica:

> News travelled fast about the young, radical and dynamic university teacher who could spellbind his audience, and 'knew all about Africa'. The dependency theorists at the university had just started an intensive public campaign and Rodney became the most sought-after lecturer at black, middle-class gatherings as well as at Rasta meetings in the slums of Kingston. He was dangerous [to the established order], primarily because he described the dependence *problematique* in racial terms – a subject which had never been brought up in the Latin American debate ...
>
> (Blomström and Hettne 1984: 108)

Blomström and Hettne (1984: 147) go on to note how, in comparison to the bulk of dependency theory writings: 'Rodney ends up far to the left on all axes.' More significantly, in writing on Africa from a dependency perspective, Rodney was a significant agent in the dissemination of dependency beyond its Latin American roots (Kay 1989). Blomström and Hettne (1984: 108–9) point out that:

> Emancipation was to Rodney not so much a matter of economic rights as it was a right to cultural identity: in this context his thoughts are more reminiscent of African ideas, of *African personality* and *Négritude*, than of Latin American dependency theory. Consequently, his contacts with the Rastafari movement were well developed, which in turn, the [Jamaican] government considered to be a greater danger than communist agitators.

In particular, *How Europe Underdeveloped Africa* highlighted the roles of the slave trade and subsequent predatory colonial capitalism in producing the simultaneous flow of wealth to Europe and impoverishment of Africa. Rodney pictured African social and commercial relations being fragmented and distorted as they became increasingly orientated to Western markets. For Rodney (1972: 37) therefore:

> The question as to who and what is responsible for African underdevelopment can be answered at two levels. Firstly, the answer is that

> the operation of the imperialist system bears major responsibility for African economic retardation by draining African wealth and by making it impossible to develop more rapidly the resources of the continent. Secondly, one has to deal with those who manipulate the system and those who are either agents or unwitting accomplices of the said system.

The latter point relates to the agency of African (and other postcolonial) elites as intermediaries, a theme that crops up in Rodney's other writings on African and Caribbean histories, development and underdevelopment (e.g. Rodney 1967, 1970, 1981). Yet during his time in Tanzania, Rodney celebrated some of the claims and putative achievements of the Arusha Declaration (proclaiming a strategy of socialism and self-reliance) in charting an alternative to the usual neo-colonial pattern in Africa. Here Rodney was undoubtedly caught up in, and his judgements shaped by, the revolutionary optimism of the times and the sense of an alternative to the neo-colonialism that he felt characterised most of Africa and the Caribbean. However, *How Europe Underdeveloped Africa* transcends that moment. By insisting that African underdevelopment was not a natural condition and that African cultures and histories were rich and not themselves the source of the continent's contemporary relative underdevelopment, Rodney's book would prove inspirational among a broad readership. In particular, for many black folk in the diaspora (hence in Europe, the Caribbean and the Americas), Rodney's book was part of a wave of expression, mobilisation and radicalisation in the 1970s.

Robert Shenton (then Professor of History at the University of Toronto) ended his review of *How Europe Underdeveloped Africa* (subsequently expanded and reprinted as an introduction to later editions of Rodney's book) with the following words inviting further questions on Africa's trajectories:

> Dr Rodney's book is important because rather than just presenting history he seeks to make it. It is doubly important because Rodney, through the use of a popular, polemical style seeks to reach those who in his eyes are truly capable of making history. My own reservation is that the use of this style sometimes cloaks the need for analysis with popular polemic and creates the impression that all the important questions regarding Africa's past and present development have been asked and answered. And this clearly is not the case.
>
> (Shenton 1975: 150)

Rodney's book is clear about the genesis of African underdevelopment. As Rodney intended, it was perhaps most effective as a polemic and an invitation to think historically and comparatively about African underdevelopment. In his original preface to the text, Rodney (1972) explains that: 'the purpose has been to try and reach Africans who wish to explore further the nature of their exploitation, rather than to satisfy the "standards" set by their oppressors and their spokesmen in the academic world' (1983 edn: 8). Rodney was not much interested in convention, theoretical abstraction or academic honours. His commitment to activism underlay his return to Guyana and ultimately led to his early death.

Lewis (1998: 256) ends his book-length study of Rodney's thought and impact by returning to his place in a radical Caribbean tradition:

> Rodney's intellectual legacy, as a historian and as an activist, forms an important part of the unfolding Caribbean intellectual tradition. … The central feature underlying Rodney's contribution to this tradition was his positive awareness of himself as a person of African descent in the Caribbean, the link he forged with Africa and the intellectual agenda that emerged in relation to the major challenges of decolonisation.

It is interesting to reflect on how Rodney's thought would have evolved had he not been killed. What would Rodney have said about 'globalisation', neo-liberalism and new configurations of Empire? How would his politics and writings have dealt with the eclipse of state socialism, the relative decline of Third Worldism and the rise of Asian (and other so-called 'newly industrialised') economies? Many things have changed since Rodney's days. Yet, more than three decades after its publication, *How Europe Underdeveloped Africa* has not lost its polemical capacity to incite reactions and to radicalise some of its readers. A major conference in Guyana in June 2005 marked the twenty-fifth anniversary of his death and celebrated his emancipatory vision (www.rodney25.org//index.htm). Meanwhile, issues of Africa's past and present (under)development remain open-ended questions and contests.

Major works

Rodney, W. (1967) *West Africa and the Atlantic Slave-Trade*, Nairobi: East Africa Publishing House.

—— (1969) *The Groundings with my Brothers*, London: Bogle-L'Ouverture Publications.

—— (1970) *A History of the Upper Guinea Coast 1545–1800*, London and New York: Monthly Review Press.

—— (1972) *How Europe Underdeveloped Africa*, London: Bogle-L'Ouverture Publications (republished 1983).
—— (1981) *A History of the Guyanese Working People, 1881–1905*, Kingston and London: Heinemann Educational Books.

Further reading

Blomström, M. and Hettne, B. (1984) *Development Theory in Transition. The Dependency Debate and Beyond: Third World Responses*, London: Zed.
Kay, C. (1989) *Latin American Theories of Development and Underdevelopment*, London and New York: Routledge.
Lewis, R.C. (1998) *Walter Rodney's Intellectual and Political Thought*, Kingston: The Press, University of the West Indies.
Shenton, R. (1975) 'Review of Walter Rodney, How Europe Underdeveloped Africa', *Canadian Journal of African Studies* 9(1): 146–50.

James Sidaway

WALT WHITMAN ROSTOW (1916–2003)

Rostow's most famous book, *The Stages of Economic Growth*, is subtitled *A Non-Communist Manifesto* – an appropriate motto for both his academic and political works. Among the 'pioneers in development' of the 1950s with his theory of economic 'take-off', Rostow was by far the most influential because he saw his academic work as a political mission and under Presidents Kennedy and Johnson he held high government posts. That gave him direct influence on US policy towards the Third World and also led him to become one of the most controversial political figures of the 1960s.

Rostow was born in New York on 7 October 1916 to a Russian immigrant father. He graduated from high school in New Haven, Connecticut in 1932, aged fifteen, and was awarded a scholarship to study economics and history at Yale. In 1936 the highly gifted Rostow won a Rhodes Scholarship and experienced his first stay in Europe at Oxford from 1936 to 1938. Following his return, he wrote his PhD thesis on 'British Trade Fluctuations, 1868–1896', graduating from Yale in 1940. Having worked briefly as an economics instructor at Columbia University, Rostow was called up for military service when the USA entered the Second World War, serving from 1941 to 1945 in the Office of Strategic Services (the forerunner of the CIA) in London, analysing aerial photographs for the planning of strategic bombing. In 1945/6, he became the Harmsworth Professor of American History at Oxford, but decided to return to political work in 1947. He became Assistant to **Gunnar Myrdal**, head of the UN Economic Commission for Europe, based in Geneva. Yet not even the highly prestigious UN job could hold him for long. He resigned in 1949 and became Visiting Pitt Professor of American History at Cambridge (England). It was not until 1950, aged only thirty-three, when he was appointed to the chair of economic history at the Massachusetts Institute of Technology (MIT), a position which he held until 1961, that he more or less settled down.

For Rostow, these were the decisive academic years, marked by the founding of the Center of International Studies (CENIS) at MIT in 1951. Its head was Max Millikan, whom he knew from his Yale days. With the Korean War raging, CENIS was to develop strategies against the spread of Communism. The circle of colleagues in Cambridge and Boston, where the MIT and Harvard University are separated only by the Charles River, included such highly acclaimed people as Paul Samuelson, Robert Solow, Paul Rosenstein-Rodan, Everett Hagen, **Charles Kindleberger**, Benjamin Higgins, Wilfred Malenbaum, Lucian Pye, Robert Baldwin, Richard Eckaus and Daniel Lerner. They constituted an illustrious collection of the pioneers in development – economists, sociologists and political scientists, all of whom would deserve inclusion in a Who's Who of development theory. At CENIS, he collaborated in two projects on the Soviet Union and China, which he summed up in 1955 in the essay, 'Marx Was a City Boy, or Why Communism May Fail'. Even more important was *A Proposal* (Rostow and Millikan 1957), which presented the perception that development policy can be a political instrument in the East–West conflict. From 1956 to 1958, Rostow did the groundwork for *The Stages of Economic Growth* (1960), which is based on a series of lectures he gave in Cambridge in late 1958.

The third stage of Rostow's life began in 1960 with his appointment to John F. Kennedy's election campaign team. He took leave of absence from MIT, and after Kennedy's election was named as Deputy Special Assistant to the President for National Security Affairs in January 1961. The following autumn he became Chairman of the Policy Planning Council at the State Department. Together with Paul Schlesinger, John K. Galbraith and his brother Eugene, Rostow formed the Charles River clique, Kennedy's liberal brains trust. During this time, Rostow wrote many memoranda in which development policy was conceived as a new field of US foreign policy. After Kennedy's assassination in November 1963, Rostow continued to work for Lyndon B. Johnson. In May 1964 he became the US member of the Inter-American Committee on the Alliance for Progress with the rank of ambassador. At the beginning of 1966, Rostow returned to the White House as National Security Adviser, and held this office until January 1969, i.e. at the height of the Vietnam War. Nixon's election as President ended Rostow's political career, which had brought him increasing criticism from the liberal public because he was seen as one of the

Administration officials mainly responsible for the escalation of the war. Rostow's old faculty at MIT refused to let him return to his chair, and an attempt to win an appointment at Harvard also failed. But in February 1969, he was appointed to the specially created chair of Jr. Professor Emeritus for Political Economy at the Lyndon B. Johnson School of Public Affairs at the University of Texas, in Austin, where he ended the fourth and now purely academic stage of his life. He dropped out of the public eye and wrote many academic books. He died in Austin on 13 February 2003, aged eighty-six.

Rostow was a convinced liberal with a missionary-like zeal which was expressed equally in his commitment to development and his anti-communism. Economic growth and the modernisation of society, according to the CENIS theory, were to prevent the spread of communism. From his studies of China and the Soviet Union, Rostow was convinced that, because Marxist theory neglected agriculture, it could not master the problem of development. Therefore he argued that force of arms should be used to assert what he believed to be sensible: the stages of economic growth that he had outlined. Thus his work revealed many parallels with that of **Marx**, against whom he fought so bitterly. These related not only to his claim to have formulated a universally historic counter-concept to the Communist Manifesto, but also to his derivation from it of his demand for political action.

However, Rostow is distinguished from Marx by one thing: whereas Marx never possessed personal political power, only supplying the concepts for his successors, Rostow did have power through direct access to two US Presidents. That applied both to the period of the 'Pioneers in Development' and to the Kennedy years. Rostow succumbed to the hubris of power; he believed he could not only define the world but also change it. What influenced him? Initially, it was surely the liberal minds of his parents and the anti-communism of his Russian immigrant family. In addition, there was the academic influence of Oxford and Cambridge, Keynesianism, the influence of the German Historical School, and finally the context of MIT and Harvard with their unique group of luminaries.

His 'Stages' theory was so influential because of its simplicity. Like Marx, he distinguished five stages through which all countries have to pass. These are: (l) the traditional society; (2) the preconditions for take-off; (3) the take-off; (4) the drive to maturity (self-sustained growth); and (5) the age of high mass consumption. The most important are stages 2–4, because they mark the transition from traditional to modern society (Rostow 1960). Rostow claimed that, in emphasising that economic change was the result of human will, he was not formulating merely a theory of growth but one of societal development in general.

The building blocks in Rostow's theory are the modernisation-promoting use of science and technology, the sharp increase in the savings and investment rates in order to achieve continual growth, the role of the innovative entrepreneur and the concept of key sectors. This is a crude synopsis of elements which can be found among prominent contemporaries: big push (Rosenstein-Rodan), spurt (**Gerschenkron**), linkage concept and key sectors (**Hirschman**), the role of the entrepreneur (Schumpeter), and stages theory (Fourastie).

Whether it constitutes a theory at all rather than merely a taxonomy of economic stages is one of the fundamental criticisms levelled at Rostow. It is also said that, notwithstanding his own claim, he took too little account of social and political factors. However, this objection is wrong if one looks at other works in which he emphasises the decisive role of new societal forces, whereby nationalist sentiment *vis-à-vis* more advanced countries is recognised as a motivating force for modernisation. Rostow is also criticised for his fixation on the Anglo-Saxon path, which ignores the top-down modernisation implemented in countries such as Germany, Japan and the Soviet Union. In empirical terms, much does not tally; for instance, no take-off can be verified in France or Austria-Hungary. Even Rostow's description of the USA as a model case is only partly right because, despite its mass consumption, it is not a welfare state. The relative decline of countries such as Britain can certainly not be explained at all, and countries such as Turkey, Argentina and India have never emerged from their alleged take-off stage. In that respect, Rostow suffers the fate of many global theorists because there are always many objections in respect of empirical details. Like other representatives of the Historical School, neo-classical authors deny that his work has the character of theory because the theoretical stringency of deductively obtained models can never be achieved in an inductive way.

The highpoint of criticism was reached at the Constance Conference of the International Economic Association in 1960, which Rostow edited in 1963 under the title *The Economics of Take-off into Sustained Growth*. The main critics were Kuznets and Solow. The former found fault with the lack of an empirical basis and the latter questioned whether Rostow's work had the character of theory at all. It was not until 1978, in *The World Economy*, that he provided the empirical proof for his stages theory. He originally intended his *How It All Began* (1975) to be the introduction to that. Basically, his enormous final research programme was nothing other than an attempt to put straight his earlier critics.

The controversy over take-off marked the core of the problem. Only if the existence of take-off could be proved, did it make sense to research the preconditions, and only then was the theory correct that the automatism of self-sustaining growth followed take-off. Also, only then did the normative conclusion make sense that one could orchestrate this connection and promote it externally. Less noticed but much more important were the political consequences of Rostow's stage theory in reformulating American foreign policy which, against the background of the Korean War, was fixated on the military dimension of the East–West conflict. Rostow believed that the USA should assume the leadership in a new international partnership programme for world economic growth. He aimed to broaden the understanding of Containment and make clear that the East–West conflict also had to be conducted in the countries of Asia, the Middle East, Africa and Latin America because they were in the second stage of the Rostowian model, where the preconditions for take-off are laid. If the preconditions were not in place, the economy would come to crisis. The communists then would have the chance to take over power, as had already happened in China and was in the offing in Vietnam and elsewhere. Therefore the process had to be supported from outside – by means of development assistance. Rostow calculated how much capital would be required to bring the investment rates of the countries involved up to the critical level.

The partners in the assistance programmes would be the new elites in the countries mentioned, whose nationalistic strivings are expressed in the wish for economic and social modernisation. At least in the initial phase, a strong state component would be essential; if necessary, the military could certainly become a suitable partner. This is very much how things were actually handled from the 1960s onwards, with support given to every autocratic regime provided that it was robustly anti-communist and promised development. The Alliance for Progress was to contribute to 'political maturity'.

Previously, Rostow had believed it necessary to convince the US foreign policy elite that development policy was definitely in their national security interest. The Republicans criticised Rostow vehemently for his programme: in fact he had to justify himself to Congress against charges that he was seeking to water down American security policy and placing too much emphasis on a planned economy. Conservative economists such as Peter Bauer and Milton Friedman even said that they suspected him of socialism because of his budgetary approach. However, Rostow and his fellow combatants were able to assert their views and convince Kennedy. To be sure, they were helped by Khrushchev's political offensive from the Sputnik shock in 1957 to the Cuban missile crisis of 1962, which signalled a substantial growth in Soviet power.

The establishment of development policy at the beginning of the 1960s, the announcement of the first development decade, the founding of the USAID, the Alliance for Progress, the Peace Corps and the OECD's

Development Assistance Committee, and the reorientation of the World Bank from post-war reconstruction aid to financing development – all of which meant putting the new foreign policy strategy on an operational basis – reflect Rostow's influence. But a small flaw was overlooked amid the hubris of power: despite all efforts, reality could not be shaped in the manner proposed in the Rostowian stages theory. That again links him with **Marx**. Vietnam, the chosen model, became a trauma. Because the Vietcong prevented the creation of the preconditions for take-off in South Vietnam, the war was driven ahead ever more, due not least to Rostow's advisory position under Johnson. It was not until the change to the Nixon administration that the conservative realist, Henry Kissinger, was able to stop the bustling activity of Rostow, the liberal missionary, by accepting the American defeat in Vietnam.

What is left of Rostow today? In the 1970s, he became a classic figure of controversy – not only because of his role in the Vietnam War, but also because he had dared to provide a counter-concept to Marx. He never mentioned the subjects of dependency, the world market, terms of trade or colonialism. He liked assistance from outside, the World Bank, direct foreign investment, multinationals and military advisors. But for Vietnam, he could have taken his place among the honoured school of the development pioneers, could have returned to the MIT without his reputation having the taint of warmonger and fanatical anti-communist. What have remained are his concepts: take-off, preconditions of growth, self-sustained growth, the age of mass consumption. From today's viewpoint, against the background of the major disaster of failed and rogue states, we must recognise that his demand for the political preconditions for take-off is more topical than ever. He was one of the decisive theorists able not only to dream up development policy and justify its necessity, but also use his position at the levers of power to ensure its practical introduction. Whether US security policy interests and combating communism were the crucial motive, or whether he knew how to package his developmental engagement cleverly in political terms, as realistic critics ascribe to him, is an open question. Either way, the countless Vietnamese and the counted American victims of the Vietnam War are also part of his appraisal. The title of his last autobiographical book, published posthumously, is *Concept and Controversy: Sixty Years of Taking Ideas to Market*.

Major works

Rostow, W.W. (1960) *The Stages of Economic Growth: A Non-Communist Manifesto*, Cambridge: Cambridge University Press, 2nd edn 1971.

—— (1971) *Politics and the Stages of Growth*, Cambridge: Cambridge University Press.

—— (1975) *How It All Began: Origins of the Modern Economy*, London: Methuen.
—— (1978) *The World Economy: History & Prospect*, London: Macmillan.
—— (1980) *Why the Poor Get Richer and the Rich Slow Down: Essays in the Marshallian Long Period*, London: Macmillan.
—— (1985) *Eisenhower, Kennedy and Foreign Aid*, Austin, TX: University of Texas Press.
—— (1990) *Theorists of Economic Growth from David Hume to the Present: With a Perspective to the Next Century*, New York: Oxford University Press.
—— (2003) *Concept and Controversy: Sixty Years of Taking Ideas to Market*, Austin, TX: University of Texas Press.
Rostow, W.W. and Millikan, M.F. (1957) *A Proposal: Key to an Effective Foreign Policy*, New York: Harper; repr. 1976, Westport, CT: Greenwood Press.

Further reading

Baran, P.A. and Hobsbawm, E.J. (1961) 'The Stages of Economic Growth', *Kyklos* 14(2): 234–42.
Cornwell, R. (2003) 'Walt Rostow: Vietnam War Super-hawk Advising Presidents Kennedy and Johnson', *The Independent*, 17 February, London (www.independent.co.uk).
Economist, The (2003) 'Walt Rostow: An Adviser in the Vietnam War, Died on February 13th, Aged 86', Vol. 366(8312), 22 February: 101 (www.economist.co.uk).
Fishlow, A. (1965) 'Empty Economic Stages?', *Economic Journal* 75(297): 112–25.
Foster-Carter, A. (1976) 'From Rostow to Gunder Frank: Conflicting Paradigms in the Analysis of Underdevelopment', *World Development* 4: 167–80.
Hodgson, G. (2003) 'Walt Rostow: Cold War Liberal Adviser to President Kennedy Who Backed the Disastrous US Intervention in Vietnam', *The Guardian* 17 February, London (www.guardian.co.uk).
Kindleberger, C.P. and Di Tella, G. (eds) (1982) *Economics in the Long View: Essays in Honour of Walt Whitman Rostow. Vol. 1: Models and Methodology. Vol. 2: Applications and Cases, Part I. Vol. 3: Applications and Cases, Part II.* London: Macmillan.
Meier, G.M. (ed.) (1987) *Pioneers in Development*, Second Series, New York: Oxford University Press.
Meier, G.M. and Seers, D. (eds) (1984) *Pioneers in Development*, New York: Oxford University Press.
North, D.C. (1958) 'A Note on Professor Rostow's "Take-Off" into Self-Sustained Economic Growth', *The Manchester School of Economic and Social Studies* 26: 68–75.
Ohlin, G. (1961) 'Reflections on the Rostow Doctrine', *Economic Development and Cultural Change* 9(4): 648–55.
Pearce, K.C. (2001) *Rostow, Kennedy, and the Rhetoric of Foreign Aid*, East Lansing: Michigan State University Press.
Rosowsky, H. (1965) 'The Take-off into Sustained Controversy', *Journal of Economic History* 25(2): 271–5.
Rostow, W.W. (ed.) (1963) *The Economics of Take-off into Sustained Growth: Proceedings of a Conference Held by the International Economic Association*, New York: S. Martin's Press.

Ulrich Menzel

E.F. (FRITZ) SCHUMACHER (1911–77)

Fritz Schumacher's (1973) book *Small is Beautiful: A Study of Economics as if People Mattered* was undoubtedly a landmark in encouraging a redefinition of the complex relationships between people, environment, technology and development. Schumacher challenged orthodoxies and raised issues that are just as relevant today, with present concerns about such matters as climate change, pollution and sustainability.

It had a truly global impact, with over 700,000 copies sold in numerous languages. As Peter Lewis (1974) commented in the cover notes of the paperback edition,

> A book of heart and hope and downright commonsense about the future … The basic message in this tremendously thought-provoking book is that man is pulling the earth and himself out of equilibrium by applying only one test to everything he does: money, profits and therefore giant operations. We have got to ask instead, what about the cost in human terms, in happiness, health, beauty and conserving the planet?

Over a quarter of a century after his premature death in 1977, Schumacher is still universally associated with this book and the concept of 'intermediate technology' which he popularised. Yet, both this book and his lesser-known, more theoretical second volume (Schumacher 1977), represented the culmination of a lifetime of deep thought and indefatigable commitment, that were stimulated by Schumacher's journeys and personal encounters, his wide reading and his impressive knowledge and understanding.

Ernst Friedrich (Fritz) Schumacher was born on 16 August 1911 in Bonn, Germany, the son of a charismatic and well-connected professor of economics. He followed in his father's footsteps by studying law and economics, initially at Bonn University. Then, in October 1930, he was awarded a Rhodes scholarship to study at Oxford. But even before he took up his place, at the age of eighteen and on his first visit to England, he was introduced to John Maynard Keynes, who invited the young Fritz to attend his prestigious seminars in Cambridge. England of the 1930s was not an easy place for a young German to live, and in any case he didn't like Oxford, so that he requested an additional year of his scholarship to be based at Columbia University in New York. At Columbia, he was much more settled and developed a good rapport with the outspoken professor of banking, Parker Willis, who, recognising the twenty-two year old's ability and potential, invited Fritz to give a course of lectures and seminars in 1933.

A year later he returned to Germany, but the deteriorating situation there prompted Fritz and his young wife, Muschi Petersen, to leave for London in 1937. Through his many contacts, Fritz was offered a small labourer's cottage on Lord Brand's estate at Eydon, Northamptonshire, where he lived during the wartime period, working first as a farm labourer, and then, from 1942, at the Oxford Institute of Statistics. A period spent during 1940 in an internment camp at Prees Heath, Shropshire, experiencing harsh physical and emotional conditions, had a particular impact on Schumacher's personal development. Four months after entering the camp, he was elected camp leader by his fellow inmates, and it was there that he began to appreciate the relevance of **Marx**'s ideas and the importance of understanding the political and social dimensions of economic issues. As his daughter, Barbara Wood, writes in her biography of her father,

> When he was released from Prees Heath, he came out not as a man emerging from the hard conditions that go with the deprivation of liberty, but as though he was returning from a stimulating seminar, on fire with new ideas and visions. He left Prees Heath invigorated …
>
> (Wood 1984: 113)

Schumacher became a British citizen in April 1946, and in November 1949 was offered the post of Economic Adviser to the National Coal Board (NCB), which was to be his workplace until early retirement in 1971. In the meantime, and following visits to Burma (1955) and India (1961 and 1962), in 1965 he teamed up with his NCB colleague George McRobie, to establish the Intermediate Technology Development Group (ITDG).

With his background in economics, Fritz Schumacher was well aware of the literature and understood the arguments sufficiently well to feel confident in challenging orthodox viewpoints. His early work focused on the question of unemployment and the interface between labour and technology. In pre-war Germany he was concerned about the six million unemployed, and he talked about what he called 'Fritz's World Improvement Plan', an incentive scheme for manufacturers to employ labour rather than machinery. In 1943, he wrote a pamphlet for the left-wing Fabian Society on *Export Policy and Full Employment*, and he later worked closely with Sir William Beveridge (the architect of the British welfare state), in producing the report on *Full Employment in a Free Society* (1944). These experiences, coupled with his reading of Marx and Lenin, engendered a deep concern in Schumacher for morality, and the nature of work that people did and how they felt about it. In the early 1940s, he said that he regarded Lenin as 'more exciting and illuminating than anything I know' (Wood 1984: 137). At that time, Schumacher was intolerant of religion, believing that intelligence and Christianity were incompatible.

However, his views on religion gradually changed. Undoubtedly a key landmark in his personal development was the three months he spent in Burma in 1955 as a UN-funded Economic Adviser on secondment from the NCB. He became fascinated with the Buddhist approach to economics, which makes a distinction between 'renewable' and 'non-renewable' resources. Here too there was a link with his longstanding interest in employment. In a chapter on 'Buddhist Economics' in *Small is Beautiful*, he comments:

> The very start of Buddhist economic planning would be a planning for full employment … While the materialist is mainly interested in goods, the Buddhist is mainly interested in liberation … The keynote of Buddhist economics, therefore, is simplicity and non-violence … From an economist's point of view, the marvel of the Buddhist way of life is the utter rationality of its pattern – amazingly small means leading to extraordinarily satisfactory results.
>
> (Schumacher 1974: 47)

Schumacher was also fascinated with the views of **Gandhi**, who believed that 'the poor of the world cannot be helped by mass production, only by production by the masses' (*ibid.*: 128).

The visits to Burma and India also fired his interest in technology. In the context of the latter, Schumacher believed that 'the biggest single collective decision that any country in the position of India has to take is the choice of technology' (*ibid.*: 175). Rather than unquestioningly transferring technology from rich to poor countries, he argued that what was particularly needed was *intermediate technology*, that was both superior to outdated and primitive technology and also simpler, cheaper and freer than the technology of the rich. He was sceptical of sophisticated technology, and, in particular, its dubious role in alleviating poverty: 'Can we develop a technology which really helps us to solve our problems – a technology with a human face?' (*ibid.*: 123).

At a conference of eminent economists in Cambridge in September 1964, he presented a well-argued paper on the economic motivation for intermediate technology, which caused uproar and proved to be the talking point of the entire conference.

Fritz Schumacher was also deeply concerned with the quality of human life, and the need to improve the lot of the unemployed, the poor and other marginalised elements of economies and societies. Just as technology needs to have 'a human face', so, he argued must economics demonstrate a greater awareness of the human dimension, *as if people mattered*. Schumacher suggested that,

> Economics and the standard of living can just as well be looked after by a capitalist system, moderated by a bit of planning and redistributive taxation. But culture and, generally, the quality of life, can now only be debased by such a system.
>
> (*ibid*.: 217)

He also questioned the role of economics in clarifying the meaning and nature of development in the context of both rich and poor countries, observing that, 'Economic development is something much wider and deeper than economics, let alone econometrics. Its roots lie outside the economic sphere, in education, organisation, discipline and, beyond that, in political independence and a national consciousness of self-reliance' (*ibid*.: 171).

Schumacher criticised the focus of much economic writing; 'We do not approach economics primarily from the point of view of people; we approach it from the point of view of the production of goods, and the people as a kind of afterthought' (McRobie 1981: 6). He frequently extolled the virtues of what he termed 'non-violent economics', and in 1960 wrote a much-quoted article for the Sunday newspaper, the *Observer*, in which he said,

> A way of life that ever more rapidly depletes the power of earth to sustain it and piles up ever more insoluble problems for each succeeding generation can only be called violent. It is not a way of life that one would like to see exported to countries not yet committed to it.
>
> (*Observer*, 21 August 1960)

Schumacher was also fascinated with the nature of the workplace and the work experience, and in *Small is Beautiful* he writes about 'New patterns of ownership', in which large companies might be broken down into smaller worker-friendly subsidiaries, where all workers can feel a sense of ownership and receive rewards that are commensurate with their efforts, as reflected in production and profitability. He believed that, 'people [should] have a chance to enjoy themselves while they are working, instead of working solely for their pay packet and hoping, usually forlornly, for enjoyment solely during their leisure time' (Schumacher 1974: 16–17).

Environmental issues were a lifelong passion with Schumacher. Soon after establishing the family home at Caterham, Surrey in 1950, he became a keen gardener and an active member of the Soil Association, experimenting with organic methods in his own back garden. On a broader scale, he argued that, 'In agriculture and horticulture, we can interest ourselves in the perfection of production methods which are biologically sound, build up soil fertility, and produce health, beauty and permanence. Productivity will then look after itself' (*ibid.*: 16–17). He was a keen supporter of the Soil Association and referred to its objectives regularly in his writing and lectures. In 1970 he became the Association's President.

He was also concerned about the disproportionate use of non-renewable resources by the richer countries, referring to the case of the USA, 'the 5.6% of the world population which live in the USA require something of the order of 40% of the world's primary resources to keep going' (*ibid.*: 98). Large powerful countries and organisations are also responsible for damaging the environment.

Fritz Schumacher was a brilliant orator and was regularly invited to speak on many occasions around the world, his frequent absences from the NCB being generously condoned by his boss, Lord Robens. Schumacher visited the President of Peru, then **Julius Nyerere** of Tanzania, whose Arusha Declaration on 'African Socialism' greatly impressed him. Other high-level visits were made to Zambia and South Africa, and in 1977, he gave a six-week coast-to-coast lecture tour in the USA and was invited to meet President Jimmy Carter at the White House.

Schumacher's final lecture, delivered in Caux, Switzerland on 3 September 1977, the day before he collapsed and died, was typical. Reproduced in McRobie (1981), the lecture, 'Technology for a Democratic Society', had the key message that both highly industrialised and developing countries of the world must begin to develop technologies that are more in harmony with people and the environment, and are less dependent on non-renewable resources (McRobie 1981: 1). His University of London lectures on 'Crucial Problems of Modern Living' were highly popular with students and initiated much debate. In these, he called for the raising of consciousness, or an 'inner development', so that the necessities for the world's survival could be worked out carefully and put into practice. As he later wrote, 'Everywhere people ask: "What can I actually *do*?" The answer is as simple as it is disconcerting: we can, each of us, work to put our own house in order' (Schumacher 1974: 249–50).

Schumacher's message has been carried forward in so many ways. The ITDG and Soil Association have gone from strength to strength, together with Schumacher UK ('promoting human scale sustainable development') based in Bristol, the E.F. Schumacher Society in Great Barrington,

Massachusetts (USA), Schumacher College in Devon (UK), and the New Economics Foundation in London (UK), to name but a few. The fact that these organisations exist today, over a quarter of a century after Schumacher's death, are themselves a testament to the power and relevance of his ideas. Global warming, pollution, the depletion of non-renewable resources, the alleviation of poverty, sustainable development, the nature and effects of different technologies, organic farming and, not least, the quality of life, all remain high on international and national agendas in the twenty-first century. This is surely indicative both of the significance of Schumacher's inspirational message, and also that he was definitely well ahead of his time.

Acknowledgement

The author is grateful to Andrew Scott, Policy and Programmes Director of the Intermediate Technology Development Group, for his valuable advice in researching this chapter.

Major works

Schumacher, E.F. (1973) *Small is Beautiful: A Study of Economics as if People Mattered*, London: Blond and Briggs (paperback, London: Abacus, 1974).
—— (1977) *A Guide for the Perplexed*, London: Jonathan Cape.
—— (1979) *Good Work*, London: Harper and Row.

Further reading

ITDG (2003) *Small is Working: Technology for Poverty Reduction*, Paris: UNESCO/ITDG/TVE.
Kirk, G. (ed.) (1982) *Schumacher on Energy*, London: Jonathan Cape.
Kumar, S. (ed.) (1980) *The Schumacher Lectures*, London: Blond and Briggs.
McRobie, G. (1981) *Small is Possible*, London: Jonathan Cape.
Schumacher, E.F. (1943) *Export Policy and Full Employment*, London: Fabian Society, Research Series No. 77.
Scott, A. (1996) 'Appropriate Technology: is it Ready for – and Relevant to – the Millennium?', *Appropriate Technology* 23(3): 1–4.
Smillie, I. (2000) *Mastering the Machine Revisited: Poverty, Aid and Technology*, London: Intermediate Technology Publications.
Wood, B. (1984) *Alias Papa: A Life of Fritz Schumacher*, London: Jonathan Cape.

Tony Binns

DUDLEY SEERS (1920–83)

'The world is inconveniently large to cover in one lifetime', wrote Dudley Seers in the Preface to his most important work, completed shortly before his death, *The Political Economy of Nationalism* (1983). Not a gesture of self-importance, his listing of over thirty-five countries in which he had worked was meant to underscore several unique features of his development work: the belief in a steady engagement with real situations as a basis for reliable knowledge and policy formulation; the notion that knowledge is largely influenced by one's experiences and historical context; and an acute inclination to question established truths, including his own. The fact that those countries in which he worked happened to be small or medium-sized (e.g. Colombia, Sri Lanka, with some in the early years of radical transformation including Cuba, Chile, and Portugal) made him aware of issues that were to recur throughout his life, including dependency and inequality. He was above all an independent thinker working within a structuralist framework that conformed neither to neo-classical nor to neo-Marxist paradigms.

Seers's interest in empirical data went back to his early days as a statistician at Oxford in the 1940s. His contributions in this area were substantial, including on topics such as social development indicators (1972, 1977); the articulation between political-economic problems, planning models, and national accounting systems (see his contribution to a volume in honour of one of his collaborators, **Hans W. Singer** (Seers 1976a)); and the insistence on reliable statistics on key aspects of a country's socio-economic and cultural life that could realistically inform planning 'targets for key resources and styles of consumption' (1976b: 12). Seers's interest in statistics, however, was characteristically self-reflexive; he asserted outright that numbers are artifacts and acknowledged that underlying any accounting system there lies a view of the world. As **Richard Jolly** (1992: 494) put it, it was in the areas of national planning and statistics that Seers was 'deeply radical, perhaps more constructively radical than in any other area of his work'. Today we would call Seers a 'constructivist', believing that statistics are not value neutral, since they narrate power-laden stories about the world. He urged planners to develop statistics that could serve the cause of the poor effectively.

Seers is best known for two crucial interventions in development debates: the inadequacy of neo-classical economics to non-industrialised countries, and his critique of growth as a standard of development. Beginning in the early 1960s, he published a handful of wonderfully titled papers on these topics which became among the most celebrated of the time.

These interests were influenced by his formative years at the Economic Commission for Latin America (ECLA, or CEPAL in Spanish, 1957–61). These were the years when CEPAL's founder **Raúl Prebisch**'s core–periphery model of underdevelopment was being developed to result, in the 1960s, in the famous dependency perspective. Prebisch's challenges to neo-classicism and emphasis on structural factors did not go unnoticed by Seers.

'The Limitations of the Special Case' (1963) made the argument that the economic doctrines and preoccupations developed for industrial economies were largely inapplicable to developing countries because the frame of reference was fundamentally different – indeed, these doctrines applied to a 'highly special' case (1963: 4–5); what was needed was a framework that did not render invisible the defining features of less-developed economies (e.g. entrenched land tenure systems, economic and political dependency). Less noticed in Seers's work has been his pioneering epistemological analysis. At a time when appeal to paradigms and research programmes was barely thinkable or still to come, Seers's analysis constituted a lucid statement on how economic doctrines sought hegemony and replaced one another, underscoring some of the central practices through which scientific discourses are deployed, including the use of textbooks and graduate training. He found his fellow economists conveniently unaware of this fact, save for exceptions such as his mentor, Professor Joan Robinson. Marxist economics was vulnerable to the same critique. In 'The Birth, Life, and Death of Development Economics' he added 'professional convenience and career interests' (1979: 709) among the factors conditioning the hegemony of particular economic theories.

That Seers found more in common with the Latin Americans who emphasised structural factors than with those who applied allegedly universal economic models (1981) was at the root of his critique of economic growth as goal and yardstick of development. Instead, he proposed a view aptly encapsulated in an oft-quoted passage from 'What Are We Trying to Measure?':

> The questions to ask about a country's development are therefore: What has been happening to poverty? What has been happening to unemployment? What has been happening to inequality? If all three of these have become less severe, then beyond doubt this has been a period of development for the country concerned.

Otherwise, rapid rates of growth notwithstanding, 'it would be strange to call the result "development" ' (1972: 24). With time, his proposal would influence such important transformations in development thinking as the 'redistribution with growth' approaches of the 1970s and landmark policy developments in the 1980s and 1990s such as the basic needs approach, adjustment with a human face, and human development indicators (pioneered by **Hans Singer**, **Richard Jolly**, **Hollis Chenery**, **Paul Streeten**, **Mahbub ul Haq** and **Amartya Sen**). Seers also often referred to the social, cultural and political requirements that could make development meaningful. His death cut short the development of this framework, a task that remains pertinent to this date.

During the last decade of his life, Seers's work took a significant turn in two respects: the application of his development framework to Europe, emphasising nationalism, and a radicalisation of his views on the Third World. The result was an integrated framework which (a) encompassed the entire world and focused on worldwide problems, respecting specificities; and (b) subsumed development economics into an interdisciplinary field incorporating all important dimensions of development (1977, 1979, 1983). He even suggested that dependency theory could be more pertinent to many European situations than neo-classical theories; this was the case when thinking about Europe's own peripheries (e.g. southern Europe) or underdeveloped regions within core countries, e. g. southern Italy or the Scottish highlands (Seers, Schaffer and Kiljunen 1979; De Brandt, Mándi and Seers 1980). His disappointment with internationalism – the increased interdependence of nations which he saw as equivalent to super-power domination – was accompanied by the observation that the Cold War international system had become untenable.

These were the years of the Club of Rome reports on the planet's limits to growth (e.g. Meadows *et al.* 1972), to which, Seers felt, both capitalism and socialism failed to provide answers. Adding to this the persistence of nationalism, in his last book he postulated a third path to development, based on a judicious form of nationalism and partially self-sufficient regional blocs. Self-sufficiency in basic food staples, technology and culture were the basis of any nationalist and regional development strategy. Changing the structures of demand (particularly patterns of consumption) was essential in this regard, which called for a curb on the influence of transnational corporations in these three areas. Seers's framework resembled positions advocated by radical Third World intellectuals, particularly **Julius Nyerere**'s concept of self-reliance and **Samir Amin**'s strategy of de-linking. Today we could restate Seers's thesis in terms of selective de-linking and selective re-engagement, quite different from what came to prevail in the Thatcher–Reagan–Bush neoliberal era.

Seers's anticipatory mind was also in evidence in his discussion, often in passing, of topics rarely addressed during his time, such as race, religion and culture in development. He led five interdisciplinary country missions (particularly for the International Labour Organisation's World Employment Programme), each lasting between four and six weeks and each resulting in a major conceptual and policy report. He occupied many prominent positions with academic organisations, the United Nations, and non-governmental organisations such as the Society for International Development (the most influential development NGO at the time). Institutionally, he will remain most well known as the first director of the Institute of Development Studies at Sussex (1967–72). From its inception, the Institute became the most exciting place in the developed world for young students eager to pursue enlightened careers in the development field; he also helped provide safe haven at the Institute for scholars fleeing military dictatorships from South America and Eastern Europe, who added to the vibrant ambiance of the place.

Dudley Seers had been born on 11 April 1920, 'a white, English, Christian male', as he was prone to say at the beginning of his courses to warn students about the biases in his works.

> To have been brought up in the family of an executive of General Motors, educated at a preparatory school and Rugby, then at Cambridge, and to have served as an officer of the Royal Navy [1941–1945] and in the civil service at home and abroad, is to be a lifetime captive of these institutions, even (perhaps especially) when one is reacting against the attitudes they attempt to instil.
>
> (Seers 1983: 11)

His profound awareness of the historicity of all knowledge – which today goes under the rubric of 'positionality' – was exemplary for his moment. Lest this sound as though he was a postmodernist *avant la lettre*, it should be remembered that his acute self-reflexivity went side by side with a deep commitment to some of the grand narratives of modernity, certainly egalitarianism and a humanism infused with the best traditions of critical thought. He did not shy away from proposing practicable solutions to what he saw as the pressing problems of the day; his pragmatism was informed by a realistic diagnosis of power and a distrust of grandiose development objectives common in the literature of the time (witness his ironic and uncompromising critique of the Brandt Report (1980) on North–South relations, which emerged out of Seers's analysis as a paradigmatic instance of a discourse run amok).

Seers's prescient mind could be found wanting on a number of issues; he was, after all, a child of his age. Even if he was always willing to reinvent it, he believed in development perhaps because, as Professor **Singer** (1983) put it, he was 'a practical visionary' who had 'a knack for getting his ideas and visions translated into practice'. He was not an iconoclast in the sense of, say, Ivan Illich among his more radical contemporaries; he was critical of many aspects of the Enlightenment tradition but did not call for a systematic overhaul. However, he fulfilled a tremendously important role as one of the few real *enfants terribles* – or, more precisely, critical consciences – of the development age, and he was respected by Marxist and neo-classical theorists alike, despite the fact that both were equally the target of his acerbic pen.

By the time he died on 21 March 1983, he had become an inevitable reference in the development field. He was, and remains, one of the finest minds and kindred spirits to work for the wellbeing of what was called, in what he characterised as 'the odiously euphemistic language of the UN', the Third World. Many of his works are as relevant today as they were when they appeared, with a few pertinent adaptations. The critical examination of orthodox economics ('deconstruction', in today's jargon) is as pressing today as at the time of 'The Limitations of the Special Case', given the greater hypertrophy of econometrics and the hegemony of neo-liberal approaches. Nationalisms and regional blocs capable of countering the present imperial power have, if anything, become more salient issues today; even some contemporary attempts at harnessing multiculturalism would not be inconsistent with his vision for the European case. There is much he would decry in what is happening today – the self-serving internationalism of market forces, the pernicious effect of global media on the world's cultures, and the seeming capitulation of international lending institutions to the dictates of a world run by, and for the benefit of, large transnational corporations. His development framework has by no means become part of the mainstream. This does not mean that his work has ceased to be relevant, perhaps it is now even more so; or perhaps one could say with greater pertinence that he remains, even today, a pioneer of development thought.

Acknowledgement

I would like to thank Sir Richard Jolly for his clarifying and detailed comments, and Wendy Harcourt for her constructive suggestions on an earlier draft.

Major works

De Bandt, J., Mándi, P. and Seers, D. (eds) (1980) *New Trends in European Development Studies*, New York: St Martin's Press.

Seers, D. (1964) 'The Limitations of the Special Case', In Martin, K. and Knapp, J. (eds), *The Teaching of Development Economics*, Chicago: Aldine Publishing Company, pp. 1–27 (originally published in *Oxford University Institute of Economics and Statistics Bulletin* 25, 1963).

—— (1969) 'The Meaning of Development', *International Development Review* 11(4): 1–6.
—— (1972) 'What Are We Trying to Measure?', *Journal of Development Studies* 8(3): 21–36.
—— (1976a) 'The Political Economy of National Accounting', in Cairncross, A. and Puri, M., *Employment, Income Distribution and Development Strategy: Problems of the Developing Countries. Essays in Honour of H.W. Singer*, New York: Holmes and Meier Publishers, pp. 193–209.
—— (1976b) 'A New Look at the Three World Classification', *IDS Bulletin* 7: 8–13.
—— (1977) 'The New Meaning of Development', *International Development Review* 19(3): 2–7.
—— (1979) 'The Birth, Life and Death of Development Economics', *Development and Change* 10: 707–19.
—— (1980) 'North–South: Muddling Morality and Mutuality', *Third World Quarterly* 2(4): 681-93.
—— (ed.) (1981) *Dependency Theory: A Critical Reassessment*, London: Frances Pinter.
—— (1983) *The Political Economy of Nationalism*, Oxford: Oxford University Press.
Seers, D. and Joy, L. (eds) (1971) *Development in a Divided World*, Harmondsworth: Penguin.
Seers, D., Schaffer, B. and Kiljunen, M.-L. (eds) (1979) *Underdeveloped Europe: Studies in Core–Periphery Relations*, Sussex: Harvester Press.
Seers, D. and Vasitos, C. (eds) (1980) *Integration and Unequal Development: The Experience of the EEC*, New York: St Martin's Press.

Further reading

Brandt Commission (1980) *North–South: A Programme For Survival*, London: Pan.
Jolly, R. (1992) 'Dudley Seers (1920–1983)', in Philip Arestis and Malcolm Sawyer (eds), *A Biographical Dictionary of Dissenting Economists*, Aldershot: Elgar, pp. 491–9.
Meadows, D.H., Meadows, D.L., Randers, J. and Behrens, W. (1972) *The Limits to Growth*, London: Pan.
Singer, H. (1983) 'Tribute to Dudley Seers', in *The New Internationalist: The People, The Ideas, The Action in the Fight for World Development*, 125 (July), Oxford: New Internationalist Publications (no page number).
Times, The (1983) 'Professor Dudley Seers: Leading Overseas Development Economist', *The Times* (London), 23 March: 12.
United Nations History Project, www.unhistory.org.
Ward, M. (2004) *Quantifying the World: UN Contributions to Statistics*, New York: United Nations. (Analysis of the pressures and biases underlying the UN's role in development statistics.)

Country reports and major missions

International Labour Office (1970) *Towards Full Employment: A Programme for Colombia*, Geneva: ILO (mission led by Seers).
—— (1971) *Matching Employment Opportunities and Expectations: A Programme of Action for Ceylon*, Geneva: ILO (mission led by Seers, 2 vols).

—— (1981) *First Things First: Meeting the Basic Needs of the People of Nigeria*, Geneva: ILO.
Seers, D. (1990) in Trosky, S.M. (ed.), *Contemporary Authors*, Vol. 129: 395–6, Detroit, MI: Gale Research Inc.
Seers, D., Bianchi, A., Jolly, R. and Nolff, M. (1964) *Cuba: The Economic and Social Revolution*, Chapel Hill, NC: University of North Carolina Press.

Arturo Escobar

AMARTYA KUMAR SEN (1933–)

Amartya Kumar Sen is a leading public intellectual of our time. Born in 1933 and brought up in Dhaka (now in Bangladesh), Sen's childhood was marked by encounters with key members of the Bengali intelligentsia, and by the horrors of the Great Bengal Famine and the Partition of India. Sen continues to acknowledge the presence of these events in his life and work, large parts of which have been concerned to understand how a lack of substantial freedoms leads to a lack of 'functionings', or the ability to live a different or a better life.

Sen is best known in development studies for his work on poverty and famines, and on the measurement of 'human development', but the citation for his 1998 Nobel Memorial Prize for Economic Science also salutes his work on social choice theory. In addition, he is known to readers of the *New York* and *London Reviews of Books* for well-crafted essays on environmental sustainability, religious nationalism and violence in India, and the poetic and political careers of Rabindranath Tagore. In key respects, Sen is a descendant of the 'Bengal Renaissance' which began in the early nineteenth century. His rapid rise to prominence, and the breadth and quality of his published work, call to mind forebears like Rammohun Roy and Tagore (an early teacher), along with contemporaries like Partha Chatterjee and Partha Dasgupta. Sen was a Professor at Jadavpur University by the age of twenty-three, and later taught at the Delhi and London Schools of Economics before moving in turn to Oxford, Harvard and the Mastership of his old Cambridge College, Trinity. In addition to the Nobel Prize, which he received while at Cambridge, Sen has been awarded India's highest honour, the Bharat Ratna. In 2004 he returned to teaching and research at Harvard.

Few can have contributed more to development studies than Amartya Sen. A version of his 1957 Prize Fellowship dissertation at Trinity College was published as his first book (Sen 1960). Thereafter, Sen published widely on savings and capital in developing countries, on surplus labour in peasant economies, on the relationship between farm size holdings and productivity in Indian agriculture, and on the isolation paradox and collective action problems. In the 1970s he began to produce the work for which he is now better known – on the measurement and meanings of human development, on the causes and prevention of famines, on gender issues and inequality, and on development as freedom. In addition, while it is a matter of pride for Sen that he has never advised governments, it is clear that his work has been shaped by involvements with the United Nations Industrial Development Organization (where his work with Partha Dasgupta and Stephen Marglin (1972) formed the basis of UNIDO's guidelines for project analysis), the International Labour Office (two of his key books (Sen 1975, 1981) grew out of work for the World Employment Programme of the International Labour Organization), and the United Nations Development Programme (which incorporated aspects of Sen's work on the evaluation of development into its Human Development Index and Human Development Reports).

Notwithstanding the diversity of Sen's work, which informs an extraordinary body of research now carried out by former students and colleagues, there is consistency in many of his leading propositions. The first of these has to do with the proper focus of economic and moral evaluations. Sen rejects the idea that individuals act according to a narrow specification of what might count as self-interest. Most of us are not rational fools, and issues like class position, gender, family influence and a sense of fairness affect the way we behave (Sen 1977). Sen also rejects the idea that different policies can usefully be evaluated in terms of the classic concerns of welfare economics or utilitarianism. '[M]aximizing the sum of individual utilities [pleasure, happiness, welfare] is supremely unconcerned with the interpersonal distribution of that sum. This should make it particularly unsuitable for measuring or judging inequality' (Sen 1973: 16). This is especially the case when it is combined with an insistence on Pareto optimality, or the idea that the utility (or welfare) of anyone should not be raised if it leads to a reduction in the utility (or welfare) of someone else. Such an insistence could lead to a situation where government action in favour of the poor, or those without substantive freedoms, is disabled to the extent that better-off social actors define their utilities in opposition to those of the poor.

Sen is staunchly opposed to this conclusion. In his view we are strongly marked by the accidents of history and geography: by the fact that some of us are born to wealthy families in Beverly Hills while others are born to landless labourers in Bangladesh. He thus agrees with John Rawls (1971) that a theory of justice as fairness would wish to ensure that all humans are endowed with a minimum set of primary goods (including education and an income) subject only to a prior rule that would guarantee equal personal liberties. For Sen, however, the Rawlsian Difference Principle is good, but not good enough. He argues that a strict equation cannot be drawn between primary goods and well-being because the former cannot always be converted into the latter. 'For example, a pregnant woman may have to overcome disadvantages in living comfortably and well that a man need not have, even when both have exactly the same income and other primary goods' (Sen 1995: 27).

It follows that Sen's own preference is for a model of economic evaluation which suggests that, 'Development consists of the removal of various types of unfreedoms that leave people with little opportunity of exercising their reasoned agency' (Sen 2000: xii). For Sen, real freedom is defined precisely in terms of certain human and civil rights that must be guaranteed for all. To live a good life we need to be fit and healthy, educated and exposed to different sources of information. But we must also be free to choose our own accounts of the good life and to participate in market exchange. The choices which free agents make will necessarily be influenced by the differences which constitute us as individual human beings, or which shape our personal circumstances. The conversion of personal incomes and resources into capabilities, achievements, well-being or real freedoms will be affected by personal heterogeneities, environmental diversities, variations in social climate, and differences within the family (after Sen 2000: 70–1).

Notions of similarity and difference in turn feed through to the empirical and practical insights that can be claimed for Sen's account of development as freedom. These begin with the language of 'functionings' and 'capabilities' that Sen uses to fashion his accounts of famine or gender discrimination, among other issues. 'Functionings' refer to the things that a person may value doing or being, and thus denote a freedom to achieve a certain lifestyle. 'Capabilities' refer to the sets of resources (physical, mental and social) that a person might command and which give rise to various 'functionings'.

Sen has used this approach to understand the dynamics of famines in conditions both of 'boom' (Bengal in 1943–44) and 'bust' (Wollo, Ethiopia in 1972–74). Sen maintains that famine deaths result from a precipitate decline in the entitlements of various persons and social groups to parts of the regional food supply. In the case of the Great Bengal Famine, then, the victims were to be found mainly among agricultural labourers, fishermen, transport workers, paddy huskers and others who faced a slackening demand for their services at a time when the demand for labour in urban Bengal was helping to push up the price of the main staple crop, rice. These groups suffered because their exchange entitlements to food were all of a sudden inconsistent with their basic needs or capability to survive. More recently, Sen has argued that 'no famine has ever taken place in the history of the world in a functioning democracy' (*ibid*.: 16). This claim is at one with the 'instrumental' defence that Sen offers for the virtues of freedom. If freedom consists in part, but by definition, in democratic pluralism, then that same system of governance ensures that the most basic economic freedoms are guaranteed by its major institutions, including contending political parties and a free press. As ever, the free flow of information is vital to this process, and to the process of economic evaluation that is thereby entailed.

Similar arguments are at work in Sen's accounts of gender discrimination and the evils of child labour. Sen has written movingly about the one hundred million women who are missing from the world today as a consequence of sex-selective abortion and the comparative neglect of female health and nutrition in childhood. While the precise figure might range from 60 million to perhaps 110 million missing women, the scale of this holocaust is horrible in the extreme. Sen accounts for it in terms of the capability deprivation of women, most of whom do not enjoy the same substantive freedoms as men. This manifests itself not just in unfair patterns of food sharing and health care within a household but also in terms of the lack of voice from which many females suffer.

Sen insists that such findings bear out the importance of an approach to social comparisons that is focused on capabilities not incomes. It also restates the importance of getting inside the 'household'. Sen follows many feminist scholars in insisting on the asymmetrical distribution of power and resources within the household (almost always to the disadvantage of women), and he buttresses this insistence with his continuing commitment to the freedoms of the individual human subject. This commitment is further apparent in his treatment of the child labour issue. Sen is aware that campaigning groups have drawn attention to the income gap that drives some parents to seek employment for their children. He insists, even so, that child labour is very often linked to slavery or bondage, and that it removes from a child the freedom to attend school. It is thus wrong in and of itself, as indeed are coercive population policies in a country like China. This is where the foundational nature of Sen's commitment to freedom reasserts itself.

Sen's work is widely admired for its crisp prose, common sense, and evident concern for social justice. It has also helped to change the way we think about famines (less emphasis on food availability decline), and development itself (less emphasis on aggregate economic performance, more on empowerment and basic needs). But it is not without its critics. Peter Nolan (1993) was an early and ungracious critic of Sen's account of the causation and prevention of famines, and David Keen (1994) has shown how freedom to choose often disappears amid war, famines or complex emergencies – a point that Sen largely misses. Other critics have charged that Sen has a poorly developed sense of power and politics (Corbridge

2002; Gasper 2004). Sen's liberalism inclines him to place great emphasis upon the power of ideas and reasoned debate in the promotion of public policy. While he is not unaware of the embedded nature of social and economic inequalities – far from it – Sen is sometimes inattentive to the political struggles that must be waged to change entrenched structures of power and to secure a greater freedom of choice for a wider range of people. This relates to his additive conception of 'freedom'. Sen's most recent account of development as freedom is committed to the view that an extension of freedom in one direction is or can always be linked to an extension of freedom in all other directions. He thus has no truck with the view that relatively benign authoritarian states in East Asia might have secured high and sustained rates of economic growth and development precisely because they suspended democratic freedoms in the first decades of structural transformation. This is not the commonly accepted wisdom, however, and it can reasonably be argued that Sen underestimates the power of such regimes to solve collective action problems (Wade 1990).

Two further criticisms should be noted. First, there is a growing technical literature on Sen's accounts of functionings, capabilities and entitlements (Qizilbash 1997; Giri 2000; Devereux 2001; Alkire 2002). This is partly concerned with the ways in which Sen defines an agent or personhood. If his rational actors are not fools, to what extent and how are they embedded in particular collectivities, cultures or systems of belief? Second, concern has been voiced about the determinacy of some of the policy suggestions that seem to flow from Sen's work. How is greater educational and health investment to be achieved? What institutions does Sen have in mind, and how would he mobilise support for social investment against defenders of the *status quo*?

In Paul Seabright's view, 'Development as freedom is curiously silent about the difficulty of devising mandates that work' (Seabright 2001: 43). Sen, however, has often pointed out that this job falls mainly to others. In any case, Sen's work has aided the development of famine early-warning systems, just as it has encouraged others to build more rounded indices of Human Development. More recently, too, Sen has used his Nobel Prize money to fund two Trusts in India and Bangladesh which do have to get to grips with institutional and political realities. The Pratichi India Trust is mainly concerned with tackling illiteracy in West Bengal, while the Pratichi Bangladesh Trust is concerned with gender inequalities and the empowerment of women. Sen continues to lead by example and by responding vigorously, but always carefully, to the criticisms that have been levelled against his work. He remains a towering and much-loved figure not just in development studies, but across the social sciences.

Major works

Drèze, J. and Sen, A.K. (1989) *Hunger and Public Action*, Oxford: Oxford University Press/WIDER.

—— (2002) *India: Development and Participation*, Oxford: Oxford University Press.

Sen, A.K. (1960) *Choice of Techniques*, Oxford: Blackwell.

—— (1973) *On Economic Inequality*, Oxford: Clarendon Press.

—— (1975) *Employment, Technology and Development*, Oxford: Oxford University Press/ILO.

—— (1981) *Poverty and Famines: An Essay on Entitlement and Deprivation*, Oxford: Clarendon Press.

—— (1982) *Choice, Welfare and Measurement*, Oxford: Blackwell.

—— (1984) *Resources, Values and Development*, Cambridge, MA: Harvard University Press.

—— (1989) 'Food and Freedom', *World Development* 17: 769–81.

—— (1995) *Inequality Reexamined*, Cambridge, MA: Harvard University Press.

—— (2000) *Development as Freedom*, New York: Anchor Books.

—— (2004) *Rationality and Freedom*, Cambridge, MA: The Belknap Press.

Further reading

Alkire, S. (2002) *Valuing Freedoms: Sen's Capability Approach and Poverty Reduction*, Oxford: Oxford University Press.

Atkinson, A. (1999) 'The Contributions of Amartya Sen to Welfare Economics', *Scandinavian Journal of Economics* 101: 173–90.

Corbridge, S. (2002) 'Development as Freedom: The Spaces of Amartya Sen', *Progress in Development Studies* 2: 183–217.

Dasgupta, P., Marglin, S. and Sen, A.K. (1972) *Guidelines for Project Evaluation (UNIDO)*, New York: United Nations.

Devereux, S. (2001) 'Sen's Entitlement Approach: Critiques and Counter-critiques', *Oxford Development Studies* 29: 244–63.

Drèze, J. and Sen, A.K. (1995) *India: Economic Development and Social Opportunity*, Oxford: Clarendon Press.

Drèze, J., Sen, A.K. and Hussain, A. (eds) (1995) *The Political Economy of Hunger*, Oxford: Oxford University Press/WIDER.

Gasper, D. (2004) *The Ethics of Development*, Edinburgh: Edinburgh University Press.

Giri, A.K. (2000) 'Rethinking Human Well-being. A Dialogue with Amartya Sen', *Journal of International Development* 12: 1003–18.

Keen, D. (1994) *The Benefits of Famine: A Political Economy of Famine and Relief in Southwestern Sudan, 1983–1989*, Princeton, NJ: Princeton University Press.

Nolan, P. (1993) 'The Causation and Prevention of Famines: A Critique of A.K. Sen', *Journal of Peasant Studies* 21: 1–28.

Nussbaum, M. and Sen, A.K. (eds) (1993) *The Quality of Life*, Oxford: Clarendon Press.

Qizilbash, M. (1997) 'A Weakness of the Capability Approach with Reference to Gender Justice', *Journal of International Development* 9: 251–62.

Rawls, J. (1971) *A Theory of Justice*, Cambridge, MA: Harvard University Press.

Seabright, P. (2001) 'The Road Upward: Review of Development as Freedom', *The New York Review of Books* XLVIII(5): 41–3.

Sen, A.K. (1970) 'The Impossibility of a Paretian Liberal', *Journal of Political Economy* 78 (1): 152–7.

—— (1977) 'Rational Fools: A Critique of the Behavioural Foundations of Economic Theory', *Philosophy and Public Affairs* 6: 317–44.

—— (1990) 'More Than One Hundred Million Women are Missing', *New York Review of Books,* 20 December: 61–6.

—— (1993) 'The Causation and Prevention of Famines: A Reply', *Journal of Peasant Studies* 21: 29–40.

—— (1996) 'Secularism and its Discontents'. in K. Basu and S. Subrahmanyam (eds), *Unravelling the Nation: Sectarian Conflict and India's Secular Identity*, New Delhi: Penguin, pp. 11–43.

Sen, A.K. and Williams, B. (eds) (1982) *Utilitarianism and Beyond*, Cambridge: Cambridge University Press.

Vanhuysee, P. (2000) 'On Sen's Liberal Paradox and its Reception within Political Theory and Welfare Economics', *Politics* 20: 25–31.

Wade, R. (1990) *Governing the Market: Economic Theory and the Role of Government in East Asian Industrialization*, Princeton, NJ: Princeton University Press.

Stuart Corbridge

VANDANA SHIVA (1952–)

Vandana Shiva was born on 5 November 1952 in the valley of Dehradun (India), nestled in the Himalayan mountain ranges. Her father was the Conservator of Forests and her mother a former Education Ministry official who had been displaced from her home when part of India became Pakistan in 1947. She trained as a physicist (her doctoral thesis was on 'Hidden Variables and Non-locality in Quantum Theory' from the University of Western Ontario) before shifting her focus to interdisciplinary research on science, technology and environmental policy. This shift was in part inspired by her involvement during the 1970s in the Chipko movement, a grassroots ecological movement in the Himalayan forest comprising thousands of supporters, mainly village women, coming out to protest against commercial logging abuses, often by interposing their bodies between the contractors' axes and the trees.

As a feminist-ecologist working in and out of India, Shiva draws on detailed historical knowledge of specific natural resource conflicts in her pioneering research on biodiversity and the environment. In the main, she argues that ecological and technological issues are intimately connected to social equity and human freedoms. She criticises the market-oriented, hyper-extractive approach to development – often legitimised by modern western scientific knowledge and biotechnology – for the over-exploitation of natural resources and the devaluation of indigenous knowledge on the environment. In turn, she proposes working towards a resource-prudent, sustainable, 'survival' (i.e. subsistence-oriented) economy based on technological choices informed by indigenous knowledge of ecological relations. Her best-known books include *The Violence of the Green Revolution: Third World Agriculture, Ecology and Politics* (1991), a critique of the agricultural 'revolution' beginning in the 1960s, sponsored by the United Nations Food and Agricultural Organization (FAO), to increase world food production by introducing high-yield cereal varieties; *Ecofeminism* (1993; with feminist sociologist Maria Mies), which combines feminism with environmentalism in its critique of the twin domination of women and nature in both the North and the South; *Biopiracy: The Plunder of Nature and Knowledge* (1997) about the way intellectual property rights on life forms represent a theft of biodiversity and indigenous knowledge; *Stolen Harvest: The Hijacking of the Global Food Supply* (2000) on the decrease in food security as a result of agricultural trade liberalisation; and *Water Wars: Privatization, Pollution and Profit* (2002) about the environmental conflict arising from the destruction of water systems as a result of development projects such as dam-building.

For Shiva, there can be no separation between intellectual work and activism. Inasmuch as her books are primarily works of advocacy, she is also an outspoken environmental activist, playing a leading role in international movements against corporate globalisation, which she sees as eroding people's basic economic rights to determining their own living conditions, and advocating instead the preservation and extension of indigenous agricultural and environmental knowledge. From **Mohandas Gandhi**, she draws on and applies the idea of *Satyagraha*, or non-violent non-co-operation to defend the people's freedoms to have access to water, seed, food and medicine. She is founder of the Research Foundation for Science, Technology and Ecology, an independent institute in India dedicated to research on ecological and social issues, as well as Navdanya, a grassroots conservation movement aimed at protecting biodiversity, particularly native seeds. Her work on conservation, ecology and the environment has reaped her many awards, including the Right Livelihood Award (known also as the Alternative Nobel Prize) in 1993, for 'pioneering insights into the social and environmental costs of the dominant development process, and her ability to work with and for local people and communities in the articulation and implementation of alternatives'. In the late 1990s, she initiated the international movement, Diverse Women for Diversity, which acknowledges the role of women from less-developed countries as seed conservators and experts in the use of medicinal plants.

An important concern in Shiva's work lies with ensuring food security, sustainable agriculture and the conservation of resources such as forests and water. Her forceful critique of the Green Revolution (mainly in the context of the Punjab, India) is well known: while the Revolution has created an infrastructure of agricultural research and development and impressive yields of grain on limited land, the new strains of plants require large quantities of fertilisers, pesticides and water (justifying massive dam-building programmes with destructive ecological consequences). The adverse effects include the evolution of pesticide-resistant pests, the creation of crop dependency and arable monoculture vulnerable to disease, the destruction of fish stocks by pesticides, a decline in food production due to soil destruction, and the encouragement of agricultural costs beyond the means of many small, independent farmers. By encouraging self-reliance and individualism, the Green Revolution has also diminished the sense of community responsibility over common resources, and instead aggravated conflict and violence – including religious strife in the case of the Punjab – throughout society.

More generally, Shiva argues that market-driven development in agriculture or forestry, spurred on by trade and investment liberalisation, is a project of exclusion that siphons resources and knowledge of the poor in the South into the global marketplace. In the context of India, for example, she argues that modernising agriculture (still the primary livelihood for three-quarters of humanity) through biotechnological fixes and globalisation strategies has encouraged a shift in production from food to export crops and thereby reduced food security; flooded the local market with imports that have wiped out local business and diversity; and paved the way for global corporations to take over the control of food processing. Globalised food production of this nature is tantamount to what Shiva calls 'food totalitarianism', something that can be challenged only by the building of an alternative 'food democracy'. By 'food democracy', Shiva refers to a way of producing and trading food that preserves diverse seed and food varieties, allows for crop species to fulfil multiple functions, incorporates cultural needs and local autonomy, and encourages human labour to engage in diverse, complex and meaningful tasks. This requires resisting the top-down approach of western development models as well as promoting local activism (such as that organised by Navdanya) in, for example, placing native seeds out of the reach of commercial plant breeders and biotechnology corporations. In a similar vein, Shiva argues too that mining and damming activities, aquafarming and industrial agriculture have depleted water resources and stripped indigenous people of communal water rights. Accordingly, she appeals for a framework for just and sustainable water use through returning democratic control of water resources to the people.

A second related arena of Shiva's work is built around her contention that not only does biodiversity have intrinsic value, but control over ecosystems should be vested in the local communities in the South who have used and protected them, and should not be hijacked by large transnational corporations driven primarily by commercial motivations. Shiva uses the term 'biopiracy' to describe what she sees as the theft of biodiversity and indigenous knowledge by the extending global reach of US-style patent laws – often spurred on by giant US companies – which encourages life patenting (i.e. granting patents on life forms). She considers patents, and more generally intellectual property rights on life forms (as propelled by the Trade-Related Aspects of Intellectual Property Rights (TRIPs) under the General Agreement on Tariffs and Trade (GATT), precursor to the World Trade Organisation (WTO)) as a tool for imperialistic control, leading to the privatisation of knowledge (i.e. knowledge cannot be transmitted without permission and licence fee collection). Based on a highly restricted concept of innovation and exclusive Eurocentric notions of property, TRIPs favour transnational corporations which appropriate ownership of genetically engineered substances through private property patents, to the detriment of the citizens, peasants, and indigenous people of the less developed world who find themselves being treated as a 'market' for 'products' developed from the very seeds which were taken away from them. Furthermore, seed-saving among farmers has become defined as intellectual property theft, while exchanging seeds among neighbours – what was previously called brown-bagging, a social act of exchange for non-profit activity – has become an infringement because 'distribution' is now covered by a patent regime. As medicine can now be patented, drugs made by Indian companies as 'generic drugs' (many times cheaper than the same retrovirals made in the US as a result of patenting) are labelled by the US pharmaceutical industry as 'piracy drugs'. In short, Shiva argues that intellectual property rights on life forms imposed by the US are leading to the 'enclosure of biodiversity, of life itself' (e.g. seeds, genes, medicines) and this has become the means by which South–North wealth transfer is taking place. In campaigning against these 'rights', Shiva insists on the freedom to 'recognize that things like seeds should be accessible to farmers, things like medicine should be accessible to those who are dying of AIDS, and no regime in the world can put profit above people's lives'. Shiva goes further to argue that as biodiversity is being converted from a local common resource into an enclosed private property, transnational companies gain a monopoly on the medical and agricultural uses of biodiversity while local knowledge systems about the environment are systematically devalued and erased, and local rights to the use of environmental resources displaced.

A third major thread which runs through Shiva's work is her abiding concern that women bear the brunt of globalised capitalist expansion, including the environmental degradation that results from such expansion. Women's productive and reproductive lives are violated by development projects that erase biological and cultural diversity in favour of export-oriented monoculture, by deforestation and toxic waste disposal that destroy the environment, and by population control strategies that target their bodies for surveillance and regulation. Together with Maria Mies in *Ecofeminism*, Shiva identifies the confluence of global capitalism, western narratives of development and patriarchy (what they call 'the capitalist patriarchal world system') as the source of oppression. They argue instead for an ecofeminist perspective which rejects capitalist development based on high technology, mass consumerism and unrestrained economic growth in favour of cultivating locally rooted, culturally diverse, self-reliant and life-affirming communities that give priority to subsistence. Shiva takes the view that women are inherently closer to nature (what she calls 'the feminine principle') than men and has written in a range of places on women's approach to forest use, food production and water conservation as providing a non-violent, inclusive alternative to more exploitative, 'male' models of development dependent on technological fixes.

In her critique of the fundamental flaws in capitalist development projects and technological fixes leading to the degradation of the environment and livelihoods, in championing local subsistence and indigenous knowledge, and in challenging 'business as usual' in both international and local organisations, Shiva is often acknowledged as a powerful, morally compelling voice for the people of the South. But her critics, including Bina Agarwal and Meera Nanda, have pointed out that her vision is compromised by a tendency to provide very simple answers to complex problems. They argue that the world's environmental crisis cannot be solely attributed to the rapacious North and its elite allies in the South, and that strategies based on local autonomy and indigenous systems of resource use alone may not meet the needs of demographic expansion or necessarily ensure the preservation of biodiversity. It is also not clear that subsistence activities are inherently as emancipatory as she claims. Her environmental philosophy has also been perceived as 'Arcadian' while her views on women's relationship with nature are over-romanticised. Others have questioned the political feasibility of her project as it is flawed by the assumption implicit in her work that 'women', or the 'poor', can be treated as a unitary entity with sufficient shared experience so as to be able to crystallise grassroots resistance as a collectivity. The tendency in Shiva's work is to ascribe to the subject a singular and ahistorical consciousness and to ignore the social heterogeneity and political tensions embedded in local societies. In eschewing questions of meaning and representation, Shiva runs the danger of constructing an 'authentic' or 'heroic' subaltern, and proposing a development project which may be inadequate to the complex tasks of transformation called for.

Major works

Mies, M. and Shiva, V. (1993) *Ecofeminism*, New Delhi: Kali for Women.
Shiva, V. (1988) *Staying Alive: Women, Ecology and Survival in India*, New Delhi: Kali for Women.
—— (1991) *Ecology and the Politics of Survival: Conflicts over Natural Resources in India*, New Delhi: Sage Publications.
—— (1991) *The Violence of the Green Revolution: Third World Agriculture, Ecology and Politics*, Penang: Third World Network.
—— (1993) *Monocultures of the Mind: Perspectives on Biodiversity and Biotechnology*, Penang: Third World Network.
—— (1997) *Biopiracy: The Plunder of Nature and Knowledge*, Cambridge, MA: South End Press.
—— (2000) *Stolen Harvest: The Hijacking of the Global Food Supply*, Cambridge, MA: South End Press.
—— (2001) *Protect or Plunder?: Understanding Intellectual Property Rights*, London, New York: Zed Books.
—— (2002) *Water Wars: Privatization, Pollution and Profit*, Cambridge, MA: South End Press.

Further reading

Agarwal, B. (1994) *A Field of One's Own*, Cambridge: Cambridge University Press.
BBC Reith Lectures (2000) 'Poverty and Globalisation', Interview with Vandana Shiva, http://news.bbc.co.uk/hi/english/static/events/reith_2000/lecture5.stm.
DeGregori, T.R., 'Shiva the Destroyer?', http://www.butterfliesandwheels.com/articleprint.php?num=17 (based on DeGregori, T.R. (2003) *Origins of the Organic Agriculture Debate*, Oxford: Blackwell).
In Motion Magazine (2004), interview with Vandana Shiva, 'The Role of Patents in the Rise of Globalization', 28 March, http://www.inmotionmagazine.com/global/vshiva4_int.html.
Jewitt, S. (2002) *Environment, Knowledge and Gender: Local Development in India's Jharkand*, Aldershot: Ashgate.
Malhotra, P. (2001) 'Vandana Shiva: The Paradigm Warrior in Pursuit of Environmental Justice', *The Financial Express* (Bombay, India), 4 June, http://www/commondreams.org/view01/0606-04.htm.
Nanda, M. (1991) 'Is Modern Science a Western Patriarchal Myth? A Critique of the Populist Orthodoxy', *South Asian Bulletin* 11(1&2): 32–61.
—— (1997) 'History is What Hurts: A Materialist Feminist Perspective on the Green Revolution and its Ecofeminist Critics', in Rosemary Hennessy and Chrys Ingraham (eds), *Materialist Feminism: A Reader in Class, Difference and Women's Lives*, London: Routledge, pp. 364–94.
Research Foundation for Science, Technology and Ecology (n.d.) Short Curriculum Vitae of Dr. Vandana Shiva, http://www.vshiva.net/vs_cv.htm.
Van Gelder, S.R. (2003) 'Earth Democracy – An Interview with Vandana Shiva', *Yes Magazine*, Winter, http://www.futurenet.org/article.asp?id=570.

Brenda S.A. Yeoh

HANS WOLFGANG SINGER (1910–)

Hans (now Sir Hans) Singer is one of the best-known and most respected development economists. His work has been recognised in awards, honorary doctorates, and a knighthood in 1994 'for services to economic issues'. He was included as one of the ten 'pioneers in development' in a World Bank book published in 1984 (Meier and Seers 1984). Six *Festschriften* have been published in his honour (Cairncross and Puri 1976; Clay and Shaw 1987; Chen and Sapsford 1997; Sapsford and Chen 1998; Sapsford and Chen 1999; Hatti and Tandon 2004). They show the depth and breadth of Singer's influence in development economics, and the esteem and affection in which he is held. Singer has produced 450 publications since his first paper appeared in 1935 (A complete list appears in Shaw 2002 and 2004). He has addressed more issues than any other development economist. The downside is that he has dispersed his efforts so widely that he has not produced the one definitive work that he carries in him.

Singer was born on 29 November 1910 in Elberfeld, now part of Wuppertal, in the German Rhineland into a strongly assimilated, largely secular, middle-class Jewish family. Two personalities in particular helped to shape his outlook on life; his father, a hard-working doctor with a large practice, who often treated the poor without charge, and the local rabbi, a highly respected member of society who gave him instructions in ethics, moral values and civic responsibility.

Entering Bonn University in 1929 with the intention of studying medicine, Singer switched to economics after attending a lecture by Joseph Schumpeter and came under his spell, and that of his economic masterpiece, *The Theory of Economic Development.* Singer was forced to leave Nazi Germany in 1933. Fortunately, on the recommendation of Schumpeter, he received a two-year scholarship (1934–6) at Cambridge University to complete his PhD work. At Cambridge, he came under another spell, that of John Maynard Keynes, precisely at the time that Keynes was producing his economic masterpiece, *The General Theory of Employment, Interest and Money*. Singer's PhD dissertation was entitled 'Materials for the Study of Urban Ground Rent', which covered the period 1845 to 1913. This showed his aptitude for analysing long-term data series. He was among the first group of students to receive doctorates in economics from Cambridge University.

Singer's first employment after university was on a major two-year (1936–8) study of long-term unemployment in the depressed areas of the United Kingdom organised by the Pilgrim Trust under the chairmanship of Archbishop William Temple, with Sir William (later Lord) Beveridge as its main adviser (The Pilgrim Trust 1938). These two personalities and the study made a strong impression on him. There followed lectureships at Manchester (1938–44) and Glasgow (1946–7) universities. During brief employment at the Ministry of Town and Country Planning in London (1945–46), Singer was involved in the calculation of compensation for owners as part of the Labour Party's intended programme for the nationalisation of development rights in urban land, drawing on his PhD work at Cambridge.

In 1947, there occurred an event that was to dramatically change the course of Singer's career. David Owen, who had worked with Singer on the unemployment study, was appointed the first head of the department of economic affairs at the newly created United Nations. Owen sought Singer's services to strengthen his new department. Singer accepted a two-year secondment, which was to last for twenty-two years (1947–69). He was assigned to work in a small section that dealt with the problems of developing countries. Simultaneously, he was appointed a professor in the graduate faculty of the New School for Social Research in New York. He welcomed the opportunity of maintaining his links with the academic world. It also gave him the opportunity to think about the theoretical framework of economic development within which his work at the UN could be placed. And he was able to publish papers in his own name, a practice not allowed at the UN, which were widely acknowledged and quoted.

Analysis of the long-run terms of trade between industrialised and developing countries is perhaps Singer's best-known contribution to development economics (Singer 1949 and 1950; UN 1949). He showed that the net barter terms of trade between primary products and manufactures were subjected to a long-term downward trend. This contradicted the belief widely held at the time that the long-term trend favoured primary products. Singer argued that economic history had been unkind to developing countries. Most of the secondary and cumulative effects of investment had been removed from the developing country to the investing, industrial, country. Developing countries had also been diverted into types of activities that offered less scope for technical progress, withholding a central factor of 'dynamic radiation' that had revolutionised society in the industrial world.

Singer emphasised that his work on the terms of trade was meant to be more a *policy guide* than a long-term projection. He advised developing countries to diversify out of primary exports through developing domestic markets and industrialisation. Subsequently, he put more emphasis on relations between types of countries than between types of commodities, and on the distribution of 'technological power' in more contemporary terms in answer to some of the criticisms of his original work (**Rostow** 1990).

Singer also emphasised the objective of 'distributive' justice, pointing to in-built, long-term inequalities between industrialised and developing countries. He has constantly argued that unless there are fundamental changes in the world economic order, divergence rather than convergence will continue between them, threatening global economic and social advancement, stability and peace. Singer found strong allies, especially **Raúl Prebisch**, who used Singer's original work (Prebisch 1950; Toye and Toye 2003). The Prebisch–Singer thesis, as it came to be known, has created a growth industry in the development economics literature. Increasingly sophisticated statistical and econometric analyses have vindicated the thesis, making it one of the very few hypotheses in economics that have stood the test of time (Sapsford and Chen 1998: 27–34).

What are some of the special attributes that have set Singer apart from other pioneering development economists? Singer came under the direct influence of *both* Schumpeter and Keynes, the two great beacons in his intellectual development. The combination of these forces has had a decisive influence on Singer's ideas: Schumpeter with his views on the importance of technology and innovation and the role of the entrepreneur; and Keynes' thesis that economics is not a universal truth applicable to all countries and conditions. They provided Singer with an enabling framework for much of his work.

Work on the terms of trade has been the fulcrum for many of the other issues Singer has taken up in order to achieve 'distributive justice' for the developing countries including: industrialisation, science and technology, human investment, planning, trade and aid, technical assistance and pre-investment activities, regional co-operation, and a new international economic order with multilateral global governance.

Long service in the United Nations during its formative years was also to benefit Singer's work. He carried out a wide range of assignments that enabled him to see at first hand conditions in the developing world, discuss with leaders and planners their development problems and aspirations, and set his own theoretical and conceptual framework against the background of concrete reality. Out of this milieu emerged Singer's 'perspectives' on development including: the mechanics of development, the role of the public sector in economic growth, a balanced view of balanced growth, the concept of pre-investment financing, the notion of human investment, the interaction of population growth and education with economic growth, his concept of 'redistribution from growth' to tackle poverty, and his view that development is 'growth *plus* change', culturally, socially and economically.

Singer was also one of the founders of the structural analysis of development. He has maintained that there can be no 'blueprint' for development (Schiavo-Campo and Singer 1970). However, a number of themes have permeated his work. He considers that the starting point should be people, not money and wealth, which gave a whole new perspective on the development process. Sustained and equitable development depended not on the creation of wealth but on the capacity of people to create wealth. Hence, Singer's insistence on the importance of the human factor in economic development, and on health, education and training, the well-being of children, and food security, science and appropriate technology, employment, income distribution and the conquest of poverty, and planning and sound institutions, all viewed in an international context in which trade and aid are conducted with distributive justice and efficiency so that all countries, developing and developed, might flourish and converge, not diverge. Many of these perspectives have since become conventional wisdoms of progressive development work.

Not being content only to formulate the theoretical underpinnings of the problems of developing countries, Singer became involved in many pioneering ventures to overcome them. A number of the organisations he helped to set up or strengthen in order to tackle Third World problems are still operating today including: the United Nations Development Programme, the United Nations Children's Fund, the UN World Food Programme, the UN Research Institute for Social Development, the UN Industrial Development Organization, and the African Development Bank. He was also indirectly involved in the creation of the International Development Association (IDA), the soft-loan arm of the World Bank. The IDA was established in part as a foil to prevent attempts to set up a Special UN Fund for Economic Development, which Singer and others tried to establish in the United Nations (Meier and Seers 1984: 296–303).

Since 1969, Singer has been a professorial fellow of the Institute of Development Studies (IDS), and professor (now emeritus), at the University of Sussex, UK. During this period his output has been prolific. He has made a significant contribution to the activities of IDS, which is now recognised as one of the world's leading institutes of development studies. He is remembered with respect and affection as a source of unlimited help and inspiration by the cohorts of students. And he has been in constant demand by governments, multilateral and bilateral aid agencies and non-governmental organisations. Singer's commitment to the case for a more just and equitable world economic order remains undiminished. His life and work serve as an inspiration for the next generation of development economists.

Major works

The following is a representative selection of Singer's many publications:

Ansari, J.A., Ballance, R.H. and Singer, H.W. (1982) *The International Economy and Industrial Development: Trade and Investment in the Third World*, Brighton: Harvester Press.

Ansari, J.A. and Singer, H.W. (1977) *Rich and Poor Countries: Consequences of International Economic Disorder*, London and Winchester, MA: Unwin Hyman.

Jolly, R. and Singer, H.W. (1972) *Employment, Incomes and Equality: A Strategy for Increasing Productive Employment in Kenya*, Geneva: International Labour Office.

Raffer, K. and Singer, H.W. (1996) *The Foreign Aid Business: Economic Assistance and Development Co-operation*, Cheltenham: Edward Elgar.

—— (2001) *The Economic North–South Divide. Six Decades of Unequal Development*, Cheltenham and Northampton, MA: Edward Elgar.

Roy, S. and Singer, H.W. (1993) *Economic Progress and Prospects in the Third World: Lessons of Development Experience Since 1945*, Aldershot and Brookfield, VT: Edward Elgar.

Schiavo-Campo, S. and Singer, H.W. (1970) *Perspectives in Economic Development*, Boston: Houghton Mifflin Co.

Singer, H.W. (1964) *International Development, Growth and Change*, New York: McGraw-Hill.

—— (1975) *The Strategy of International Development: Essays in the Economics of Backwardness by H.W. Singer*, in Cairncross, A. and Puri, M. (eds), Basingstoke and London: Macmillan.

—— (1998) *Growth, Development and Trade: Selected Essays of Hans Singer*, Cheltenham: Edward Elgar.

—— (2001) *International Development Co-operation. Selected Essays by H.W. Singer on Aid and the United Nations System*, in D.J. Shaw (ed), Basingstoke and New York: Palgrave Macmillan.

Singer, H.W., Wood, J. and Jennings, T. (1987) *Food Aid. The Challenge and the Opportunity*, Oxford: Clarendon Press.

Further reading

Cairncross, A. and Puri, M. (eds) (1976) *Employment, Income Distribution and Development Strategy: Problems of Developing Countries. Essays in Honour of H.W. Singer*, Basingstoke and London: Macmillan.

Chen, J. and Sapsford, D. (eds) (1997) 'Economic Development and Policy: Professor Sir Hans Singer's Contribution to Development Economics', *World Development*, 25(11): 1853–956.

Clay, E. and Shaw, D.J. (eds) (1987) *Poverty, Development and Food: Essays in Honour of H.W. Singer on his 75th Birthday*, Basingstoke and London: Macmillan.

Hatti, N. and Tandon, R. (eds) (2004) *Trade and Technology in a Globalizing World: Essays in Honour of H.W. Singer*, New Delhi: BPRC (India) Ltd.

Meier, G.M. and Seers, D. (eds) (1984) *Pioneers in Development*, New York: Published for the World Bank by Oxford University Press, pp. 273–311.

Prebisch, R. (1950) *The Economic Development of Latin America and its Principal Problems*, Santiago, Chile: United Nations Commission for Latin America.

Rostow, W.W. (1990) *Theorists of Economic Growth from David Hume to the Present with a Perspective on the Next Century*, New York: Oxford University Press.

Sapsford, D. and Chen, J. (eds) (1998) *Development Economics and Policy: The Conference Volume to Celebrate the 85th Birthday of Professor Sir Hans Singer*, Basingstoke and London: Macmillan.
—— (1999) 'The Prebisch–Singer Thesis: A Thesis for the New Millennium', *Journal of International Development* 11(6): 863–916.
Shaw, D.J. (2002 and 2004) *Sir Hans Singer: The Life and Work of a Development Economist*, Basingstoke and New York: Palgrave Macmillan (hardback); New Delhi: BRPC (India) Ltd (paperback).
Singer, H.W. (1949) 'Economic Progress in Under-developed Countries', *Social Research* 16(1): 236–66.
—— (1950) 'The Distribution of Gains between Investing and Borrowing Countries', *American Economic Review* 40(2): 473–85.
The Pilgrim Trust (1938) *Men Without Work: A Report Made to the Pilgrim Trust*, Cambridge: Cambridge University Press. Reprinted (1968) New York: Greenwood Press.
Toye, J. and Toye, R. (2003) 'The Origin and Interpretation of the Prebisch–Singer Thesis', *History of Political Economy* 35(3): 437–67.
UN (1949) *Relative Prices of Exports and Imports for Under-Developed Countries: A Study of Post-War Terms of Trade between Under-Developed and Industrialized Countries*, 1949/II.B.3, New York: United Nations.

John Shaw

JOSEPH STIGLITZ (1943–)

Born in Gary, Indiana on 9 February 1943, Joseph Stiglitz is reputed to have been dubbed by Paul Samuelson the best economist to have originated from the declining steel town. This is, however, the home town of the first Nobel Laureate for the discipline. Stiglitz's career has been astonishing. As a graduate student he edited Samuelson's Collected Papers. At twenty-seven, he became full professor at Yale, subsequently taking appointments at Stanford, Oxford, Princeton and Columbia. He received the J.B. Clark award for outstanding economist under the age of forty. He joined Clinton's Council of Economic Advisors, serving as Chair and a member of Cabinet. In 1997, he became the World Bank's Chief Economist, promoting the Comprehensive Development Framework with President James Wolfensohn. In the putative turn away from the discredited neo-liberal Washington Consensus, he launched the post-Washington Consensus. This helped to restore legitimacy to the Bretton Woods Institutions as more state-, people- and development-friendly. In 2001, he was awarded the Nobel Prize in Economics in part for his role in founding the New Development Economics, an element in his more general information-theoretic approach. In the volume of 2003 honouring his sixtieth birthday, his list of publications runs to almost fifty pages, including multiple editions of textbooks in economics, micro-economics, macro-economics and the economics of the public sector.

Despite these astonishing achievements, Stiglitz would have remained a significant but minor figure within development studies and economics but for an event that catapulted him to prominence. He was forced to resign from the World Bank at the instigation of Larry Summers, one of his predecessors and, at the time, US Treasury Secretary of State. He insisted that Stiglitz must go if Wolfensohn were to remain in office. The reason is that, from the outset of his appointment at the World Bank, Stiglitz had used his position to criticise the IMF. This became unacceptable once it threatened policy rather than offering rhetoric, and Stiglitz had become particularly outspoken over his opposition to privatisation in the former Soviet Union and the austere macro-economic policies and financial liberalisation imposed upon South Korea following the Asian crisis.

Close examination of Stiglitz's economics reveals paradoxes. First, he is not particularly radical in policy and theory nor are his conclusions particularly original. Close attention to market imperfections leads him to seek balance between market and state, safety nets, proper sequencing and regulation of financial liberalisation in particular and other policies in general as opposed to neo-liberal shock therapy. But from the 1980s onwards, literature critical of the Washington Consensus, especially that concerned with the developmental state and adjustment with a human face, drew similar conclusions. His emphasis on market and institutional imperfections as especially characteristic of developing countries represents a partial rediscovery of the old and classic development economics of what might be termed the pre-Washington consensus of the McNamara period. Thus, Stiglitz's novelty lies less in the conclusions that he draws than in the information-theoretic approach that he uses to obtain them, and the positions from which he has been able to promote them with considerable and rare intellectual integrity given the personal consequences.

Second, in his career, he seems to have become more radical, especially after his break with the World Bank. He has set up an Initiative for Policy Dialogue to promote alternatives to what he now terms neoclassical economics, having redefined it as the idea that markets work perfectly in order to distance himself from it. His critique of the Bretton Woods Institutions, especially the IMF, has intensified. His book, *Globalization and Its Discontents*, published in 2002, mainly concerned with a critique of the IMF and his own experiences in promoting debate over his disagreements with it, is a best-seller. He has been hailed as the champion of the poor for his stances. But there is no increasing radicalisation in Stiglitz's economics apart from the loss of a naïve innocence about the power of his own ideas. This is someone who has remained over his entire career committed to

Keynesianism at the macro level and the correction of market imperfections by capable government at the micro level. It is the intellectual environment that has changed rather than his position within it, opportunity for prominence aside.

Third, then, within mainstream (development) economics, Stiglitz has been at the forefront of the turn against neo-liberalism since the 1990s. **Marx** argued of Ricardo that, for his analytics, it was sufficient to read the first two chapters of his *Principles*. The first deals with the labour theory of value as average labour time of production in industry, and the second, inconsistent with the first by defining value at the margin of cultivation, constructs a theory of rent. All the rest is application with no further theoretical advance. While in part a put-down, Marx also viewed Ricardo as the bourgeois political economist *par excellence*. This is because he applied his principles and his own version of the labour theory of value ruthlessly to whatever economic phenomena he cared to address.

What has this to do with Stiglitz? There are some striking parallels, and contrasts, to be drawn between him and (Marx's view of) Ricardo. First, Stiglitz can also be understood as relying upon two principles, to the first of which he is obsessively committed. With the second, as will be seen, it is inconsistent although it is mainly implicit and unconscious. Stiglitz understands the capitalist economy in terms of what he himself dubs a new paradigm based on imperfect information. It deploys the idea that exchanges take place between agents who are imperfectly and asymmetrically informed. For this reason, markets are imperfect, and three potentially inefficient sorts of outcome result. Markets may clear (supply equals demand) but be inefficient (some trades beneficial to both parties do not occur). This is so if potential buyers (sellers) of lower (higher) quality goods do not wish to trade at a price reflecting a higher (lower) average quality.

Second, markets may not clear but prices not adjust in favour of the short side (price increases if excess demand and decreases for excess supply). An example is efficiency-wages where, despite surplus labour, wages fall because employers anticipate a more than compensating loss in productivity, loyalty and work intensity. Third, markets can be absent, as in health insurance for the aged or sick. Any level of premium would attract an average level of health risk that is too high as the healthier choose to risk self-provision rather than pay premiums that reflect the sicklier.

For Stiglitz, these are general principles that apply to all markets, their significance depending upon the nature of the imperfections, the goods, the agents, and so forth specific to each market. His voluminous writings reflect one application after another. But a difference from Ricardo is that Stiglitz's horizons extend beyond capitalism to deal with development and the transitional economies. Further, Stiglitz also addresses the non-economic or non-market responses to market imperfections. Thus, institutions are perceived to be a, not necessarily efficient, response to market imperfections. His theory of share-cropping summarily dismisses both customary behaviour and exploitation as explanations in favour of monitoring and incentive problems in case tenants know more than landlords over production possibilities and work effort. Share-cropping is also used to argue that informational imperfections in one market, land, can have knock-on effects on others, such as availability of credit, choice of technique and capital intensity.

But other social sciences and political economy are notable for their absence from Stiglitz's work. His point of departure is the neoclassical economics of perfectly working markets, which he carefully picks to pieces from his own perspective. His own paradigm, however, displays considerable continuities with the neoclassical approach. Like Ricardo, his method is deductive, even more explicitly drawing conclusions from axioms in the form of models. These are embedded in the optimising behaviour of individuals from which micro-foundations underpin macro-outcomes in the form of equilibrium. As with Ricardo, the logic of his deductive system is used to confront chosen topics with scant regard to methodology (deductivism wedded to naïve empiricism and falsifiability) or history of economic thought (with occasional token misinterpretations of Adam Smith as anticipating general equilibrium theory and Hayek the information-theoretic approach, for example).

Stiglitz's second analytical principle is to abandon his first and appeal to social, structural or other factors than optimising individuals. The need to do so arises in three ways. First, the deductive method always depends upon the exogenous as parameters. How are these to be explained other than to push out the boundaries further only to recreate them anew? Second, the real world has a nasty habit of revealing exploitation and power, and inventiveness and culture, not reducible to a calculus of imperfectly informed optimising agents engaged in individual acts of exchange with one another. As **Marx** said of Ricardo's theory of rent, his assumption of free exchange and flow of capital onto the land would have been incomprehensible to his contemporaries anywhere else in the world. By the same token, can share-cropping be understood as free of exploitation and cultural factors, and development or lack of it the consequence of imperfections, and independent of power and associated conflicts?

Third, Stiglitz's own experience outside the academic world of neoclassical economics leads him to recognise the failure of his logic to prevail, most notably in the acknowledged rise to power of a financial system with its own compelling interests or the irrational dogma of his opponents. Significantly, these analytically residual categories leap to prominence in his more reflective and recent contributions. *Globalization and Its Discontents* has no theory, scarcely a concept, of globalisation (other than reduction in transport and communication costs) and, by contrast, the term is effectively absent from all of his earlier work. This, and the account of his period as advisor to Clinton, is entirely free of the information-theoretic approach. But, if he is ultimately drawn to recognise social power and interests, these should surely serve as analytical starting point rather than conclusion of informationally constrained optimisation.

In short, in the intellectual assault upon neo-liberalism, Stiglitz has played the leading role from within mainstream economics. He has not promoted alternatives other than his own. His command of these, methodology, history of economic thought, and the other social sciences is limited. For some, this puts him in the vanguard, for others he holds back a more radical and wide-ranging critique of neo-liberalism. Ultimately, his lasting contribution will depend upon the extent to which others build upon, and break with, his partial and distorted restoration of the classic understandings of development.

Major works

For his own 'Memoir', see http://www-1.gsb.columbia.edu/f.PDF.

Chang, H.-J. (ed.) (2001) *Joseph Stiglitz and the World Bank: The Rebel Within*, London: Anthem Press. (A collection of his essays promoting the post-Washington Consensus.)

Stiglitz, J. (1986) 'The New Development Economics', *World Development* 14(2): 257–65. (A concise and early account of his approach to economics and development.)

—— (1994) *Whither Socialism?*, Cambridge, MA: MIT Press. (For a fuller account with application to transitional economies.)

—— (2001) 'Information and the Change in the Paradigm in Economics', available at http://nobelprize.org/economics/laureates/2001/stiglitz-lecture.pdf. (His own overview of his work, provided in his Nobel Prize Lecture.)

—— (2002) *Globalization and Its Discontents*, New York: W.W. Norton and Co. (His critique of the IMF.)

—— (2003) *The Roaring Nineties: Seeds of Destruction*, New York: W.W. Norton and Co. (For a narrative of his time on the Council of Economic Advisers.)

Further reading

For an appreciative assessment, see 'Markets with Asymmetric Information', available on the Nobel website, http://www.nobel.se/economics/laureates/2001/press.html, 'Advanced Information'.

Arnott, R. *et al.* (eds) (2003) *Economics for an Imperfect World: Essays in Honor of Joseph E. Stiglitz*, Cambridge, MA: MIT Press. (Essays in honour of his sixtieth birthday, including a full bibliography of his work.)

Fine, B. (2002) 'Economics Imperialism and the New Development Economics as Kuhnian Paradigm Shift', *World Development* 30(12): 2057–70. (For critical assessments of the information-theoretic approach as paradigm.)

Fine, B. *et al.* (eds) (2001) *Development Policy in the Twenty-First Century: Beyond the Post-Washington Consensus*, London: Routledge. (For a critique of the post-Washington Consensus.)

Fine, B. and Jomo, K. (eds) (2005) *The New Development Economics: A Critical Introduction*, Delhi, Tulikar and London: Zed Books.

Wade, R. (2001) 'Showdown at the World Bank', *New Left Review* 7: 124–37. (For his 'resignation' from the World Bank.)

Ben Fine

PAUL PATRICK STREETEN (1917–)

It would be a mistake to confine Paul Streeten to a small box labelled 'development economics' and to seek to assess him within the limits of a subdiscipline when, like Keynes, **Marx** or Adam Smith himself, he is not only a total economist but a person of many disciplines who has focused all his intelligence on the great economic issues of his day. He is one of that small band, which includes his two mentors and colleagues, **Gunnar Myrdal** and Thomas Balogh, who are not only widely regarded in Europe and North America as highly sophisticated economists, albeit dissenters, but who have also thought long and hard about poorer parts of the world.

It has been observed that Streeten's style 'did not develop, it appeared like Minerva from the head of Zeus, fully grown and armed' (Stretton 1986: 1). He burst on the scene in 1949 with articles on the theories of profit and pricing. These 'dense, laconic, lucid papers' (*ibid.*: 2) were followed during the next four years by seven more (one written in German, one translated into French) on the inadequacy of the price mechanism; exchange rates and national income; the inappropriateness of simple elasticity concepts in the theory of international trade (with Balogh); reserve capacity and the kinked demand curve; the modern theory of welfare economics; the effect of taxation on risk taking; plus a translation from German (with a thoughtful appendix) of Myrdal's (1953) *Political Element in the Development of Economic Theory*. And all the while he was tutoring across the full range of economics at Balliol College, Oxford, where he had arrived as a wounded soldier-student near the end of World War II.

For the next fifty years a steady stream of articles, books and policy reports was to emerge from the head of Zeus all beautifully written, clear, erudite, witty, modest and fully armed. For Paul Streeten has an unrivalled mastery of economics. His next essay was on 'Keynes and the Classics' (1954). Then he edited Myrdal's essays and gave expression to his concern with methodology and with values in the social sciences: 'Values are not something to be discarded, nor even something to be made explicit in order to be separated from empirical matter, but are ever-present and permeate empirical analysis through and through' (Streeten 1958: xiii). And he saw the importance of Heisenberg's Uncertainty Principle, concerning the mutual impact of the observer and the observed, with all its implications (*ibid.*: xv). It is an essay, published in the *Quarterly Journal of Economics* (1954), which he regards as his most genuinely innovative contribution but which made no impact on the profession. It remains required reading. While he continued to generate new insights as his experience widened, Streeten was to remain remarkably consistent down the years in his analytical style and focus.

'My interest in social justice', he wrote in his mid-seventies, 'dates back to my childhood' (Streeten 1994: preface). It was a childhood far removed from the quiet Balliol of his early writing and teaching. For Paul Hornig, as he was born in the last days of the Austro-Hungarian Empire, grew up in the hectic society of Vienna between the wars. His father died before he was two but his mother and his aunt moved in intellectual circles, where the young Paul met and was influenced by a wide range of thinkers, including Max Adler, Karl Popper, several psychologists and the leading Utopian socialist, Otto Bauer. As a teenager he was active in the socialist youth movement, the *Roten Falken* (Red Falcons), battling fiercely with the radical right. But all this ended abruptly in March 1938, when the Nazis marched into Vienna:

> I was on several lists and the combination of being a Jew and politically active on the left would have been enough. But though enquiries were made at our old flat, I was not given away. I believe there was an SS officer who was a friend of mine (vintage Streeten!) and who had deleted my name from the list of those to be arrested. I witnessed the hysterical city on the day of the Anschluss. I walked through the streets, and saw the armoured Nazi cars cheered by the crowds. The Viennese, reputed for their *Gemütlichkeit*, revealed faces distorted by hate as they shouted hysterically 'Heil Hitler', 'Ein Volk, ein Reich, ein Führer'. It was a deplorable and frightening sight.
>
> (Streeten 1985: 47)

If Austria in the 1930s had been turbulent, the next five years were to take the young man on a dizzying roller-coaster. 'Attempts to leave were entirely haphazard' but with the help of English friends he managed to cross the Channel, where he was looked after by a group of Christians including a family in Cambridge and two sisters in Sussex, Marjorie and

Dorothy Streeten. Arriving just before May Week, 'The transition from the turbulence, hysteria, fear and ghastliness of Vienna to the peace, tranquillity, and sunshine of the Master's lodgings in a Cambridge College ... was an extraordinary experience' (*ibid.*: 48). But more was to follow. He went to Aberdeen University, where he studied Political Economy for nearly two years until he was interned for being an enemy alien, even though he had volunteered for the Royal Air Force. For the next year he was shunted from pillar to Canada where he and his companions, some of whom later became very distinguished, were held behind layers of barbed wire, towers, armed sentries and searchlights. Like the prisoners of Robben Island twenty-five years later, they started a University at which Paul overcame his deficiency in mathematics at the feet of Hermann Bondi, cosmologist and subsequently co-creator of the steady-state theory of the universe. He learned enough to be able to view with a beady eye the subsequent work of the mathematical model builders in economics. Forty years on he was to remark with his special combination of self-deprecation and rapier wit that,

> Of course, without a thorough training in mathematics, one feels nowadays like a handloom weaver in the days after the invention of the power loom. But the thought is made bearable by the fact that the power loom seems to be weaving the Emperor's clothes.
>
> (*ibid.*: 56)

After five months behind the Canadian barbed wire, this motley crew was returned to England for another few months of cold, hungry, overcrowded internment. But eventually, as a long-distance runner, he managed to attract the attention of the authorities who were looking for men able to handle the arduous hush-hush work of the Inter-Allied Commando. Here, during training, they were told to pick English names in case they were taken prisoner behind enemy lines and shot as traitors. And thus, 'in a few seconds' the young man changed from Hornig to Streeten and assumed the cover story of his adopted family. The maiden sisters were very pleased. Subsequently he was trained for the invasion of Sicily where, in the summer of 1943, he was badly wounded. It was a watershed day. '[N]o more punting, rock climbing, skiing, running' (*ibid.*: 54). Indeed he was not expected to survive but, after hospitalisation in Cairo, he arrived in Balliol late the following year to read Politics, Philosophy and Economics. The roller-coaster ride was over.

Twenty years later he was still in Oxford, teaching and writing hard. During the 1960s two things happened to determine his future direction. The critique of 'balanced growth' by Rosenstein-Rodan, Nurkse and others became a focus of debate in the profession and Streeten found that his own thinking about Europe (Streeten 1961) was relevant. Second, **Gunnar Myrdal** asked him to collaborate in the research and writing of *Asian Drama*. 'And underlying these two tributaries, there was my ... interest in the world community and my objections to the nation state and nationalism as a form of heresy' (*ibid.*). And so the die was cast.

In 1968 *Asian Drama* was published and Streeten co-edited *The Crisis of Indian Planning* (Streeten and Lipton 1968). Three years later he co-edited *Commonwealth Policy in a Global Context* (Streeten and Corbet 1971). This was followed by a small gem, *Aid to Africa* (Streeten 1972a). A generation later it is still worth reading. This was accompanied by *The Frontiers of Development Studies* 'on the problems of development in which rich and poor countries co-exist and where the presence and policies of the rich crucially affect the development efforts and prospects of the poor' (Streeten 1972b). In Oxford in 1973 (at the invitation of the later-notorious publisher, Robert Maxwell), he became founding editor of the journal, *World Development*, which was to become influential as a monthly forum for debate. 'Our goal', he wrote in what might almost be his testament, 'is to learn from one another, regardless of nation or culture, income, academic discipline, profession or ideology. We hope to set a modest example of enduring global co-operation through maintaining an international dialogue and dismantling barriers to communication.'[1] For the next twenty-five years a steady stream of articles, reviews and 'essays in biography' sprang from the editorial pen. It was not until 2003, in his mid-eighties, that he stepped down as chairman of the editorial board.

In June 1976 the International Labour Organisation held a seminal world employment conference in Geneva, where a strategy to 'satisfy the most basic needs of all people in the shortest possible time' was endorsed (ILO 1977). Shortly after this, Streeten was asked by **Mahbub ul Haq**, then director of policy planning at the World Bank, to lead the Bank's strategic thinking about 'Basic Needs'. He did a brilliant job and, in his dual role of missionary and interpreter, converted everybody within the Bank so effectively that the strategy of Basic Needs is widely perceived to be that institution's major contribution to thinking about development (Streeten 1981). But even then his questioning mind could see the unanswered questions, the ambiguities. In a paper delivered in *apartheid* South Africa in 1984, he observed that 'It is not at all clear whether the basic needs approach mobilizes the power of the poor to improve radically their situation or whether it reinforces the oppressive existing order' (Streeten 1984). But he stated pointedly that, 'the freedom to define one's needs is itself a basic need' (*ibid.*: 3).

During the 1980s, Paul Streeten (then based at Boston University) was at his most effective as scholar-statesman. After his success at the World Bank, he was asked by the ILO to be joint leader of a Mission to Tanzania to help devise a basic needs oriented development strategy for that country (ILO 1982). Coming at a moment of great crisis when President **Julius Nyerere**'s idealistic economic and social policies, with their consistent goal of slaying the postcolonial dragons of *Umaskini, Ujinga na Maradhi*,[3] seemed to have come unstuck, the report was a masterly document. And Streeten's fingerprints were all over it. While expressing deep empathy for the vision of one of Africa's great leaders, the report examined, unflinchingly, what had gone wrong and what could be done to put it right. From what happened subsequently in Tanzania, it would seem that the report was heard. A few years later the IMF and the Indian Council for Research on International Economic Relations convened a seminar in Mumbai (Streeten 1988) to tackle the hot topic of structural adjustment in Asia. Once again Streeten was to play a crucial role as enabler and mediator. Observing that his criterion for a good seminar, which it had been, was to meet again one old friend and to make one new one, he went on to edit the book and write the summary of the discussion with its characteristic conclusion:

> Perhaps the most important general lesson that emerged was that there are no general lessons, and that each case has to be treated separately and on its merits. This pragmatic approach made it possible to discuss policies without getting bogged down in ideological or dogmatic positions.
>
> (*ibid.*: 17)

Streeten was now in his seventies, an age when most people are beginning to slow down, but his greatest book was still to come. Five long lectures (plus appendices), delivered in Italy, form the core of *Thinking about Development* (Streeten 1995). It is a potent distillation, with even now new insights, of a life-time focused on the great global questions of the past sixty years. It is full of what **Michael Lipton** calls 'Streetenesque subtlety' (Lipton 1977: 214). As Peter Timmer of Harvard has said, he is truly 'one of our wisest scholars'.[3] While the lectures in Italy were focused on 'the eradication of poverty in the world', a small book (Streeten 1994) published in Denmark at much the same time focused more particularly on unemployment. But he was not yet finished. In 2001, in his mid-eighties, Streeten published *Globalisation: Threat or Opportunity?*, where he was able to clarify the two contradictory aspects and to consider the implications of a phenomenon with which he had been wrestling all his life.

Paul Streeten is truly a world economist, not limited by the blinkers of the hidden assumptions of those confined to any one nation state or region; someone who is able to see problems both from the inside, at the grassroots, and also from an overall perspective. Drawing on his mastery of economics, his wide reading in many other subjects, and on his own vast experience, Paul Streeten has increased the understanding and enhanced the capacity of scholars and policy-makers around the globe.

Notes

1 *World Development*: Notes for contributors.
2 Poverty, Ignorance and Disease in kiSwahili, Personal observation, Dar es Salaam, July 1963.
3 Cover comment on Paul Streeten, *What Price Food? Agricultural Price Policies in Developing Countries*, Cornell University Press, Ithaca, 1987.

Major works

Streeten, P. (1949) 'The Theory of Profit', *Manchester School*, vol xvii, no.3. For a detailed bibliography of his work (up to 1991) see Streeten 1995: 357–80.

—— (1954) 'Keynes and the Classics', in K. Kurihara (ed.), *Post-Keynesian Economics*, New Brunswick, NJ: Rutgers University Press.

—— (1954) 'Programmes and Prognoses, Unbalanced Growth and the Ideal Plan', *Quarterly Journal of Economics*, lxviii(3) (August): 355–76; reprinted as Introduction to Gunnar Myrdal (1958) *Value in Social Theory*, London: Routledge and Kegan Paul.

—— (1955) 'Reformed Capitalism in Britain', in Robert L. Heilbroner, *The Great Economists: Their Lives and their Conceptions of the World*, British edn, London: Eyre & Spottiswood.

—— (1958) 'Introduction', in Gunnar Myrdal (ed.), *Value in Social Theory*, London: Routledge and Kegan Paul, p. xiii.

—— (1961) *Economic Integration: Aspects and Problems*, Leiden: A.W. Sythoff, chap. 5.

—— (1964) *Economic Integration: Aspects and Problems*, Leiden: A.W. Sythoff, 2nd enlarged edn, chap. 5.

—— (ed.) (1970) *Unfashionable Economics: Essays in Honour of Lord Balogh*, London: Weidenfeld & Nicolson.

—— (1972a) *Aid to Africa. A Policy Outline for the 1970s*, New York: Praeger.

—— (1972b) *The Frontiers of Development Studies*, London: Macmillan.

—— (1984) 'Basic Needs: Some Unsettled Questions', Carnegie Conference Paper no. 8, University of Cape Town.

—— (1985) *An Autobiographical Fragment*, Balliol College Annual Record, Oxford: Balliol College, Oxford.

—— (1987) *What Price Food? Agricultural Price Policies in Developing Countries*, Ithaca, NY: Cornell University Press.

—— (ed.) (1988) *Beyond Adjustment: The Asian Experience*, Washington, DC: IMF.

—— (1994) *Strategies for Human Development: Global Poverty and Unemployment*, Copenhagen: Handelshojskolens Forlag, preface.

—— (1995) *Thinking About Development*, Cambridge: Cambridge University Press.
—— (2001) *Globalisation: Threat or Opportunity?*, Copenhagen: Handelshojskolens Forlag.
Streeten, P. and Corbet, H. (eds) (1971) *Commonwealth Policy in a Global Context*, London: Frank Cass.
Streeten, P. and Lipton, M. (eds) (1968) *The Crisis of Indian Planning: Economic Policy in the 1960s*, London: Oxford University Press.
Streeten, P., with others (1981) *First Things First: Meeting Basic Human Needs in the Developing Countries*, New York: Oxford University Press for the World Bank.

Further reading

Balogh, T. (1963) *Unequal Partners*, 2 vols, Oxford: Basil Blackwell.
International Labour Office (1977) *Employment, Growth and Basic Needs: A One-World Problem*, New York: Praeger.
—— (1982) *Basic Needs in Danger: A Basic Needs Oriented Development Strategy for Tanzania*, Addis Ababa: ILO.
Lipton, M. (1977) *Why Poor People Stay Poor: A Study of Urban Bias in World Development*, London: Temple Smith.
Myrdal, G. (1953) *The Political Element in the Development of Economic Theory*, London: Routledge and Kegan Paul.
—— (1958) *Value in Social Theory: A Selection of Essays on Methodology*, ed. Paul Streeten, London: Routledge & Kegan Paul.
—— (1968) *Asian Drama: An Inquiry into the Poverty of Nations*, New York: Random House.
Stretton, H. (1986) 'Paul Streeten: An Appreciation', in Sanjaya Lall and Frances Stewart, F., *Theory and Reality in Development*, London: Macmillan.

Francis Wilson

JAMES TOBIN (1918–2002)

James Tobin is one of the few economists working primarily in the mainstream of that discipline who has gained wide respect among development scholars, politicians and activists. Paradoxically, perhaps, he is best known for the proposal in 1972 to tax speculative electronic financial transactions and that, thanks to the recent international campaign, now bears his name. By contrast, his simultaneous contribution to measuring human welfare in broader terms than by per capita GNP alone is now less well remembered.

James Tobin, Nobel Laureate and widely regarded as the USA's foremost Keynesian economist of the twentieth century, was born in Champaign, Illinois, on 5 March 1918, where he grew up with his younger brother, attending neighbourhood schools and then the University High School in Urbana, Champaign's twin city. There, according to the autobiography that he delivered on the occasion of his Nobel Prize award in 1981, he obtained a 'marvellous' education from the university's student teachers and their trainers and was able to play in the university basketball team, 'fulfilling ambitions that had seemed beyond reach in my childhood' (Tobin 1981: 1). His father, Louis, a journalist and publicity director for University of Illinois athletics, had a profound influence on his development: 'My father also happened to be an intellectual, as learned, literate, informed, and curious as anyone I have known. Unobtrusively and casually, he was my wise and gentle teacher' (Tobin 1981: 1).

Tobin also attributes to his father the decision to apply to Harvard rather than the local University of Urbana-Champaign (where he had expected to land up studying law). Success in the entrance exams saw him head for Cambridge, Massachusetts, in late 1935, leaving the Midwest for the first time. There he spent six years, four of them as an undergraduate, in a period which he described as Harvard's 'golden age' in Economics before and during World War II. Among his most influential teachers he counts Joseph Schumpeter, Alvin Hansen, Seymour Harris, Edward Mason, Gottfried Haberler and Wassily Leontief, while young faculty and fellow graduates included Paul Samuelson, Lloyd Metzler, Paul Sweezy, John Kenneth Galbraith, Richard Musgrave and Richard Gilbert, all of whom later became prominent economists.

Following US naval service after Pearl Harbour, Tobin returned to complete his PhD at Harvard in 1946–7 on one of his abiding interests, the theory and statistics of the consumption function, by then persuaded that he would follow an academic career. Shortly after returning, he met, and in September 1946 married, Elizabeth (Betty) Fay Ringo, a former student of Samuelson. A postdoctoral fellowship enabled Tobin to remain at Harvard – with a stint at the Deptartment of Applied Economics in Cambridge (UK) – until 1950. Thereafter, he moved to an associate professorship in economics at Yale University in New Haven, Connecticut, gaining a full professorship in 1955 and assuming the named Sterling Professorship of Economics two years later. There he directed the Cowles Foundation for Research in Economics from 1955 to 1961, chaired the Economics Department from 1974 to 1978 and remained a distinguished figure until retirement in 1988 and subsequently as emeritus professor until his death on 11 March 2002, just six days after his eighty-fourth birthday.

Tobin's career was punctuated by a series of visiting appointments at other universities. However, the absence from Yale of which he was most proud was to serve on President John F. Kennedy's Council of Economic Advisors in Washington, DC from 1961 to 1962 alongside fellow luminaries like Robert Solow and Kenneth Arrow. Although he soon returned to academia, he continued as a consultant to the Council for several more years and the experience influenced much of his subsequent work (Tobin 1966, 1974, 1987; Tobin and Wallis, 1968; Tobin and Weidenbaum 1988;

Perry and Tobin 2000). Tobin's enduring interest in financial markets and investment decision-making is reflected in another string of books (Hester and Tobin, 1967a, 1967b, 1967c; Tobin 1970, 1982; Tobin and Golub 1998) and journal articles. His proposal for a tax on speculative foreign exchange transactions originated from this strand of work (Tobin 1974).

Altogether, Tobin published some 400 journal articles, the most important of which were subsequently republished in a four-volume set of *Essays in Economics* (1972, 1975, 1982 and 1996a), reflecting the principal thematic phases of his output. A fifth set, embracing his later work on global markets and finance, was published posthumously (Tobin 2003). Unlike most academic economists, Tobin was as comfortable – and forthright – penning topical potboilers for newspapers as in preparing scholarly treatises.

His reputation in the development world rests principally on three contributions, none of which was directed specifically at developing countries at the time. Indeed, he was very much a liberal American international economist in the Keynesian tradition. The first pertinent contribution reflects this directly, namely his sustained critique of the monetarist dogma popularised by Milton Friedman in Chicago and later adopted in part by President Reagan. Among those impressed by Friedman's monetarism were a group of young Chilean economics students who subsequently undertook its 'purest' implementation under Augusto Pinochet's military dictatorship after 1973. Combined with the political repression, this caused much hardship to the poor but heralded what was widely known internationally as Chile's 'economic miracle', one of the inspirations for the international financial institutions in formulating their neoliberal policies to tackle the 1980s debt crisis.

Tobin's other two most notable development-oriented interventions date from the early 1970s. 'Is Economic Growth Obsolete?' (Nordhaus and Tobin 1972) initiated what became a sustained critique of using GNP (per capita) as the principal measure of human welfare when, in fact, it is an indicator of marketed production. This idea, and the several deductions and additions required to approximate a more useful indicator, the Measure of Economic Welfare (MEW), quickly entered economic textbooks and informed a generation of economics undergraduates. To those, myself included, learning their craft amid the daily realities of the global South, where most people engaged at least partially in non-marketed production and trade, this made eminent sense and represented an all-too-rare attempt to relate conventional economic theory to familiar conditions. This line of critique spawned experiments with compound indicators to avoid the limitations of reliance on a single variable. Ultimately the Human Development Index (HDI) was introduced by the United Nations Development

Programme (UNDP) in 1990 and has become the most widely used index for international comparisons of development as well-being.

Tobin's most widely known contribution to development thinking is the 'Tobin Tax' proposal for what he called 'sand in the wheels' of international financial markets to reduce volatility and the profits from short-term speculative transactions relative to longer-term productive flows. Many years later, Tobin (1996b: x) observed that his early effort on the subject (Tobin 1974, 1978) 'did not make much of a ripple. In fact, one might say that it sank like a rock. The community of professional economists simply ignored it.' That fate he attributed to the inherent objection of most economists and bankers to any market interference, accompanied by claims that it would drive financial markets to offshore tax havens while not reducing exchange rate fluctuations and speculative attacks. However,

> Most disappointing and surprising, critics seemed to miss what I regarded as the essential property of the transactions tax – the beauty part – that this simple, one-parameter tax would automatically penalize short-horizon round trips, while negligibly affecting the incentives for commodity trade and long-term capital investments. A 0.2% tax on a round trip to another currency costs 48% if transacted every business day, 10% if every week, 2.4% if every month. But it is a trivial charge on commodity trade or long-term foreign investments.
>
> (Tobin 1996b: xi)

Tobin viewed the tax as the principal objective, with use of the funds thereby raised for multilateral purposes 'as a by-product'. However, the resurgence of interest that commenced at the World Social Summit in Copenhagen in 1994 and which is attributable to the extreme currency speculation of the mid-1990s, regards this revenue as no less important. Even a tax rate of only 0.1 per cent (half of Tobin's original proposal), could raise between US$50 and 300 billion annually, thereby matching existing levels of official development assistance (War on Want n.d.). By 1996, the IMF's chief economist was also quite positive; the campaign has gathered momentum since, led in part by the development NGOs, War on Want and Oxfam, while several countries, notably Canada, France and Belgium, support the plan and/or have introduced enabling legislation. It is important to note, though, that supporters' motives vary: some, like Tobin, focus principally on the need for reduced financial market volatility and the efficiency of the proposed tax; many development lobbyists see stability and the funds accruing as being equally important; while anti-globalisation protesters hope that it will retard financial globalisation.

Whether tactically or genuinely, Tobin (1996b; Patomäki 2001: 124–5) has expressed uneasiness at the political agendas of some of the new protagonists of the tax. Today, concerns over some practical implementation problems and the loss of governmental sovereignty thereby implied and a lack of political will remain the major obstacles to its implementation. However, Tobin (1996b, 1996c) rejected most of these claims, pointing out, for instance, that the tax would restore to governments some economic sovereignty already lost to international financial markets. Moreover, Patomäki (2001) argues that universal agreement is no longer a prerequisite. His important contribution also takes a broader view of financial markets as co-responsible for the widening of global disparities and the current global financial institutional arrangements as inadequate and inefficient. Hence Patomäki advocates the tax on the additional grounds of global-scale distributive justice, the shared ideal of democracy, and human emancipation.

James Tobin's robust commentaries on contemporary economics also embraced the international arena, as illustrated by the conclusion to a 1993 paper on the challenges facing nation states in our increasingly interdependent world:

> Our economies are increasingly interdependent. Our opportunities are increasingly worldwide. Our maladies are contagious and intertwined. The leaders of the G7 countries have yet to show appreciation of the seriousness of the problems they jointly face, let alone enough imagination and initiative to seek joint remedies. In comparison with their predecessors who confronted the tasks of world economic recovery after World War II and responded with the Marshall Plan, the Bretton Woods institutions and the GATT, our present leaders are pygmies.
>
> (Tobin 1996d: 189)

Tobin's reputation spawned invitations to deliver a bewildering array of named lectures worldwide. Between 1967 and 1996, he also received no fewer than twenty-one honorary doctorates (mainly LL Ds and Doctorates of Humane Letters), from some of the USA's leading universities and colleges (including Syracuse, Illinois, Dartmouth, Swarthmore, New York, Colgate, Harvard and Wisconsin-Madison) and three in Europe. He became a member of the National Academy of Sciences in 1972 and in 1981 was awarded the Nobel Prize in Economic Science in recognition of his analysis of financial markets and their relations to expenditure decisions, employment, production and prices.

These honours, his prolific published output and their influence will ensure that his reputation endures. However, outside the USA at least, he will probably best be remembered for a modest proposal in 1972, long deprecated but ultimately resurrected in the form of a global campaign for the Tobin Tax about which he claimed to feel uneasy. As for the man himself, fellow economist, Paul Krugman (2002), paid him this tribute:

> He was a great economist and a remarkably good man; his passing seems to me to symbolize the passing of an era, one in which economic debate was both nicer and a lot more honest than it is today ... and I mourn not just his passing, but the passing of an era when economists of such fundamental decency could flourish, and even influence policy.

Major works

Hester, D.D. and Tobin, J. (eds) (1967a) *Studies of Portfolio Behavior*, New York: Wiley.
—— (eds) (1967b) *Risk Aversion and Portfolio Choice*, New York: Wiley.
—— (1967c) *Financial Markets and Economic Activity*, New York: Wiley.
Nordhaus, W. and Tobin, J. (1972) 'Is Economic Growth Obsolete?', *National Bureau of Economic Research, Fifth Anniversary Colloquium V*, New York: NBER (an earlier version appeared as *Cowles Foundation Discussion Papers 319* in 1971).
Perry, G.L. and Tobin, J. (2000) *Economic Events, Ideas and Policies: the 1960s and After*, Washington, DC: Brookings Institution Press.
Tobin, J. (1966) *National Economic Policy*, New Haven, CT: Yale University Press.
—— (1970) *Asset Accumulation and Economic Activity: Reflections on Contemporary Macroeconomic Theory*, New Haven, CT: Yale University Press (reprinted by University of Chicago Press, 1982).
—— (1972) *Essays in Economics Vol. 1: Macroeconomics*, Chicago, IL: Markham (reprinted by MIT Press, 1987).
—— (1974) *The New Economics a Decade Older*, Princeton, NJ: Princeton University Press.
—— (1975) *Essays in Economics Vol. 2: Consumption and Econometrics*, Amsterdam: North-Holland (reprinted by MIT Press, 1987).
—— (1978) 'A Proposal for International Monetary Reform', *Eastern Economic Journal* 4: 153–9.
—— (1982) *Essays in Economics, Vol. 3: Theory and Policy*, Cambridge, MA: MIT Press.
—— (1987) *Policies for Prosperity; Essays in a Keynesian Mode*, Brighton: Wheatsheaf Books.
—— (1996a) *Essays in Economics, Vol. 4: National and International*, Cambridge, MA: MIT Press.
—— (1996b) 'Prologue', in Ul Haq, M., Kaul, I. and Grunberg, I. (eds), *The Tobin Tax: Coping with Financial Volatility*, New York and London: Oxford University Press.

—— (1996c) 'A Currency Transactions Tax: Why and How', *Open Economies Review* 7: 493–9.
—— (1996d) *Full Employment and Growth: Further Keynesian Essays on Policy*, Cheltenham and Lyme, NH: Edward Elgar.
—— (2003) *World Finance and Economic Stability: Selected Essays of James Tobin*, Aldershot: Edward Elgar.
Tobin, J. and Golub, S.S. (1998) *Money, Credit and Capital*, Boston, MA: Irwin/ McGraw-Hill.
Tobin, J. and Wallis, W.A. (1968) *Welfare Programs: An Economic Appraisal*, Washington, DC: American Enterprise Institute.
Tobin, J. and Weidenbaum, M. (eds) (1988) *Two Revolutions in Economic Policy: the First Economic Reports of Presidents Kennedy and Reagan*, Cambridge, MA: MIT Press.

Further reading

Krugman, P. (2002) 'Missing James Tobin', *New York Times*, 3 December (also at www.pkarchive.org/column/031202.html, accessed on 9 February 2005).
Patomäki, H. (2001) *Democratising Globalisation: the Leverage of the Tobin Tax*, London: Zed Books.
Tobin, J. (1981) Autobiography (www.nobelprize.org/economics/laureates/ 1981/tobin-autobio.html, accessed 9 February 2005).
War on Want (n.d.) *The Tobin Tax: Win–Win for the World's Poor* (Briefing), London: War on Want (available at http://www.tobintax.org.uk).

David Simon

MAHBUB UL HAQ (1934–98)

> [The] UNDP and the United Nations system as a whole owe Mahbub a big debt of gratitude. Perhaps more than any other individual, he changed forever the way we think about development.
>
> (Dijibril Diallo 1998: 2)

When he passed away on 16 July 1998, the Pakistani economist, Mahbub Ul Haq ('Loved by God' as the name means in Arabic), had achieved international acclaim as a scholar-administrator who had pioneered new visions of human development and initiated the UNDP's *Human Development Report* series. Ul Haq served as Chief Economist in the Pakistan Planning Commission (1957–70) and as Director of the World Bank's Policy Planning Department (1970–82). Mahbub's influence on global development policy debates was also tangible; he was the Chairman of the North–South Roundtable (1979–84), an eminent adviser to the Brandt Commission (1980–2) and a Governor of the IMF (1985) and also the World Bank (1988). Mahbub served on the governing boards of numerous international institutions and think-tanks, including the Earth Council, the World

Commission on Culture and Development and the Institute of International Economics (Speth 1998). UN Secretary General, Kofi Annan, described Ul Haq's untimely death from pneumonia as 'a loss to the world' (cited in HDC 1998: 1).

Having studied Economics at Government College in Lahore, Ul Haq moved to Cambridge University where he studied for a Masters degree and began a longstanding friendship with **Amartya Sen**. In the memorial lecture for Ul Haq on 15 October 1998, Sen recalled the discussions they had after arriving at Cambridge about conventional economics and its potential contribution in India and Pakistan. Sen explained that they both needed to learn Economics to be heard 'but not use it much … who really wants to know what determines the price of toothpaste?' (Sen cited in Rosenfield 1998: 1). At twenty-one, Ul Haq left Cambridge and continued his education studying for a PhD at Yale, from where he returned to Pakistan to work for the Federal Government as Chief Economist of the National Planning Commission until 1970. Working on the formulation and implementation of Pakistan's five-year development plans, Mahbub later declared, represented 'happy days. My sights were set, my horizon was clear, and there was no hesitancy in my views about economic development' (Ul Haq 1976: 3).

Ul Haq set to work in 1960, while at Harvard, on *The Strategy of Economic Planning: A Case Study of Pakistan* (Ul Haq 1963). The book attempted to express themes of poverty and economic development in Pakistan and was concerned with the potential barriers to progress: the deeply unequal pattern of landholding, the pervasive illiteracy and the 'warped development' that favoured a privileged minority (Ul Haq 1973: 1). In April 1968, Mahbub made a speech in Karachi about the twenty-two 'industrial family groups' that had come to warp, skew and dominate national economic and political affairs. Not surprisingly, the speech created major shockwaves in Pakistan and beyond. Although **Julius Nyerere** wrote to Ul Haq with a letter of appreciation and Indira Gandhi used sections of Mahbub's speeches in her own policy presentations, Dr Haq later recalled that '[t]he academic community in the Western world reacted in shocked disbelief: one of its own products had suddenly gone berserk' (Ul Haq 1976: 8). Keen to avoid doctrinaire economics and planning philosophies, Amartya Sen has noted that this book was 'informed by a general recognition that while a poor economy may take a very long time to become a rich country through GNP growth, the conditions of human living can be changed much more rapidly through intelligent policy making' (Sen cited in Rosenfield 1998: 1). Calling for properly targeted social intervention, the book began a longstanding commitment to interrogate GNP as a measure of progress, focusing on non-economic methods of securing positive change and improving the quality and targeting of development policy. In reference to this work, Ul Haq later reflected that '[t]hough I have written much else since then my detractors have seldom allowed me to forget my original writings, perhaps believing that the evolution of ideas is an unforgivable sin' (cited in HDC 1998: 2).

Ul Haq's writings and speeches about inequality and economic growth brought him to the attention of World Bank President, Robert McNamara. During his tenure at the World Bank (1970–82), Ul Haq was credited with making a major contribution to the Bank's philosophy of development economics. Upon his death in 1998, the President of the World Bank, James Wolfensohn, wrote in a letter to Mahbub's wife Khadija that 'probably more than anyone else [Ul Haq] provided the intellectual impetus for the Bank's commitment to poverty reduction in the early 1970s' (cited in HDC 1998: 1). As Director of the Bank's Policy Planning Department, Ul Haq set out to steer more attention towards poverty alleviation programmes, nutrition, water supply, education, social welfare and increased allocations for small farm production. According to one tribute, through this work Mahbub was able to 'help sensitise a cold commercial lending institution to the concerns of the poor in the Third World' (Jabbar 1998: 2). Drawing upon his involvement with the Bank, Ul Haq wrote *The Poverty Curtain: Choices for the Third World* (1976), which highlighted the neglect of human resources in development planning and was a seminal study that provided an important precursor for the Bank's later development of the basic needs and human development approaches of the 1980s (Ul Haq and Burki 1980; HDC 1998). In the book, the 'seven sins' committed by the 'priesthood of development planners' are highlighted. These included playing 'numbers games', constructing excessive economic controls, being constantly preoccupied with 'investment illusions', the addiction to 'development fashions', the divorcing of planning and implementation and the growing 'mesmerization' of planners with high growth rates in GNP.

During the early 1970s, Mahbub shifted his attention towards the intellectual self-reliance of the 'Third World' and became more concerned at the simplicity of some intellectual dialogues about development at the international level (Ul Haq 1976). This feeling was sharpened by his experience of the UN Conference on the Human Environment in Stockholm in 1972 and a growing awareness of North–South differences of environmental perspective: 'what were we really doing sitting through endless seminars and conferences where our own voice was neither solicited nor heard?' (Ul Haq 1976: 84). Along with **Samir Amin** and others who had shared these concerns and had also been present at Stockholm, Ul Haq founded the Third World Forum, an action group of around 100 leading intellectuals from the South, which first met in Santiago (Chile) and formalised its own constitution in Karachi in 1975.

In the early 1980s, Mahbub left the World Bank to return to Pakistan to work as Minister of Finance, Planning and Commerce (1982–88). Under General Zia Ul Haq's regime (no relation), Mahbub is credited with a major acceleration in social spending and with instigating significant tax reforms, new initiatives for poverty reduction, economic deregulation and an increased emphasis on human development (HDC 1998). Nonetheless, in reflecting some years later on his time in cabinet, Ul Haq spoke of his lack of independence and the fact that he was 'not able to do very much' (cited in Brazier 1994: 4). As he put it in an interview in 1994: 'I did accomplish some things but I was part of a very élitist system dominated by landlords in the Assembly, by élitist groups in the government, and by the Army, which would not let any trade-off take place between military and social expenditure' (cited in Brazier 1994: 4). It could be argued that Mahbub's experience of these restrictions and 'trade-offs' between the economic and social realms of development were to have a big impact on his ideas in the years ahead and particularly when his attention turned to the measurement and conceptualisation of 'human development'.

In 1989 Ul Haq and his wife and intellectual partner, Khadija, moved to New York where he became Special Adviser to the UNDP administrator (1989–95). In doing so, he was reunited with **Amartya Sen** and also came together with Frances Stewart, **Richard Jolly** and Meghnad Desai to prepare the annual *Human Development Reports* (HDRs). According to Sen, Mahbub explained that '[w]e need a measure of the same level of vulgarity as the GNP – just one number – but a measure that is not as blind to social aspects of human lives as the GNP is' (cited in Rosenfield 1998: 2). The Human Development Index (HDI) is now a central and recognised tool for the UNDP and although Sen admits to initially finding the HDI a bit 'coarse', it helped bring about a 'major change in the understanding and statistical accounting of the process of development' (Sen, cited in HDC 1998: 2). Critics of the HDI would come back to this issue of the coarse and oversimplified nature of the Index and have continued to highlight the costliness and numerous errors of HDRs as well as the poor quality of the underlying data. The initiation of the HDRs has been seen by some as a way for Ul Haq (who was their chief architect) to atone for his failure with human development issues in Pakistan (Brazier 1998: 4). According to Amartya Sen, however, the special achievement of the HDRs is that they bring 'an inescapably pluralist conception of progress to the exercise of development evaluation' (Sen 2000: 18).

Following his retirement from the UNDP, Mahbub initiated a new series of HDRs in South Asia published by the Human Development

Centre (HDC) he established in Islamabad. In his final years, Ul Haq began to deepen and further his interest in human security issues and in the regional crisis of governance in Asia. Through the HDC in Islamabad, Mahbub was able to champion 'an earnest debate on restoring a better balance between arms and the people in South Asia' (Ponzio 1998: 10). Taking its lead from the UNDP's 1995 HDR on *Gender and Human Development*, the South Asia Human Development Report of 2000 (HDC 2000) opens with Ul Haq's often-cited injunction that 'Human Development, if not engendered, is fatally endangered' (HDC 2000: 1). At one of the last meetings of the HDC attended by Mahbub, he quoted Bernard Shaw in telling colleagues '[y]ou see things that are, and ask why? I dream of things that never were and ask why not?'. Over several decades, Ul Haq regularly raised the question of cuts in global military spending and consistently connected this to the possibility of improved social provision, human development and welfare. This was somewhat ironic given that he once served under a military dictatorship in Pakistan in the 1950s and 1960s. The goal of his diplomatic efforts in the 1990s was a desire for a more aggressive stance on disarmament, suggesting that the UN system needed new rules of intervention for the superpowers in order to avoid multilateral imperialism: '[o]therwise we're back to the whole philosophy of colonialism which was "the natives can't handle it, let's go in and teach them" ' (cited in Brazier 1994: 2).

In the foreword to Ul Haq's *Reflections on Human Development* (Ul Haq 1995), **Paul Streeten** recalls how he met Mahbub while working for the British Overseas Development Administration (ODA), arguing that dealing with 'hardened British civil servants' gave Ul Haq a resilience, a healthy scepticism and also an awareness that development did not always involve the mobilisation of additional resources but that it could also be achieved through a reallocation of existing resources. Interestingly, the book notes the unholy alliance of 'shock therapists' and Bretton Woods institutions which impedes the provision and redistribution of basic needs. Nonetheless, Mahbub agreed with many of the Bank's core tenets of economic liberalism, placing an emphasis on privatisation and a belief in market competition. This comes through clearly in a speech he gave about Africa's future shortly before he died (Ul Haq 1998: 2). Having been dealt a 'cruel hand' by history, African countries, he argues, should develop new technical skills and follow the Indian example, becoming 'a major supplier of computer software and electronic goods to Europe' (Ul Haq 1998: 3).

Amartya Sen has recalled Mahbub's general 'impatience with theory' (Sen 2000: 22) but also notes the intense pragmatism that he brought to the UNDP's formulation of 'human development' and his 'open minded approach, his scepticism and his perpetual willingness to listen to new suggestions' (Sen 2000: 23). Ul Haq regularly advocated an integrated UN system with a strong human development message and in some of his final writings, speeches and interviews he talked of how the World Bank had 'seriously misled policy makers' in focusing attention 'on the symptoms, not the causes', noting that 'to ignore the poor upstream and to count them endlessly downstream is merely an intellectual luxury' (Ul Haq 1997: 1). This was one indulgent sin committed by the 'priesthood of development planners' that Mahbub considered unforgivable.

Major works

Ul Haq, M. (1963) *The Strategy of Economic Planning: A Case Study of Pakistan*, Karachi: Oxford University Press.

—— (1976) *The Poverty Curtain: Choices for the Third World*, New York: Columbia University Press.

—— (1995) *Reflections on Human Development*, New York and Oxford: Oxford University Press.

—— (1999) 'Human Rights, Security and Governance', in *Worlds Apart: Human Security and Global Governance*, London: IB Tauris.

Ul Haq, M. and Burki, S.J. (1980) *Meeting Basic Needs: An Overview*, Washington, DC: World Bank.

Further reading

Brazier, C. (1994) 'The New Deal: Interview with Mahbub ul Haq', *New Internationalist*, 262(12): 1–4, http://www.newint.org/issue262/, accessed 4 July 2004.

Desal, N. (1998) 'Heartbeat: Tribute to Development Economist Mahbub Ul Haq – Obituary', *UN Chronicle*, 22 September: 1. Copy available at http://www.findarticles.com/p/articles/mi_m1309/is_3_35/ai_54259329.

Diallo, D. (1998) 'From the Editor in Chief', *Choices: The Human Development Magazine*, 7(4): 2.

Human Development Centre (HDC) (1998) 'A Tribute to Dr Haq', http://www.un.org.pk/hdc/Tribute/, accessed 7 July 2004.

—— (2000) *Human Development in South Asia 2000: The Gender Question*, Mahbub ul Haq Human Development Centre, Delhi, Oxford: Oxford University Press.

Jabbar, J. (1998) 'Two South Asian Stalwarts: Tribute [Nikhil Chakravarty and Mahbub ul Haq]', *Himal*, 11(8), copy available at http://www.himalmag.com/Aug98/stalwart.htm/, accessed 4 July 2004.

Minna, M. (1999) 'Tribute to Mahbub ul Haq', speech given at the Mahbub ul Haq commemorative conference 'People and Poverty: Human Development into the Next Millennium', Ottawa, Canada, 13 October, http://www.acdi-cida.gc.ca/cida_ind.nsf/, accessed 4 July 2004.

Ponzio, R. (1998) 'Disarmament and Human Development: The Legacy of Dr Mahbub ul Haq', *Bonn International Centre for Conversion: Bulletin*, 1 October, 9: 1–2.

Rosenfield, S.S. (1998) 'The not so Dismal Economist', *Washington Post*, 23 October, http://www.wright.edu/~tdung/mahbub.html, accessed 8 July 2004.

Sen, A. (2000) 'A Decade of Human Development', *Journal of International Development* 1(1): 17–23.

Speth, J.G. (1998) 'UNDP Mourns Loss of Mahbub Ul Haq, Economist and Development Visionary: Administrator James Gustave Speth Praises Haq's Unwavering Commitment to Social Justice', UNDP Press Release, http://www.undp.org/dpa/pressrelease/releases/P980717e.html/, accessed 4 July 2004.

Ul Haq, M. (1973) 'System is to Blame for the 22 Wealthy Families', *The London Times*, 22 March, copy available at http://www.un.org.pk/speeches, accessed 4 July 2004.

—— (1997) 'Poverty is Cancer, Not Flu', introductory remarks at the Special Event on Poverty Eradication arranged by UNDP, 20 May 1997, http://www.un.org.pk/hdc/speeches/ accessed 3 July 2004.

—— (1998) 'Does Africa have a Future?', Opinion, *Earth Times News Service*, http://www.meltingpot.fortunecity.com/lebanon/254/ulhaq.htm, accessed 9 July 2004.

United Nations Development Programme (UNDP) (1990) *Human Development Report 1990: Concept and Measurement of Human Development*, New York: UNDP.

Marcus Power

ERIC R. WOLF (1923–99)

As an anthropologist, Eric Wolf's unique contribution to the theory of development stems from his critical approach, insisting on the inseparable linkages between the local and the global. This helps to overcome vital impasses and difficulties in development thinking.

Born in Vienna in 1923, Wolf was the son of a 'multicultural' couple. His father, a secular Jew from Austria, had been in Siberia as a prisoner of war during World War I. There he met his wife, from a Jewish family that had been exiled for their participation in the 1905 revolution. His mother's tales of Siberia, as well as his father's accounts of travels in Latin America, enhanced the cosmopolitan bent of the family and may well have fostered Wolf's later interest in cultural anthropology and particularly in Latin America. In 1933, the year the Nazis came to power in Germany, Wolf moved to the Sudetenland (then Czechoslovakia) with his family. In 1938, with Nazi expansion into Austria and Czechoslovakia, Eric Wolf was sent to school in England. Ironically confined to an internment camp for enemy aliens in early 1940, he met the sociologist, Norbert Elias, who raised his interest in social science. After arriving in New York City, Wolf 'discovered' cultural anthropology after stints in biology, political science, economics and sociology. After serving in the US Army, he gained his BA from Queens College in 1946 and became a PhD student at Columbia. There he belonged to a group of students sharing a similar background as veterans and Marxists who later went on to become quite eminent cultural anthropologists. Studying under Julian Steward, the 'founding father' of 'cultural ecology' or 'cultural materialism' from 1947, Wolf commenced field research in Puerto Rico among poor coffee growers and gained first-hand knowledge of their risky and hard everyday life. After completing his PhD in 1951, Wolf held various teaching positions and conducted research in Mexico and South Tyrol. A professor at the University of Michigan from 1961, he became increasingly committed to political issues: in 1965, together with fellow-anthropologist, Marshall Sahlins, he originated the idea of and participated in the first teach-in against the Vietnam War on the University of Michigan campus. Under the impact of this war, Wolf wrote his famous *Peasant Wars of the Twentieth Century*. In this period, together with Joseph C. Jorgensen, he also led an investigation of the American Anthropological Association's Ethics Committee on the misuse of ethnographical research for counter-insurgency in Southeast Asia (cf. Jorgensen and Wolf 1971). A Distinguished Professor at the City University of New York (CUNY) from 1971, Wolf, encouraged by his second wife and fellow-anthropologist Sydel Silverman, shifted his research towards world systems theory and the development of capitalism. This resulted in *Europe and the People Without History* (1982). Besides teaching at CUNY, he continued research into the worldwide spread of capitalism and its consequences for the rise of new forms of power and domination. These were also the themes of his last book (Wolf 1999). Eric Wolf died in the same year.

During the early years of Eric Wolf's academic career, development thinking was virtually possessed by various modernisation theories. For all their differences, these theories shared 'dualistic' concepts: 'tradition', seen as an obstacle to development, had to be modernised away to achieve 'rational, progressive modernity'. Arguably the 'corporate community' was considered the most formidable of these traditional obstacles. Populated by a 'backwards-oriented peasantry', such communities were pictured as regulated by immutable tradition that did not allow for individual development and success (see **Walt Rostow**).

One of Wolf's early contributions (1957) showed that this 'ancient traditional' corporate community was a red herring. Referring to Mesoamerica and Central Java, Wolf unveiled such communities as, in fact, products of European expansion, thus anticipating critical debate by ten to twenty years. In *Peasants* (1966), he argued that the peasantry is not an amorphous mass beholden by tradition, but has its own structured forms of social, economic and political organisation, set between 'tradition' and 'modernity', and quite capable both of change as well as of stubborn persistence. Responding to US involvement in Vietnam, Wolf (1969) examined the reasons why peasants emerged as the most revolutionary force of the twentieth century. He had a substantial impact on the debate on the role of the peasantry in history as well as the modern, especially the underdeveloped world (see Shanin 1971).

As an anthropologist, Wolf grappled with a problem of cultural-relativist scientific fiction that is still current today: the idea of bounded, more or less self-contained community. In contradistinction to this fiction, Wolf insisted on the intimate and 'dialectical' interrelation between local communities or 'small populations' and complex, or 'total societies' (Cole and Wolf 1999: 3). The task was to unveil the close interconnections between processes evolving on a regional, national or even global level. It was from this vantage point that Wolf contributed decisively to social science development theory. Rather than attributing pristine, 'traditional' conditions to the local, while 'modern' would be the big world somewhere outside, for Wolf, peasant communities, as well as individual persons and households, constitute interfaces between local environments and overarching social connections, reaching from regions right up to the global nexus of the capitalist world market.

In *Europe and the People Without History*, his vital contribution to development theory, Wolf posits that 'the world of humankind constitutes a manifold, a totality of interconnected processes' (1982: 3). He stresses that we all live in one huge and complex ecosystem, but also insists that throughout history, there have been 'connections everywhere' (*ibid.*: 3). Therefore, the constructions of modernisation theories are, in fact, mere mythology, based on the false assumption of discrete, solitary cultures or communities. US history, the penultimate model and goal of modernisation theory, also was 'a complex orchestration of antagonistic forces' rather than 'the unfolding of timeless essence' (*ibid.*: 5). Therefore, history did not follow an inexorable trajectory. Rather, there had been alternatives. Any idea, then, of human history or social development as a triumphant march to progress was deeply misconceived. Yet as Wolf noted, such conceptions underlie the splintered-up array of disciplines that since the mid-nineteenth century deal with 'inquiry into the nature and varieties of human kind' (*ibid.*: 7).

Against widespread reification, Wolf reconstructed what happened with the *People Without History* in the course of European expansion since the fifteenth century. Above all he was concerned with the role that non-Europeans had played in the making of today's world and thus in shaping our ideas about development. In reconstructing historical process both at the global and at the micro levels, he also hoped to overcome the blind spots of historical and developmental vision that he saw inherent in the relations of domination established on a world scale during the last 500 years. In this conception, 'both the people who claim history as their own and the people to whom history has been denied emerge as participants in the same historical trajectory' (*ibid.*: 23). Accordingly, the idea of differential stages of development among societies is mistaken. Modernisation theory was fundamentally wrong precisely in its fundamental tenets that necessarily discriminate between different peoples and regions in the world: the fundamental changes of modernity:

> affected not only people singled out as the carriers of 'real' history but also the populations anthropologists have called 'primitives' ... The global processes set in motion by European expansion constitute *their* history as well. There are thus no 'contemporary ancestors,' no people without history.
>
> (*ibid.*: 385)

The book title, *Europe and the People Without History*, is therefore deeply ironic. Wolf first sketches the situation of the world around 1400 CE as one of manifold interconnectedness, by trade, by relations between pastoralists and agriculturalists, and by war. Western Europe developed from a marginal peninsula of the Eurasian land mass into a 'hub of wealth and power' (*ibid.*: 123) by processes that set in around 800 CE, and from the sixteenth century onwards, England and the Netherlands were able to harness commerce and war to a grand strategy of expansion that was to transform the world. Wolf explores four exemplary cases: the implantation of Iberian colonies in the Americas, the North American fur trade, the transatlantic slave trade, and the establishment and expansion of European control in South and East Asia.

By giving centre stage to the intricate relationships of agency, Wolf manages to overcome the interrelated views of the colonised as basically backward and as mere victims and objects of colonial power. Rather, social structures pertaining in 'Indian' communities in Spanish America or at the Guinea coast in West Africa during the age of the slave trade were shaped deeply by contemporary relationships of domination, exchange and exploitation in the context of an expanding world market. Moreover, Africans or American Indians were actively involved in shaping these conditions, even though the trading system was centred around Western Europe and precious metals flowed from America towards the Iberian Peninsula and East Asia. While peasant communities in the Spanish colonies resulted from complex relationships and conflict between colonial administrators and peasants, the famous League of the Iroquois underwent fundamental changes in reorienting towards the fur trade and warfare in an age of intense British–French rivalry around the Great Lakes; further, the lifestyles of

Prairie Indians resembled today's standard image of 'Indian' life only in the wake of deep-going transformations through the spread of trade, horses and firearms. The transatlantic slave trade that transported millions from West and Central Africa to the plantations of the 'New World' (itself a very Eurocentric construct – Ed.) also built on previously existing institutions in African societies which were then fundamentally transformed by its consequences. In particular, centralised military and commercial powers were enormously enhanced in the Guinea Coast kingdoms that thrived on the slave trade.

To date, the historical process that has been common to all societies reviewed here has culminated in the advent and universal expansion of capitalism. In contradistinction to authors such as Wallerstein or **A.G. Frank**, Wolf refers the capitalist mode of production strictly to the domination of capital at the point of production, not just to the spread of market relations around the world. Wolf's detailed account of the specific constellation that conditioned the industrial revolution and the advent of capitalism in England reveals an 'unusual development' (*ibid.*: 268) with immediate global repercussions. Overseas regions were at various times converted into raw material bases for the burgeoning British cotton industry. Again, Wolf stresses the ambivalent dimension of agency: besides suffering oppression, the slave population in the Cotton South also developed 'its own store of experience and modes of coping' which it 'would pass … on across the generations' (*ibid.*: 281). During the twentieth century, novel social relationships were spawned by capitalism's 'tendency … to expand in search of new raw materials … and … of cheap labor to process them' (*ibid.*: 302). For example, the commercialisation of rubber caused vastly divergent transformations, from one stage of a successive chain of fundamental changes touched off by capitalist expansion in the Amazon rainforest, to the recruitment of Indian labourers to Malaya to work the rubber plantations. The worldwide recruitment of labour for widely varying concerns caused outcomes including migration, proletarianisation, and ethnic segmentation. For Wolf, 'working classes are not "made" in the place of work alone' (*ibid.*: 360), underscoring the need for multidisciplinary approaches, integrating cultural factors. Here as in his other works, Wolf combines rich and diverse empirical data with a conceptual rigour to forge a vitally creative theoretical contribution.

Major works

Cole, J.W. and Wolf, E.R. (1999) *The Hidden Frontier: Ecology and Ethnicity in an Alpine Valley*, Berkeley, Los Angeles and London: University of California Press (originally, New York and London: Academic Press 1974).

Jorgensen, J. and Wolf, E. (1971) 'Anthropology on the Warpath: An Exchange', *New York Review of Books* 8 April and 22 July 1971.
Wolf, E.R. (1957) 'Closed Corporate Peasant Communities in Mesoamerica and Central Java', *Southwestern Journal of Anthropology* 13 (Spring): 1–18.
—— (1966) *Peasants*, Englewood Cliffs, NJ: Prentice Hall.
—— (1969) *Peasant Wars of the Twentieth Century*, New York: Harper and Row.
—— (1982) *Europe and the People Without History*, Berkeley, Los Angeles and London: University of California Press.
—— (1999) *Envisioning Power: Ideologies of Dominance and Crisis*, Berkeley and Los Angeles: University of California Press.
Wolf, E.R. with Silverman, S. (2001) *Pathways of Power: Building an Anthropology of the Modern World*, Berkeley and Los Angeles: University of California Press.

Further reading

Schneider, J. (2002) 'Eric R. Wolf', *American National Bibliography* online: http://www.anb.org/articles/14/14-01125.html.
Schneider, J. and Rapp, R. (eds) (1995) *Articulating Hidden Histories: Exploring the Influence of Eric R. Wolf*, Berkeley and Los Angeles: University of California Press.
Shanin, T. (ed.) (1971) *Peasants and Peasant Societies*, Harmondsworth: Pelican.

Reinhart Kößler and Tilman Schiel

PETER WORSLEY (1924–)

Peter Worsley, the British social anthropologist, was born in Birkenhead (Cheshire) in 1924. He began to read English at Cambridge University but then joined the British Army during the Second World War, serving as an officer in Africa and India. Returning to Cambridge after the war, he switched his studies to the field of anthropology. After a short spell in East Africa he returned to Britain, taking a Master's degree at Manchester University where he focused on a critique of Meyer Fortes's work. He subsequently worked as a researcher with S.F. Nadel at the Australian National University, completing a dissertation on the Aborigines in 1954. Back in Britain he returned to Manchester, joining Max Gluckman there and starting his famous work on cargo cults. His career, however, was to switch from social anthropology to sociology when he took up an early post in that discipline at Hull University. He then became the first Professor of Sociology at Manchester University, a post he held from 1964 to 1982.[1]

Worsley's intellectual career, however, cannot be understood in purely academic terms. At around the time of the Battle of Stalingrad he joined the British Communist Party, along with other intellectuals of his generation. Also in common with them, he left the CPGB in 1956 after the Soviet invasion of Hungary. This 'political' side of Worsley's intellectual career is rarely made explicit but it helps to explain, for example, why the Preface to his book, *The Third World*, contains an acknowledgement to labour historians E.P. Thompson and John Saville. Their humanist, democratic, and 'English' vision of **Marx**ism was to inform all his subsequent work. It never became sectarian, pedantic or merely scholarly. The negative side of Worsley's political commitment was a succession of Cold War establishment vetoes on his work as an anthropologist, especially refusals of posts and access to fieldwork sites. This partly explains his conversion to sociology, although, as we shall see, Worsley never stopped being a social anthropologist at heart.

The Trumpet Shall Sound (Worsley 1957) was Worsley's first and still best- known work, virtually creating the area of 'cargo cult' studies in social anthropology. He set these millenarian social movement cults in Melanesia in a broad historical and comparative context for the first time. Against the prevailing 'nativist' interpretation that saw these movements as traditional and frankly irrational flights of fancy, Worsley stressed their proto-nationalist character and the protest they articulated against racial and colonial oppression. In a methodological appendix to the riveting story of the cargo cults, Worsley argues that:

> The concept of 'nativism' ... fails to account for many cult elements ... Where old beliefs and practices are retained or revived, they have a new content ... And there is little that is reversionary or perpetuative in the concept of expelling the Europeans ... The movements are thus forward-looking, not regressive, the order of the future being the *reversion* of the present
>
> (Worsley 1957: 275–6)

Worsley not only writes an extremely effective (and affecting) account of the Melanesian cargo cults but also sets them within their anti-colonial context against the prevailing traditional anthropological focus of the day. Many generations of students were 'converted' by reading *The Trumpet Shall Sound* to a passionate commitment to development and anti-colonialism.

In 1964 Worsley published *The Third World* (Worsley 1964) that certainly popularised if not started the circulation of a term that had originated in French development thinking. It dealt with the colonial relationship of the majority world with the West, the rise of nationalism, the importance of populism, the structure of the new emerging post-colonial states, and the 'positive neutralism' of the Third World in the Cold War, then at its height. Peter Worsley, against the dominant modernisation perspective in social and political theory, did not see 'democracy' versus 'totalitarianism' as the main global divide but, rather, one 'between those movements and organisations oriented to radical social change and those resisting it' (Worsley 1964: 355), primarily located in the USA. In terms of the politics emerging in the post-colonial world he argued, against the orthodox Marxists of the day, for the importance of populism, a cross-class movement that 'has been infinitely greater than has been recognised and the potential under-estimated' (Worsley 1964: 170). Despite his own recent communist past, Worsley argued that the Third World challenges both 'late capitalism' and 'late communism' (Worsley 1964: 271) and offers even to transcend itself. Against the dominant materialism of both 'bourgeois' and 'radical' development thinkers alike, Worsley declared that: 'People who consider ideas and ideals unimportant ... are ignorant of the real world' (Worsley 1964: 271).

Sometimes it is an author's lesser-known work that is more revealing in a way. Not having the cult status of *The Trumpet Shall Sound* or *The Third World*, these works have not received the attention they deserve but they are perhaps even more relevant to Peter Worsley as development thinker. The first to be considered is *Two Blades of Grass* (Worsley 1971), an edited collection on the role of rural co-operatives in agricultural modernisation. Worsley sets the theme of co-operation in the broad context of the struggle against inequality and the possessive individualism of contemporary capitalism. He also develops a theme that is central in his later writings, namely the 'populist' dimension of democracy in the Third World, an aspect 'that so much structural political sociology ignores, concentrating as it does on the formal elements in democracy' (Worsley 1971: 10). He also rightly sees the state being more active than in the advanced industrialised countries 'because capital accumulation will be inadequate if left to the free play of the market' (Worsley 1971: 15–16). Yet the state most often intervenes to assist those who already have power and it is 'traditional ties of kinship – be it those of neighbourhood, ethnicity or caste – that most often work against the requirements of strict economic rationality' (Worsley 1971: 23).

The mid-1970s were a period in which many Western intellectuals made pilgrimages to communist China and enthused on the Cultural Revolution and agrarian reform in that country. Not many people know Worsley's *Inside China* (Worsley 1975) but it provides great insight into his own politics. As an ethnographic study, the book works well and is more balanced than some others appearing at the time. Worsley is sympathetic to the socialist transformation experiment going on in China at the time but bemoans 'the virtual absence of "development theory" and of social science as we know it' (Worsley 1975: 98) in China and even acknowledges that 'bourgeois scholars have often been right about very many things, big and small, and have contributed enormously in areas where the contribution of Marxism has been virtually invisible' (Worsley 1975: 199). While acknowledging the 'pragmatic' Chinese role in world politics, Worsley still concludes that 'there is no reason to assume that China has abandoned her commitment to world socialist revolution' (Worsley 1975: 247). Perhaps more jarring is Worsley's somewhat naive statement that: 'I have no idea how common the death penalty is but I should think it very rare. Nor do I believe that there are many concentration camps' (Worsley 1975: 215).

In the mid-1980s, Worsley came back into development theory with the magnificent monograph, *The Three Worlds* (Worsley 1984), focused on culture and world development. While in one way it was simply *The Third World* twenty years on, its focus was quite different. Reflecting his much earlier engagement with British labour historians like Edward Thompson and John Saville, Worsley had two of his three core chapters dealing with the 'undoing of the peasantry' and the 'making of the working class' respectively. Ethnicity and nationalism, traditional concerns of his, formed the third leg of this great piece of work. The rise of agribusiness and the development of a strong urban working class described in *The Three Worlds* reflected work he was then beginning on international labour studies. Worsley also, if belatedly, moved beyond his previous British Africa/Asia focus to engage with the rapidly transforming Latin America with its strong indigenous intellectual and analytical traditions. But, the main point of this work was that Worsley identified 'culture' as the main missing ingredient in the traditional sociology of development (and Marxist sociology too if he had wanted to develop the theme). Taking his cue from the British cultural theorist, Raymond Williams, as much as from anthropology, Worsley reintroduced culture into the analysis of world development.

Worsley's latest book, *Knowledges* (Worsley 1997), returns to his early interest in language, thought and culture more generally. His basic argument is that knowledge takes many forms and is by nature plural. Extremely diverse in its topics – from Oceanic navigation techniques to Disneyland, and from aboriginal Australian botanical classification to the secularisation of Christmas – this is an ambitious and challenging book. Conventional Western distinctions between science and culture are, in the course of their discrete narratives, deconstructed and undermined. In the era of resurgent US imperialism, Worsley effectively shatters the premise of Western cultural superiority. Worsley is seemingly returning to his vocation as cultural anthropologist and revisiting the fieldwork carried out among the aboriginal people of 'Groote England' in Australia in the early 1950s. In a worthy reprise of *The Trumpet Shall Sound*, Worsley engages with contemporary debates on the science/culture and knowledge/belief distinctions with an original and radical intervention. As with all Worsley's work, it is supremely accessible.

Peter Worsley is a theorist of development but in Britain he is probably best known for his edited sociology textbooks that served to educate a whole generation of students in the 1970s and 1980s (Worsley 1970). There was a breadth and accessibility to these texts that completely surpassed the competition for a whole generation. Then there was his popular text on *Marx and Marxism* (Worsley 2002) that systematised the open **Marx**ist method underpinning his own particular brand of social anthropology. For Worsley, Marxism was anything but an abstract philosophy and his engagement with the influential contemporaneous 'French school' of anthropology (Claude Meillassoux, Maurice Godelier and Emmanuel Terray) was tangential at best. Rather, and quite characteristically, Worsley portrayed Marxism as a political philosophy that had been put to use in various parts of the world with varying degrees of success. That open, flexible and pragmatic approach to social and political theory was reflected in his continuous commitment to interdisciplinarity, really a post-disciplinarity where terms like anthropology, sociology and social anthropology never took on a life of their own and whose disputes were never allowed to dominate the problems of development.

With any thinker there are, finally, invariably certain silences or lacunae. The strengths of an 'understated' Marxism mean that Worsley did not engage in the important (if perhaps 'theological') Marxist debates in the 1970s and 1980s on issues of development and underdevelopment. Likewise, Worsley shows no interest whatsoever in the 'post' debates – post-colonialism, post-structuralism and post-modernism – that were well underway in the mid-1980s when *The Three Worlds* appeared. Also one of his strengths – a groundedness in the British communist labour historian tradition – was his distinctly Anglocentric perspective in terms of sources and debates, however global his own view was. Thus while **Andre Gunder Frank**'s dependency approach is flagged up already in the 1960s work, *The Three Worlds* of 1984 fails even to mention the classic *Dependency and Development in Latin America* by **Fernando Henrique Cardoso** (and Enzo Faletto) that had already appeared in English in 1979. Notwithstanding these ethnocentric tendencies (even stronger in the earlier works) Worsley does 'see' globalisation coming over the horizon in the early 1980s when he refers to 'a new "transcendental" global level' (Worsley 1984: 317) and articulates the globalisation and culture debates before they even happen.

Note

1 This section draws on Keith Hart's note on Peter Worsley for the *Biographical Dictionary of Anthropology*, Routledge, forthcoming.

Major works

Worsley, P. (1957) *The Trumpet Shall Sound: A Study of 'Cargo' Cults in Melanesia*, London: McGibbon and Kee, 2nd edn 1968.
—— (1964) *The Third World*, London: Weidenfeld and Nicolson, 2nd edn 1967.
—— (ed.) (1971) *Two Blades of Grass: Rural Co-operatives in Agricultural Modernisation*, Manchester: Manchester University Press.
—— (1975) *Inside China*, London: Allen Lane.
—— (1984) *The Three Worlds: Cultural and World Development*, London: Weidenfeld and Nicolson.
—— (1997) *Knowledges: Culture, Counter-culture, Subculture*, New York: New Press.

Other works

Worsley, P. (ed.) (1970) *Introducing Sociology*, Harmondsworth: Penguin.
—— (1987) *The New Introducing Sociology*, Harmondsworth: Penguin.
—— (2002) *Marx and Marxism*, London: Routledge.

Ronaldo Munck

ABOUT THE CONTRIBUTORS

Rita Abrahamsen is Senior Lecturer in the Department of International Politics, University of Wales, Aberystwyth, where she teaches African and Postcolonial Politics. She is the author of *Disciplining Democracy: Development Discourse and Good Governance in Africa* (Zed Books, 2000).

V.N. Balasubramanyam (Baloo) holds the Chair in Development Economics, University of Lancaster. His publications include *Multinational Enterprises and the Third Word, The Economy of India, International Transfer of Technology to India, Meeting the Third World Challenge* (with A.I. MacBean), and *Conversations with Indian Economists.* He has edited several books on international trade and investment and published articles on trade, international investment and development issues.

Jo Beall is Director of the Development Studies Institute (DESTIN) at the London School of Economics (LSE). A political sociologist, she has expertise in development policy and management and social development and her specialist research interests include the intersection of formal and informal institutions in local governance and urban development. She is author of *Funding Local Governance, Small Grants for Democracy and Development* (IT Publications, 2004) and co-author of *Uniting a Divided City: Governance and Social Exclusion in Johannesburg* (Earthscan, 2002).

Anthony Bebbington is Professor and Director of Research in the School of Environment and Development at the University of Manchester. Previously he was at the University of Colorado at Boulder and has also worked at Cambridge University, the World Bank, the Overseas Development Institute and the International Institute for Environment and Development. His research in Latin America has addressed NGOs and rural social movements; poverty and rural livelihoods; agricultural development; and the links between development interventions and political economy.

Tony Binns has been Ron Lister Professor of Geography in the University of Otago, Dunedin, New Zealand since October 2004. Prior to this, he was based at the University of Sussex, Brighton, UK. Tony has worked in the field of

Development Studies for over thirty years, with particular experience in Africa, where he has had a longstanding interest in food production systems, rural change and community-based development.

John Brohman is Associate Professor of Geography and a member of the Latin American Studies Program at Simon Fraser University in Burnaby, BC, Canada. His general research interests concern theories, strategies and practices of development, and the principal geographic area of his recent research has been Latin America. This research focuses on regional development and planning, rural development and agrarian reform, community participation and popular movements, and neoliberalism and alternative globalisations.

Sylvia Chant is Professor of Development Geography at the London School of Economics, specialising in issues of gender and development. Her most recent books are *Mainstreaming Men into Gender and Development: Debates, Reflections and Experiences* (with Matthew Gutmann) (Oxfam, 2000), and *Gender in Latin America* (in association with Nikki Craske) (Latin America Bureau/Rutgers University Press, 2003). She is currently undertaking comparative research on the 'feminisation of poverty' in Asia, Africa and Latin America.

Reginald Cline-Cole, a geographer by training, is with the Centre of West African Studies, University of Birmingham, UK. His recent publications include *West African Worlds: Paths through Socio-Economic Change, Livelihoods and Development* (Pearson, 2005, co-edited with Elsbeth Robson) and *Contesting West African Forestry* (Ashgate, 2000, co-edited with Clare Madge).

John Connell is Professor of Geography at the University of Sydney. He has written numerous books and articles about development in the Pacific region, most recently *Urbanisation in the Island Pacific* (with John Lea) (Routledge, 2002).

Stuart Corbridge teaches at the London School of Economics and the University of Miami. He is the author, with John Harriss, of *Reinventing India* (2000), with Sarah Jewitt and Sanjay Kumar of *Jharkhand: Environment, Development and Ethnicity* (2004), and with Glyn Williams, Manoj Srivastava and Rene Veron of *Seeing the State: Governance and Governmentality in Rural India* (2005).

Christopher Cramer is Senior Lecturer in Development Studies at the School of Oriental and African Studies (SOAS), University of London. His research and teaching interests include the economics of Africa, rural labour markets, primary commodity processing, and the political economy of violent conflict. His book, *Why War? Making Sense of Violence, War and Development*, will be published in 2005.

Vandana Desai is Senior Lecturer in Geography at Royal Holloway, University of London. Her research interests are in the area of community participation, low-income housing, the role of non-governmental organisations, gender and social transformation in the context of globalisation. Her recent books include *The Companion to Development Studies* (2002) and *Doing Development Research* (2006) both co-edited with Robert Potter; and *The Introduction to Displacement* (2001) co-authored with Jenny Robinson.

Arturo Escobar was born in Colombia, where he studied chemical engineering in the hope of helping his country to 'develop'. After working for a number of years in the areas of food and nutrition – including a master's degree and a year with Colombia's Department of National Planning – he moved on to study the political economy of development in the late 1970s and early 1980s, and from there to the cultural analysis of development in the 1980s and 1990s. Over the years, he has worked with a number of NGOs and social movements on alternative development and taught and lectured in this area chiefly in the United States but also in Latin America, Europe and Africa. He is currently Kenan Distinguished Professor of Anthropology and Director of the Institute of Latin American Studies at the University of North Carolina, Chapel Hill and Research Associate at the Instituto Colombiano de Antropología e Historia, ICANH in Bogotá.

Ben Fine is Professor of Economics at the School of Oriental and African Studies, University of London. Recent books include (with A. Saad-Filho) *Marx's Capital* (4th edn, Pluto, 2003), *The World of Consumption: The Material and Cultural Revisited* (Routledge, 2002), *Social Capital versus Social Theory: Political Economy and Social Science at the Turn of the Millennium* (Routledge, 2001); *Development Policy in the Twenty-First Century: Beyond the Post-Washington Consensus*, with C. Lapavitsas and J. Pincus (eds) (Routledge, 2001). In preparation is a two-volume book on the shifting relationship between economic history and economic theory (with D. Milonakis) as part of broader research on 'economics imperialism'.

Tim Forsyth is senior lecturer in Environment and Development at the London School of Economics. He specialises on aspects of environmental governance under conditions of rapid industrialisation, with special reference to contested environmental science and expertise, civil society, and public–private partnerships in Asia. He is the author of *Critical Political Ecology: The Politics of Environmental Science* (London and New York: Routledge, 2004), and *International Investment and Climate Change* (London: Earthscan, 1999), and is the editor of the *Encyclopedia of International Development* (London and New York: Routledge, 2005).

Gary Gaile is Professor of Geography at the University of Colorado, Boulder and Executive Director of the Developing Areas Research and Teaching (DART) Programme. He is also Co-Chair of the Association of American Geographer's Developing Areas Specialty Group. He earned his PhD in Geography in 1976 from UCLA. His research interests focus on Africa and include food security, rural–urban linkages, and micro-enterprise credit.

W.T.S. (Bill) Gould is Professor of Geography at the University of Liverpool, and currently Director of its Masters Programme in Population Studies and its Undergraduate Programme in International Development Studies. His research and teaching interests are broadly in population/development relationships, especially in Africa, and more specifically in the demographic and development impacts of HIV/AIDS.

Robert Gwynne is Reader in Latin American Development at the School of Geography, Earth and Environmental Sciences, University of Birmingham. In recent years, he has also been Visiting Professor at the Catholic University of Chile. His research interests focus on industrialisation in the developing world and on the impacts of neoliberalism and globalisation on regional development in Latin America. He is the author of *Industrialisation and Urbanisation in Latin America* (Routledge, 1985); *New Horizons? Third World Industrialization in an International Framework* (Longman, 1990); co-author of *Alternative Capitalisms: Geographies of Emerging Regions* (Arnold, 2003); and co-editor of *Latin America Transformed: Globalization and Modernity* (Arnold, 2004).

John Harriss is Professor of Development Studies at the London School of Economics. He was previously the founding Director of the Development Studies Institute in the School, and earlier the Dean of the School of Development Studies at the University of East Anglia. His current research interests are in democratisation, civil society and governance. He is the author of *Reinventing India* (with Stuart Corbridge, Polity Press, 2000); *Depoliticising Development: The World Bank and Social Capital* (Anthem Press, 2002); and (with Kristian Stokke and Olle Tornquist) editor of *Politicising Democracy* (Palgrave, 2004). He is a Managing Editor of the *Journal of Development Studies*.

Cristóbal Kay is Associate Professor in Development Studies and Rural Development at the Institute of Social Studies, The Hague, The Netherlands. His main research interests are in the fields of rural development and theories of development and underdevelopment, mainly within the Latin American context. His most recent books include *Disappearing Peasantries? Rural Labour in Africa, Asia and Latin America* (co-editor) and *Latin America Transformed: Globalization and Modernity*, second edition (co-editor, 2004).

Reinhart Kößler is Professor of Sociology at the University of Münster. Besides sociology of development, he has researched and published widely on social theory, culture of work and industrialisation, the post-colonial state, nationalism and ethnicity and more recently, culture and politics of public memory. His main regional interest has for some time been southern Africa.

John P. Lea, who was born in South India during the last years of the 'Raj', was educated in England and has spent most of his academic career at the University of Sydney, Australia. His work on urbanisation and development in southern Africa,

Australasia and the Pacific has focused on housing, urban planning and governance issues at the resource frontier.

Cathy McIlwaine is a Senior Lecturer in the Department of Geography, Queen Mary, University of London. Her research focuses on issues of poverty, survival strategies, gender and urban violence, mainly in Latin America. Her most recent publications include *Encounters with Violence in Latin America*, with Caroline Moser (Routledge, 2004) and *Challenges and Change in Middle America*, with Katie Willis (Prentice Hall, 2002). She is currently researching the livelihood strategies of Colombians living in London.

Henning Melber holds a PhD in Political Science and a *venia legendi* in Development Studies at Bremen University. He has been a Senior Lecturer in International Politics at Kassel University (1982–92), Director of The Namibian Economic Policy Research Unit (NEPRU) in Windhoek (1992–2000) and is currently Research Director at The Nordic Africa Institute in Uppsala, Sweden. He has published widely on a range of African Studies related subjects, mainly Namibia, but also Southern Africa, racism and of late NEPAD (New Programme for Africa's Development). He is currently a Vice-President of the European Association of Development Research and Training Institutes (EADI).

Ulrich Menzel, born 1947 in Düsseldorf, Germany, teaches International Relations and Comparative Politics at the Technical University of Braunschweig and is Managing Director of its Institute of Social Sciences. He has published twenty-six books and many articles on the subjects of International Relations, International Political Economy and Development Theory. Further information about the author can be obtained at www.tu-bs.de/~umenzel/.

Ronaldo Munck is Strategic Director for Internationalisation and Social Development at Dublin City University, having previously held chairs in sociology at the University of Liverpool and of Durban-Westville. He has written widely on Latin America and on development issues more generally, including *Contemporary Latin America* (Palgrave, 2002) and *Critical Development Theory: Contributions to a New Paradigm* (co-editor) (Zed Books, 1999). His most recent publication is *Globalisation and Social Exclusion: A Transformationalist Perspective* (Kumarian Press, 2004).

Dani W. Nabudere is a senior researcher and executive director of the Afrika Study Centre, Mbale, Uganda. He studied at Lincoln's Inn, where he was called to the Bar in 1963. Later, he spent six years as Associate Professor of Law at the University of Dar es Salaam, Tanzania, where he was able to observe at close quarters the effects of the *Ujamaa* strategy between 1973 and 1979. Subsequently, he was a minister of Culture and Community Development in the post-Amin interim government in Uganda. He is currently promoting the establishment of the

Marcus Garvey Pan-Afrikan Institute dedicated to research and mainstreaming of African indigenous knowledge and wisdom.

Anders Närman was, until his sudden death on 15 November 2004, Associate Professor of Human and Economic Geography, Göteborg University. His research interests included various aspects of Development Geography. Educational matters were a central part of his research in Kenya and Tanzania. Until his death, he was involved in research on conflict in Uganda, the Greater Horn of Africa, and on regional development in Sri Lanka. These were collaborative projects, with the universities of Dar es Salaam, Makerere and Kelaniya respectively.

Michael Parnwell is Reader in South East Asian Geography in the Department of East Asian Studies at the University of Leeds. He has specialised on the development process in South East Asia since the late 1970s. His PhD was on return-migration in North East Thailand, and he continued to take an interest in Thailand during the 1980s and early 1990s, focusing on processes of regional development, rural industrialisation and extended metropolitanisation. More recently he has worked on the human impact of deforestation in Sarawak, Malaysia, small-scale industries and urban sustainability in Sulawesi, Indonesia, sustainable development in South East Asia, and on coping mechanisms and sustainable livelihoods in Thailand, Vietnam and the Lao PDR.

Richard Peet is Professor of Geography at Clark University in Worcester, Massachusetts. He has been editor of *Antipode: A Radical Journal of Geography and Economic Geography*. His recent publications include *Modern Geographical Thought* (Oxford: Blackwell, 1998), *Theories of Development* (with Elaine Hartwick, New York: Guilford, 1999, 2002) and *Unholy Trinity: The IMF, World Bank and WTO* (London: Zed Books, 2003). He is interested in issues of globalisation and development, global governance and economic policy, culture, consciousness and ideology, and the possibility of post-neoliberal societies.

Marcus Power is a Lecturer in Human Geography in the University of Durham. His research interests are focused on Southern Africa and the Lusophone world and are concerned with geopolitics, post-colonialism and development. His recent publications include *Rethinking Development Geographies* (London: Routledge, 2003).

Sarah Radcliffe teaches at the Department of Geography, University of Cambridge. Her research focuses on the development trajectories and state–society relations in Ecuador, Peru and Bolivia, and examines the gendered and ethnic outcomes of social change. Her forthcoming book is *Multiethnic Transnationalism: Indigenous Development in the Andes* (Duke University Press).

Jonathan Rigg is a geographer based at the University of Durham. He has worked on rural and environmental change in Southeast Asia since the early 1980s. He is the author of *Southeast Asia: The Human Landscape of Modernization and Development* (London: Routledge, 2003) and has recently completed a manuscript on livelihoods and marketisation in Laos. He is currently working on aquatic food production in peri-urban areas of Bangkok, Hanoi, Phnom Penh and Ho Chi Minh City and on wider processes of de-agrarianisation and de-peasantisation.

Barbara Rugendyke is a Senior Lecturer in Geography at the University of New England, in Armidale, NSW, Australia. Barbara teaches about development issues and her current research interests include the sustainability of ecotourism in Southeast Asia and she is currently writing a book about the impacts of the advocacy activities of development NGOs.

M.A. Mohamed Salih (PhD Economics and Social Studies, University of Manchester, UK) is Professor of Politics of Development at both the Institute of Social Studies, The Hague and the Department of Political Science, University of Leiden in the Netherlands. His latest books include, *African Democracies and African Politics* (Pluto Press, 2001), *African Political Parties: Evolution, Institutionalisation and Governance* (co-edited) (Pluto Press, 2003) and *Africa Networking: Information Development, ICTs and Governance* (co-edited) (International Books, 2004).

Roberto A. Sánchez-Rodríguez completed his BA in Architecture from the National Autonomous University of Mexico (UNAM) and his PhD in Urban and Regional Planning from the University of Dortmund, Germany. He is currently a Professor of Environmental Studies at the University of California, Riverside, and Director of The University of California Institute for Mexico and the United States (UCMEXUS). His research interests are in development studies, sustainable development, trade and the environment, and the interactions between urbanisation and global environment change.

Tilman Schiel (born 1943) studied Cultural Anthropology at the University of Heidelberg (MA) and Sociology at the University of Bielefeld (Dr.rer.soc. and 'Habilitation'). Until his retirement in 2004 he worked as an itinerant scientist at several universities and research institutions. He still is member of the board of the Starnberg Institute for the Study of Global Structures, Developments and Crises.

John Shaw was associated with the UN World Food Programme for over thirty years, latterly as economic adviser and chief of its Policy Affairs Service before his retirement. He has also served as consultant to the Commonwealth Secretariat, FAO and the World Bank. He is currently on the International Editorial Board of the journal *Food Policy* and continues to write on development and food security issues.

James Sidaway is Reader in Globalisation at Loughborough University. He previously taught at the National University of Singapore, the Universities of Birmingham and Reading in the UK and has been Visiting Professor at the

Universities of Seville and Groningen. His research interests include geographies of development, geopolitics and the history and sociology of geography.

David Simon is Professor of Development Geography in the Centre for Developing Areas Research, Royal Holloway, University of London. His particular research interests include development theory and policy; the development–environment interface; urbanisation and urban–rural interaction; transport and geopolitics. He has extensive research experience in sub-Saharan Africa and tropical Asia. He is currently development/social science editor of the *Journal of Southern African Studies*, and co-editor of *The Peri-Urban Interface: Approaches to Sustainable Natural and Human Resource Use* (Earthscan, 2005) and of *Aquatic Ecosystems and Development: Comparative Asian Perspectives* (Kluwer, 2006).

Rana P.B. Singh is Professor of Cultural Geography at Banaras Hindu University, Varanasi, India. He has been involved in mass awakening programmes and heritage planning in Varanasi region for the last twenty years as promoter, collaborator and organiser. On these topics he lectured at various centres in America, Europe, East Asia and Australia. His publications include over 140 papers and 30 books on these subjects.

Morris Szeftel is Senior Lecturer in Politics at the University of Leeds. He has also worked and taught at universities in the United States, South Africa and Zambia. He has published on African politics, problems of democratisation, development and debt, and on corruption in Africa. Until recently he was Editor of the *Review of African Political Economy*.

Jan Toporowski is Research Associate in London University's School of Oriental and African Studies; Official Visitor in the Faculty of Economics and Politics, University of Cambridge; and Research Associate in the Research Centre on the History and Methodology of Economics, University of Amsterdam. After studying economics at Birkbeck College and Birmingham University, he worked in fund management and international banking, and has published widely on money, finance and economic development.

Juha I. Uitto has worked for over two decades on development and environment issues both in international organisations and academia. He holds an MSc in Geography from the University of Helsinki in his native Finland and a PhD in Social and Economic Geography from the University of Lund in Sweden. He is currently Senior Evaluation Advisor at the Global Environment Facility unit of the United Nations Development Programme in New York, USA.

Ton van Naerssen is a Senior Lecturer in development geography and co-ordinator of the Master's programme for Globalisation and Development at the Nijmegen School of Management (NSM), Radboud University of Nijmegen, the Netherlands. His current fields of interest include development theories, the involvement of civil society in urban development and international migration. He has carried out research mainly in Southeast Asia. His most recent books are (both as co-editor) *Healthy Cities in Devloping Countries: Lessons to be Learned* and *Asian Migrants at the European Labour Market* (Routledge, forthcoming).

Michael Watts is Class of 1963 Professor of Geography at the University of California, Berkeley where he has taught development studies for over twenty-five years. He has just finished a book on American Empire (*Afflicted Powers*, Verso, 2005) and is working on another on petro-politics in Nigeria. He is currently at the Center for the Study of the Advanced Behavioral Sciences at Stanford.

Katie Willis is Senior Lecturer in Geography at Royal Holloway, University of London. Her main research areas are gender, development and migration, with a particular focus on Mexico, China and Singapore. She is co-editor of *International Development Planning Review* and the author of *Theories and Practices of Development* (Routledge, 2005).

Francis Wilson was trained in Physics (at the University of Cape Town) and Economics (Cambridge). He has been teaching in the School of Economics at UCT for the past thirty-eight years. In 1975 he founded SALDRU, the Southern Africa Labour and Development Research Unit which he directed until 2000 when he started Data First, a university resource unit For Information Research & Scientific Training. His main work has been in labour (gold mines; migrant; farms), South African history, data collection and in poverty about which, to find out more, he directed the second Carnegie Inquiry into Poverty and Development in Southern Africa during the 1980s.

Ben Wisner flew around the Caribbean in a small plane with Jim Blaut and learned about peasants and tropical soils from him, and misses him sorely. Wisner is now research fellow at DESTIN, LSE and also at the Benfield Hazard Research Centre, University College London, as well as teaching as Visiting Professor of Environmental Studies at Oberlin College at the northern end of the underground railway, in Ohio, USA.

Lakshman Yapa hails from Sri Lanka, where he took his first degree at Peradeniya University (then the University of Ceylon at Peradeniya). He holds a PhD in Geography from Syracuse University in New York. He is currently Professor of Geography at Pennsylvania State University and Director of the service learning project entitled, Rethinking Urban Poverty: Philadelphia Field Project. His recent publications have addressed critical perspectives on scarcity, poverty and poverty discourses in Sri Lanka and the USA.

Brenda S.A. Yeoh is Associate Professor, Department of Geography, National University of Singapore and Principal Investigator of the Asian MetaCentre at the University's Asia Research Institute. Her research foci include the politics of space in colonial and post-colonial cities; and gender, migration and transnational communities. Her most recent books include *Toponymics: A Study of Singapore Street Names* (2003, with Victor R. Savage), *Theorising the Southeast Asian City as Text* (2003, with Robbie Goh), *The Politics of Landscape in Singapore: Construction of 'Nation'* (2003, with Lily Kong), and *State/Nation/Transnation: Perspectives on Transnationalism in the Asia-Pacific* (2004, with Katie Willis).

Alfred Babatunde Zack-Williams is Professor of Sociology at the University of Central Lancashire, Preston, UK. He was formerly Senior Lecturer in Sociology at the University of Jos, Nigeria. Among his recent publications are: *Africa in Crisis: Challenges and New Possibilities*, (ed. Pluto Press, 2002); *Africa Beyond the Post-colonial* (ed. with Ola Uduku, Ashgate, 2004); *The Politics of Transition: State, Democracy & Economic Development in Africa* (ed. with G. Mohan and James Currey, 2004). He is an editor for the *Review of African Political Economy*.

INDEX

POLITICS: THE BASICS

3rd edition

Stephen D. Tansey

This highly successful introduction to the world of politics has been fully revised and updated to explore the key issues of the 21st century. The new edition builds on the reputation for clarity and comprehensive coverage that has made previous editions essential reading for students of politics. The third edition of *Politics: The Basics*:

- introduces all the key areas of politics, explaining all the basic ideas and terms, making it an ideal text for propsective undergraduate students and the general reader
- is clearly and accessibly written, making use of boxes, figures and tables to illustrate key issues
- has a wider international focus and includes a variety of case studies and examples
- contains brand new material on postmodernism, terrorism, information technology, globalization and the media
- features an appendix which gives guidance to a variety of useful political sources, including books, newspapers and the Internet as well as information on politics courses and associations.

> "Tansey reveals an admirable breadth of knowledge and an ability to present this in most helpful ways."
>
> *Talking Politics*

0–415–30329–X

Available at all good bookshops
For ordering and further information please visit www.routledge.com